Mogens Henze · Poul Harremoës
Jes la Cour Jansen · Eric Arvin

Wastewater Treatment

Biological and Chemical Processes

Third Edition

With 193 Figures and 88 Tables

Springer

Series Editors:
Prof. Dr. U. Förstner Arbeitsbereich Umweltschutztechnik
Technische Universität Hamburg-Harburg
Eißendorfer Straße 40
21073 Hamburg, Germany

Prof. Robert J. Murphy Dept. of Civil Engineering and Mechanics
College of Engineering
University of South Florida
4202 East Fowler Avenue, ENG 118
Tampa, Fl 3362-5350, USA

Prof. Dr. ir. W.H. Rulkens Wageningen Agricultural University
Dept. of Environmental Technology
Bomenweg 2, P.O. Box 8129
6700 WV Wageningen, The Netherlands

Authors:
Prof. Mogens Henze Technical University of Denmark
Prof. Poul Harremoës Dept. of Environmental Science and Engineering
and Prof. Erik Arvin Building 115, DK-2800 Lyngby, Denmark
Dr. Jes la Cour Jansen Dept. of Water and Environmental Engineering
LTH, P.O. Box 118
22100 Lund, Sweden

Translated from the Danish Spildevandsrensning, biologisk og kemisk published by Polyteknisk Forlag, Lyngby, Denmark

ISBN 3-540-42228-5 Springer-Verlag Berlin Heidelberg New York

Library of Congress Cataloging-in-Publication Data
Spildevandsrensning, biologisk og kemisk. English
Wastewater treatment : Biological and chemical processes / Mogens Henze ... [et al.]. -- 3rd ed.
p. cm. -- (Environmental engineering)
Includes bibliographical references and index.
ISBN 3-540-42228-5
1. Sewage--Purification. I. Henze, M. II. Title. III. Environmental engineering (Berlin, Germany
TD745 .S6413 2002   628.3--dc21   2001041097

This work is subject to copyright. All rights are reserved, whether the whole or part of the material is concerned, specifically the rights of translation, reprinting, reuse of illustrations, recitation, broadcasting, reproduction on microfilm or in other ways, and storage in data banks. Duplication of this publication or parts thereof is permitted only under the provisions of the German Copyright Law of September 9, 1965, in its current version, and permission for use must always be obtained from Springer-Verlag. Violations are liable for prosecution act under German Copyright Law.

Springer-Verlag Berlin Heidelberg New York
a member of BertelsmannSpringer Science+Business Media GmbH

httpp://www.springer.de

© Springer-Verlag Berlin Heidelberg 2002
Printed in Germany

The use of general descriptive names, registered names, trademarks, etc. in this publication does not imply, even in the absence of a specific statement, that such names are exempt from the relevant protective laws and regulations and therefore free for general use.

Typesetting: Dedicated PrePress, Lyngby, Denmark
Cover-design: Struve & Partner, Heidelberg
Printed on acid-free paper   SPIN: 10842593   30 / 3020 hu - 5 4 3 2 1 -

Preface

'Art is about guessing the correct solution. Science is about producing tables in which we can find the correct solution, and engineering is about consulting the tables'.

Poul Henningsen
(Danish architect
and writer)

This book has been written for engineers who wish to familiarize themselves with the tables and part of the science behind the tables. You won't learn to build wastewater treatment plants by reading this book; the idea is to create a scientifically based platform from which to work.

The book has been written for M.Sc. and Ph.D. students, consulting engineers and process engineers at wastewater treatment plants. Drawings have been produced by Birte Brejl and Torben Dolin. Translation of the 2nd Danish edition by Nancy M. Andersen was made possible by a grant from the CowiConsult Foundation.

Thank you.
Lyngby, Spring 1995
Mogens Henze

Preface to 3rd edition

The 3rd edition has been generally revised and updated. Our understanding of complex biological processes has increased, thus topics like biological phosphorus removal, and hydrolysis and modelling have been rewritten, together with many other details. The 2nd edition of the book has been translated into Chinese and Polish, and a translation into Malay is underway. Thus it seems that the book fulfils its mission: to give a detailed information on wastewater treatment at a high theoretical level for use in teaching and engineering worldwide.

Lyngby, February 2000
Mogens Henze

Table of Contents

1. Wastewater, Volumes and Composition 11
   - 1.1. The volumes of wastewater 11
     - 1.1.1. Measurements. 11
     - 1.1.2. Statistics 13
     - 1.1.3. Estimates 16
     - 1.1.4. Population Equivalent and Person Load 24
     - 1.1.5. Prognoses 25
   - 1.2. Wastewater components 27
     - 1.2.1. Domestic wastewater/municipal wastewater 27
     - 1.2.2. Variations 35

2. Characterization of Wastewater and Biomass 43
   - 2.1. Suspended solids. 43
     - 2.1.1. Settleable solids 45
   - 2.2. Organic matter. 45
   - 2.3. Nitrogen 57
   - 2.4. Phosphorus 58
   - 2.5. Alkalinity (TAL). 59
   - 2.6. Sludge volume index etc. 59
   - 2.7. Respiration rate of sludge 60

3. Basic Biological Processes 65
   - 3.1. The biology in biological treatment plants 65
     - 3.1.1. The organisms. 65
     - 3.1.2. Selection 68
   - 3.2. Conversions in biological treatment plants 72
     - 3.2.1. Biological growth 73
     - 3.2.2. Hydrolysis. 74
     - 3.2.3. Decay 75
     - 3.2.4 Storage 76
   - 3.3. Aerobic heterotrophic conversion of organic matter 77
     - 3.3.1. Reactions, aerobic conversions. 77
     - 3.3.2. Yield constant, aerobic heterotrophic conversions . . 79
     - 3.3.3. Nutrients, aerobic heterotrophic conversions. 83
     - 3.3.4. Kinetics, aerobic heterotrophic conversions 84
     - 3.3.5. Heterotrophic micro-organisms, aerobic conversions . . 85
     - 3.3.6. The influence of the environmental factors, aerobic heterotrophic conversions. 85
   - 3.4. Nitrification 89
     - 3.4.1. Reactions by nitrification. 90
     - 3.4.2. Alkalinity 92

		3.4.3. Kinetics, nitrification 93

              3.4.3. Kinetics, nitrification 93
              3.4.4. The influence of the environmental factors on nitrification . . 93
    3.5. Denitrification 100
              3.5.1. Reactions by denitrification 101
              3.5.2. Yield constant by denitrification 101
              3.5.3. Nutrients, denitrification 103
              3.5.4. Alkalinity 103
              3.5.5. Kinetics, denitrification 103
              3.5.6. The influence of the environmental factors, denitrification. . 105
    3.6. Biological phosphorus removal 109
              3.6.1. Microorganisms 109
              3.6.2 Reactions, biological phosphorus removal 111
              3.6.3. Yield constant, biological phosphorus removal 113
              3.6.4. Alkalinity 113
              3.6.5 Kinetics, biological phosphorus removal 114
              3.6.6. Environmental factors, biological phosphorus removal . . . 114
    3.7. Anaerobic processes 116
              3.7.1. Reactions, anaerobic processes 118
              3.7.2. Growth and soluble COD yield constants,
              anaerobic processes 118
              3.7.3. Nutrients, anaerobic processes 119
              3.7.4. Alkalinity, anaerobic processes 120
              3.7.5. Kinetics, anaerobic processes 121
              3.7.6. Gas production 122
              3.7.7. The influence of the environmental factors,
              anaerobic processes 123

4.   Activated Sludge Treatment Plants 131
    4.1. Mass balances, activated sludge plants 131
    4.2. Concepts and definitions of the activated sludge process 137
    4.3. Types of plants, activated sludge plants 143
              4.3.1. Activated sludge with recycle 143
              4.3.2. Single tank activated sludge plants 146
              4.3.3. Contact stabilization plants 148
              4.3.4. Biosorption plants 150
              4.3.5. Design of activated sludge processes 151
              4.3.6. Design by means of volumetric loading 151
              4.3.7. Design by means of sludge loading or sludge age 153

5.   Biofilters 157
    5.1. Biofilm kinetics 157
    5.2. Biofilm kinetic parameters 167
    5.3. Hydraulic film diffusion 169
    5.4. Two-component diffusion 172
    5.5. Filter kinetics 175
    5.6. Mass balances for biofilters 179

    5.6.1. Biofilters without recycle 180
    Biofilters with recycle 180
  5.7. Concepts and definitions 181
  5.8. Types of plants 182
    5.8.1. Trickling filters. 183
    5.8.2. Submerged filters 185
    5.8.3. Rotating discs 188
  5.9. Design of biofilters 188
    5.9.1. Design of trickling filters 189
    5.9.2. Design of discs 190
    5.9.3. Rules for other types of filter 191
    5.9.4. Design of biofilters for dissolved organic matter. 191
  5.10. Technical conditions concerning biofilters 195
    5.10.1 Aeration in biofilters 195
    5.10.2 Growth and sloughing off of the biofilm. 195
  5.11. Removal of particulate organic matter 197
  5.12. Detailed model 203

6. Treatment Plants for Nitrification 209
  6.1. Mass balances, nitrifying plants 210
    6.1.1. Separate nitrifying plants 210
    6.1.2. Combined removal of organic matter and ammonium 219
  6.2. Types of plants for nitrification 220
    6.2.1. Nitrification plants with separate sludge 222
    6.2.2. Single sludge nitrification plants 222
    6.2.3. Nitrification in two sludge treatment systems 225
    6.2.4. Nitrification plants with separate sludge in filters 226
    6.2.5. Two sludge nitrification plants in filters 227
    6.2.6. Combined biofilters and activated sludge treatment
    plants for nitrification 229
  6.3. Design of nitrifying plants 229
    6.3.1. Design of activated sludge treatment plants for nitrification . 230
    6.3.2. Optimizing operation of nitrifying plants 233
    Design of biofilters for nitrification. 234

7. Treatment Plants for Denitrification 239
  7.1. Mass balances, denitrifying treatment plants 240
    7.1.1. Separate denitrifying plant 242
    7.1.2. Combined nitrification and denitrification 247
  7.2. Types of plants for denitrification 251
    7.2.1. Denitrification plants with separate sludge 253
    7.2.2. Denitrification plants with combined sludge 254
    7.2.3. Biofilters for denitrification 258
  7.3. Design of denitrifying plants 259
    7.3.1. C/N ratio 259
    7.3.2. Oxygen/stirring 264

7.3.3. Simultaneous nitrification-denitrification. 265
                7.3.4. Nitrogen gas in settling tanks and biofilters 266
                7.3.5. Oxygen consumption 269
                7.3.6. Alkalinity. 270
                7.3.7. Design of activated sludge plants with denitrification 270
                7.3.8. Model based process design 273
                7.3.9. Design of biofilters for denitrification 277
        7.4.   Redox-zones in the biomass 279

8. Plants for Biological Phosphorus Removal 285
        8.1.   Mass balances, biological phosphorus removal plants
                with activated sludge 285
        8.2.   Plant types, biological phosphorus removal 288
                8.2.1. Biological phosphorus removal with nitrification-
                denitrification and an internal carbon source 288
                8.2.2. Biological phosphorus removal with nitrification-
                denitrification and an external carbon source 290
                8.2.3. Biological phosphorus removal with internally
                produced easily degradable organic matter 291
                8.2.4. Biological phosphorus removal without
                nitrification-denitrification 291
        8.3.   Design of biological phosphorus removal 291
                8.3.1. Easily degradable organic matter 291
                8.3.2. Design of tanks for biological phosphorus removal 293
                Optimization of plant operation, biological phosphorus
                removal 296

9. Hydrolysis/fermentation and Anaerobic Wastewater Treatment . . 299
        9.1.   Hydrolysis/fermentation 299
        9.2.   Anaerobic wastewater treatment 300
                9.2.1. Introduction 300
                9.2.2. Mass balances, anaerobic plants 301
        9.3.   Plant types, anaerobic processes 305
                9.3.1. Pretreatment of wastewater, anaerobic plants 305
                9.3.2. Plants with suspended sludge 306
                9.3.3. Anaerobic filter processes 309
        9.4.   Design of anaerobic plants 310
                9.4.1. Design of plants with suspended sludge 311
                9.4.2. Design of anaerobic filter plants 315
                9.4.3. Gas production, anaerobic processes 317
                9.4.4. Optimization, anaerobic plants 319
                9.4.5. Start-up, anaerobic plants 319
                9.4.6. Disturbances, anaerobic plants 320

10. Treatment Plants for Phosphorus Removal from Wastewater 327
        10.1.   Mass balances for phosphorus removal processes 327

    10.2. Mechanisms for chemical/physical phosphorus removal 330
        10.2.1. Precipitation 331
        10.2.2. Coagulation. 335
        10.2.3. Flocculation 341
        10.2.4. Phosphorus binding in soil 346
    10.3. Treatment plants for phosphorus removal 348
        10.3.1. Precipitants. 348
        10.3.2. Treatment processes 350
    10.4. Design of plants for phosphorus removal. 355
        10.4.1. Chemical precipitation 355
        10.4.2. Phosphorus binding in soil 361
    10.5. Operation of plants for phosphorus removal. 363

11 . Model features, calibration and application 369
    11.1. Pragmatism versus theory-based models 369
        11.1.1. Engineering craftsmanship 369
        11.1.2. Science-based determinism 370
        11.1.3. Model structure, variables, parameters and forcing input . . 370
    11.2. Model applications. 371
        11.2.1. Planning tool 371
        11.2.2. Analysis of existing plants 372
        11.2.3. Design of new plants 372
        11.2.4. Real time control of plant 372
        11.2.5. Models as research tools 373
        11.2.6. Level of aggregation 373
    11.3. Model calibration and parameter estimation. 374
        11.3.1. Model structure 374
        11.3.2. Parameter calibration, verification and estimation 375
    11.4. Treatment plant design 379
        11.4.1. Identification of problem 379
    11.5. Model for biofilm system 382
    11.6. Analysis of existing plant/pilot plant 392
        11.6.1. Identification of the problem 392
        11.6.2. Design of experimental programme 393
        11.6.3. Interpretation of results 393
    11.7. Real time control 395
    11.8. Integrated modelling. 397

    List of Symbols 401

    Index 421

1 Wastewater, Volumes and Composition

by Mogens Henze

1.1. The volumes of wastewater

Wastewater flows are not steady or uniform, but vary from one hour to another, from day to day, from month to month and from year to year. When building a treatment plant, it is important to know the volumes of wastewater and their variations, today as well as in the future. Based on the knowledge of the wastewater, the design of the treatment plant can be determined, taking into account the wastewater to be treated. In this connection measurements are useful; if such measurements do not exist, an estimate should be made. In respect of the volume of future wastewater, the development should of course be taken into account, i.e. a prognosis should be made.

1.1.1. Measurements

Measurements of the volume of wastewater will be either in the form of curves or in the form of figures (metering). Figure 1.1 shows a variation for wastewater in the influent entering a treatment plant. The curve is the sum of domestic, industrial and public institution wastewater, infiltration and exfiltration. There is no reason to consider the size of the individual contributions as the curve shows what reaches the treatment plant. In case of a projection of the volumes of wastewater and their variations, however, it is recommended to analyze the curve and the catchment area for the purpose of splitting up the sub-contributions as it will be easier to project these separately. This is briefly dealt with in Sections 1.2 and 1.5. Sampling and measurements at treatment plants are difficult. Watch out for flows of return water (such as supernatant) which are often mixed into the raw wastewater before reaching bar screen and grit chamber, hence making correct measurement on the raw wastewater difficult.

The curve shown in figure 1.1 can be used to estimate the maximum daily flow (20.000 m^3/d) and the maximum hourly flow (1100 m^3/h) on the days in question. If a sufficient number of diurnal measurements are available, two important quantities can be calculated which form part of the design of the treatment plant, i.e.

The volumes of wastewater

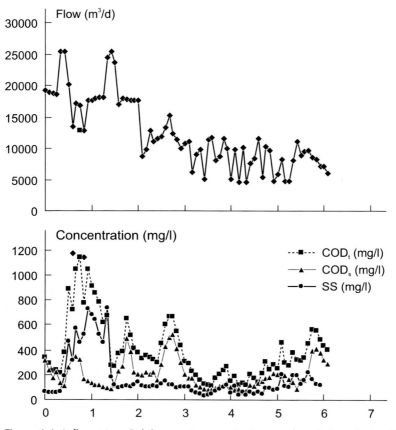

Figure 1.1. Influent to a Belgian wastewater treatment plant 7 days during 1999. The first two days shows the influence of rain on the wastewater flow and composition. The single points for the measured flow values are given as m³/d in case the flow would be the same all the day, as it was during the sampling interval /1/.

$Q_{h,max}$ the average of the maximum hourly flow rate of the individual days (m³/h)

$Q_{h,av}$ the average hourly flow rate for many days (m³/h)

The maximum hourly flow rate, $Q_{h,max}$, can be calculated on the basis of a number of maximum hourly flows rates.

The average maximum hourly flow rate, $Q_{h,max}$, is, among other things, used as a basis for the hydraulic design of sewers and ponds. The average hourly volume of water, $Q_{h,av}$, or the diurnal volume of water, $Q_{d,av}$, is for example used for the calculation of operating expenses.

1.1.2. Statistics

By processing the data statistically a more detailed picture of wastewater variations is obtained. The different volumes of water (volume, maximum hour, maximum second, etc. over a 24-hour period) will often be normally distributed or log-normally distributed. Data sets for wastewater are never ideal as there will be irregularities which may, if they are too excessive, result in a special treatment of the data.

Fractile diagrams can be an important tool in the design of treatment plants. Figure 1.2 gives an example of such a diagram. The 60 per cent fractile is frequently used as an average load and a fractile of 85-90 per cent as a maximum load.

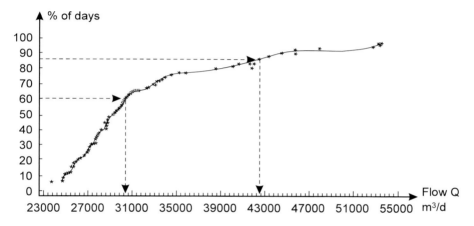

Figure 1.2. Fractile diagram for influent to Lundtofte wastewater treatment plant 1984-1989 (Denmark). Here $Q_{d,av}$ is determined as the 60%-fractile (30,400 m³/d) and $Q_{d,max}$ as the 85%-fractile (42,500 m³/d). Data from /24/.

A number of irregularities in wastewater data can be revealed visually by listing the collected data in order of time (a time series). Hence irregularities such as

– jumps
– trend (increasing or decreasing)
– variation (for example weekly or seasonal variations)

In /9/ simple methods for the examination of irregularities have been described. In figure 1.3 the influent entering the Søholt treatment plant (Denmark) is shown, listed as a time series. It appears from the figure how rain in volumes of more than 4 mm/d has an impact on the influent, and likewise a distinct weekly variation can be seen as the influent is low on Saturdays and Sundays.

By plotting a data set on logarithmic graph paper it can be determined whether the data are distributed normally or log-normally. By plotting on a piece of paper with an ordinarily divided x-axis, normally distributed data will be on a straight line. Log-normally distributed data will be on a straight line when plotted on lo-

The volumes of wastewater

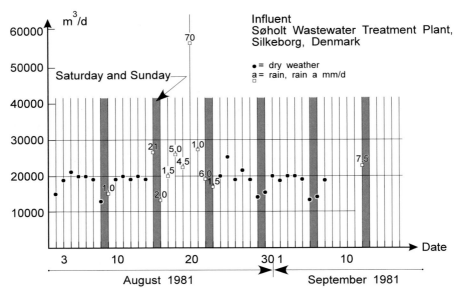

Figure 1.3. Influent to Søholt wastewater treatment plant, Silkeborg (Denmark).

garithmic graph paper with a logarithmically divided x-axis. By plotting, the average (mean value) and spread can be found. It should be noted that the mean value in a log-normal distribution does not correspond to a 50 per cent probability, but it must be calculated from the formula in Table 1.1.

Normal distribution (straight line on logarithmic graph paper with normal X-axis)			Log-normal distribution (straight line on logarithmic graph paper with logarithmic X-axis)		
Mean value,	$\overline{X}$	= f(50%)	log $\overline{X}$	= log f(50%) + 1.1513 · s^2
Spread	s	= f(84%) − f(50%)	s	= log (f(84%)) − log (f(50%))
	or			or	
	s	= f(50%) − f(16%)	s	= log (f(50%)) − log (f(16%))

Table 1.1. Determination of mean value and spread plotted on logarithmic graph paper.

Figure 1.4 shows the plot of the influent measurements for the Sjælsø treatment plant (Denmark) on a piece of logarithmic graph paper. It is seen that the maximum hours, $Q_{h,max}$, and the diurnal influent can be assumed to be log-normally distributed whereas the maximum seconds values are not on a regular straight line.

In figure 1.5 the maximum hourly flow to the Ejby Mølle treatment plant (Denmark) has been plotted for a number of dry-weather days. The points are reasonably linear, and hence they can be assumed to be normally distributed as ordinary logarithmic graph paper has been used for the plotting. The average maximum hourly flow rate, $Q_{h,max}$, can be read for 50 per cent probability

$$Q_{h,max} = 3,175 \text{ m}^3/\text{h}$$

Example 1.1

What is the spread on the maximum hourly flow to the Ejby Mølle treatment plant on a dry-weather day?

What is the percentage of the dry-weather days by which the maximum hourly flow rate is less than 3,650 m³/h?

s = f (84%) − f (50%) = 3,525 − 3,175 = 350 m³/h.

The spread, s, can be determined as the difference between the curve values for 84 per cent probability and 50 per cent probability, see Table 1.1.

By means of the curve it can be established that 90 per cent of the days the maximum hourly flow rate will be less than or equal to 3,650 m³/h.

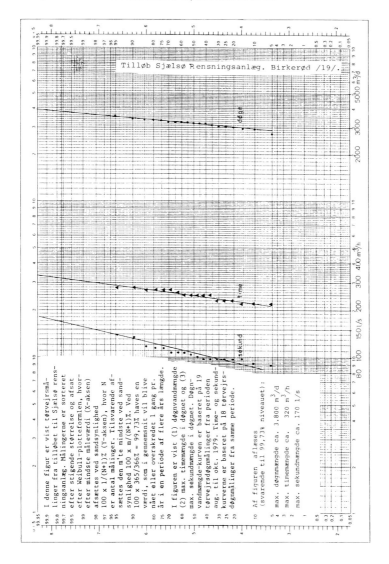

Figure 1.4. Influent to Sjælsø wastewater treatment plant, Birkerød (Denmark) /19/.

The volumes of wastewater

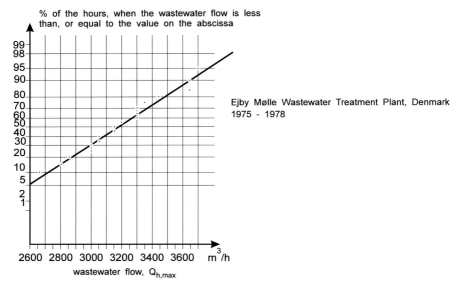

Figure 1.5. Influent Ejby Mølle wastewater treatment plant, Odense, (Denmark), data for dry-weather 24-hour periods /20/.

1.1.3. Estimates

If adequate measurements of the volumes of wastewater are not available, estimates and calculations must be made. For that purpose the wastewater is divided into parts which typically consist of

- domestic wastewater
- industrial and public institutions wastewater
- infiltration

In respect of domestic wastewater, the calculation can be made as shown in figure 1.6. The basis is a number of persons and their annual wastewater production, $Q_{yr,pers}$. Table 1.2 gives an idea of the quantity of $Q_{yr,pers}$. However, rough average figures are stated. Based on the annual volume of wastewater, the other calculations/estimates can be carried out as shown in figure 1.6.

Example 1.2

The town of Heraklion is situated on the northern coast of the Greek island of Crete. It has a fine museum with objects from Knossos which is situated in the vicinity and certainly is worth a visit.

Calculate the maximum hourly flow for domestic wastewater exclusive of infiltration and exfiltration for Heraklion on the island of Crete. The population counts 120,000 persons.

From Table 1.2 the annual volume of wastewater per capita for Greece is estimated at 60 m^3.

Country	Year	m³/(person × year) excl. infiltr.	m³/(person × year) incl. infiltr.
Albania	1977		60
Algeria	1977	40	
Australia	1981		90
Austria	1969	50	
Belgium	1969	30	
Brazil	1975		90
Denmark	1982	55	
Egypt	1977	55	
Finland	1973		210
France	1975		75
France	1976	35	
Greece	1975	60	
Italy	1970	85	
Italy	1972	80	
Norway	1978	55	
Switzerland	1969	100	
Switzerland	1976	95	
Spain	1969	50	
Spain	1977	90	
Sweden	1970	85	
Sweden	1976	75	
Sweden	1978	85	
Syria	1977	35	
The Netherlands	1970	35	
The Netherlands	1976	50	
Tunisia	1977	30	
Turkey	1977	50	
UK	1969	60	
UK	1976		70
USA	1977	140	
West Germany	1970	40	
West Germany	1976	55	

Table 1.2. Wastewater per capita (exclusive of industry). For example based on /2/,/3/,/4/,/5/,/6/,/7/,/8/.

$Q_{yr,pers} = 60$ m³/(yr · pers)

N = 120,000 persons

$Q_{yr} = Q_{yr,pers} \cdot N = (60$ m³/(yr · pers)) · (120,000 pers)

$Q_{yr} = 7.2 \cdot 10^6$ m³/yr

$Q_{d,av} = Q_{yr}/365 = (7.2 \cdot 10^6$ (m³ · yr))/(365 d/yr) = 19,700 m³/d

The hourly factor, $t_{h,d}$, is estimated at 15 h/d corresponding to an average town (see figure 1.6).

$Q_{h,max} = Q_{d,av}/t_{h,d} = (19,700$ m³/d)/(15 h/d) = 1,315 m³/h

The volumes of wastewater

In respect of industrial and public institutions wastewater, the calculation can be made as shown in figure 1.7. Here the basis will once again be an annual production of water which, in respect of industrial wastewater, will often be calculated as the product of a wastewater production per produced unit and the number of produced units per year. (For example for breweries: 0.6 m³ wastewater/hl beer, with a

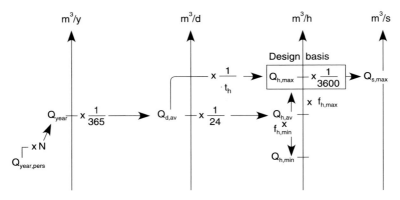

Figure 1.6. Calculation of volumes of domestic wastewater

$$\text{Households} \begin{cases} Q_{yr,pers} &= 50-100 \text{m}^3 / (\text{person} \cdot \text{yr}) (\text{see Table 1.2}) \\ t_{h,d} &= \begin{cases} 14-18 \text{ large towns} \\ 10-14 \text{ small towns} \end{cases} \\ f_{h,max} &= \begin{cases} 1.3-1.7 \text{ large towns} \\ 1.7-2.4 \text{ small towns} \end{cases} \\ f_{h,min} &= 0.2-0.4 \end{cases}$$

Q_{yr}	annual volume of wastewater, dry weather	m³/yr
$Q_{yr,pers}$	annual volume of wastewater per person	m³/yr
$Q_{d,av}$	av. volume of wastewater per day over the year, dry weather	m³/d
$Q_{h,av}$	av. volume of wastewater per hour over the year, dry weather	m³/h
$Q_{h,max}$	av. of maximum hours during the days of the year, dry weather	m³/h
$Q_{s,max}$	av. volume of water per second in an av. maximum hour, dry weather	m³/s
$Q_{h,min}$	av. of minimum hours during the days of the year, dry weather	m³/h
N	number of persons	-
$t_{h,d}$	hourly factor	h/d
$f_{h,max}$	max. hourly constant	-
$f_{h,min}$	min. hourly constant	-

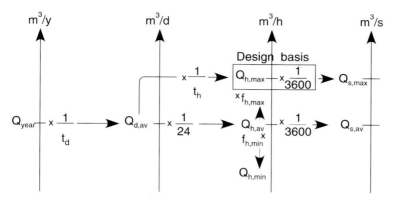

Figure 1.7. Calculation of wastewaters from industry and institutions.

Industry and institutions
$\begin{cases} Q_{yr} & = \text{most variable (see Tables 1.3 and 1.4)} \\ t_{h,yr} & = 100\text{-}365 \text{ (typical } 225\text{-}275) \\ t_{h,d} & = 4\text{-}24 \text{ (typical } 6\text{-}8) \\ f_{h,max} & = 1\text{-}6 \text{ (typical } 3\text{-}4) \\ f_{h,min} & = 0.1\text{-}0.2 \end{cases}$

Q_{yr}	annual volume of wastewater, dry weather	m³/yr
$Q_{d,av}$	av. volume of wastewater per day over the year, dry weather	m³/d
$Q_{h,av}$	av. volume of wastewater per hour over the year, dry weather	m³/d
$Q_{h,max}$	av. of maximum hours during the days of the year, dry weather	m³/h
$Q_{s,max}$	av. volume of wastewater per second in an av. maximum hour, dry weather	m³/s
$Q_{s,av}$	av. volume of wastewater per second in an av. hour over the year	m³/s
$Q_{h,min}$	av. of minimum hours during the days of the year, dry weather	m³/h
$t_{d,yr}$	diurnal factor	d/yr
$t_{h,d}$	hourly factor	h/d
$f_{h,max}$	max. hourly constant	-
$f_{h,min}$	min. hourly constant	-

production of 10^6 hl/yr, yields a volume of wastewater of $0.6 \cdot 10^6$ m³/yr). Table 1.3 gives an impression of the wastewater production from different industries whereas Table 1.4 gives an idea of the volumes of wastewater from institutions etc.

Wastewater components

Industry/production	Water consumption	Specific wastewater production	Specific pollution volumes	Contration in the effluent	Remarks
Dairies					
Milk for consumption	0.7-2.0 m³/t	0.7-1.7 m³/t	0.4-1.8 kg BOD₇/t	500-1500 g BOD₇/m³	t = tonne weighed in milk
Cheese	0.7-3.0 m³/t	0.7-2.0 m³/t	0.7-2.0 kg BOD₇/t	1000-2000 g BOD₇/m³	Caution: pH-variations discharge/emission of
Mixed production	0.7-2.5 m³/t	0.7-2.0 m³/t	0.7-2.0 kg BOD₇/t	1000-2000 g BOD₇/m³	
Slaughterhouses					
Slaughtering		3-8 m³/tp	7-16 kg BOD₇/tp	500-2000 g BOD₇/m³	tp = tonne product
				10-20 g Tot-P/m³	Caution: Strong smell, stiff hair, disinfectants
Slaughtering + meat specialities		3-12 m³/tp	10-25 kg BOD₇/tp	500-2000 g BOD₇/m³	Large variations in the water consumption
Meat specialities		1-15 m³/tp	6-15 kg BOD₇/tp	500-1000 g BOD₇/m³	depending on type of production
Breweries					
Beer and soft drinks	3-7 m³/m³*	3-7 m³/m³*	4-15 kg BOD₇/m³	1000-3000 g BOD₇/m³	m³* = product
					Caution: High pH
Canneries					
Potatoes (dry peel)	2-4 m³/t		3-6 kg BOD₇/t	1000-2000 g BOD₇/m³	t = ton raw material
Potatoes (wet peel)	4-8 m³/t		5-15 kg BOD₇/t	2000-3000 g BOD₇/m³	Caution: Flotables
Beetroots	5-10 m³/t		20-40 kg BOD₇/t	3000-5000 g BOD₇/m³	
Carrots	5-10 m³/t		5-15 kg BOD₇/t	800-1500 g BOD₇/m³	
Peas	15-30 m³/t		15-30 kg BOD₇/t	1000-2000 g BOD₇/m³	
Mixed production (vegetables)	20-30 m³/tf				tf = ton finished product
Fish	8-15 m³/t	4-8 m³/t	10-50 kg BOD₇/t	5000-10000 g BOD₇/m³	t = ton raw material
Textile industries					
The whole industry	100-250 m³/t	100-250 m³/t	50-100 kg BOD₇/t	100-1000 g BOD₇/m³	t = ton raw material
Cotton		100-250 m³/t	70-120 kg BOD₇/t	200-600 g BOD₇/m³	Caution: High water temp. extreme pH-value,
Wool		50-100 m³/t	15- 30 kg BOD₇/t	500-1500 g BOD₇/m³	chlorine gas, hydrogen sulfide gas, dangerous
Synthetic fibres		150-250 m³/t		100-300 g BOD₇/m³	chemicals (allergies)
Tanneries					
Mixed production	20-70 m³/t	20-70 m³/t	30-100 kg BOD₇/t	1000-2000 g BOD₇/m³	t = ton raw material
Hides	20-40 m³/t	20-40 m³/t	1-4 kg Cr/t	30-70 g Cr/m³	Caution: chromium, pH-variations, sludge and
Fur	60-80 m³/t	60-80 m³/t	0-100 kg S²/t	0-100 g S²/m³	hair
			10-20 kgTot-N/t	200-400 g Tot-N/m³	

Industry/production	Water consumption	Specific wastewater production	Specific pollution volumes	Contration in the effluent	Remarks
Laundries Wet washing	20-60 m³/t	20-60 m³/t	20-40 kg BOD$_7$/t 10-20 kgTot-P/t	300-800 g BOD$_7$/m³ 10- 50 g Tot-P/m³	t = tonne of washing Laundries using countercurrent wash have approx. 70% lower water consumption but the same emission of pollution (kg BOD$_7$/t) Caution: High temp.
Galvanic industries	20-200 l/m²	20-200 l/m² < 1 m³/h* max. 10 m³/h	3-30 g hm/m² 2-20 g CN/m²	*Before own treatment* approx. 150 g hm/m³ approx. 100 g CN/m³ *Before own treatment* 1-10 g hm/m³ 0.1-0.5 g CN/m³	m² = m² surface area hm = heavy metals *50% of all galvanoindustries have a flow < 1 m³/h. Caution: Solvents, cyanide, extreme pH-value, heavy metals, complex builders
Electrical circut industries	0.5-1.5 m³ / m²	0.5-1.5 m³ / m²	100-200 g Cu/m² 0- 5 g Sn/m² 0- 5 g Pb/m²	100-200 g Cu/m³ 0- 5 g Sn/m³ 0- 5 g Pb/m³	m² = m² laminate
Photolabs	0.5-1.5 m³ / m²	0.5-1.5 m³ / m²	200-400 g BOD$_7$/m²	400-700 g BOD$_7$/m³ 50-100 g EDTA/m³	m² = m² emulsion There are large variations in the pollution Caution: Damage to the skin by contact, allergic reactions
Printing houses	30-40 m³/d	30-40 m³/d	approx. 7 kg Zn/d approx. 0.04 kg Ag/d approx. 0.03 kg Cr/d approx. 0.01 kg Cd/d	170-230 g Zn/m³ 1.0-1.3 g Ag/m³ 0.8-1.0 g Cr/m³ 0.2-0.3 g Cd/m³	The expenses are based on an investigation made in the trade. The table shows an average printer with a water consumption of 30-40 m³/day Caution: Solvent, acids
Car repair/wash Cars Lorries	approx. 400 l/(Lt) approx. 200 l/(Ht) approx. 1200 l/(Ht)				Caution: Solvent Lt = Low-pressure washing Ht = High-pressure washing

Table 1.3. Industrial wastewater. Specific production and concentration. 1 kg BOD7 corresponds to approx. 0.85 kg BOD (BOD5). From /12/.

Type	Volume m³/yr	per	Ref. for example
Schools	8-10	pupil	/10/,/2/
Places of work	15-20	employee	/10/,/2/
Camping sites	25-30	pers. per day	/2/
Cottages	40-60	cottage	
Military installations	50-60	permanent resident	/10/
	15-20	employee	/10/
Hospitals	150-250	bed	/10/,/2/,/11/
Nursing homes, sanatoriums	100-150	bed	/10/
Hotels, boarding houses	60-100	bed	/10/,/2/
Restaurants etc.	100-150	employee	/10/
Swimming baths	50-60	visitor per day	/10/,/2/

Table 1.4. Volumes of wastewater from institutions etc.

Example 1.3.

Calculate the maximum hourly flow of water for a laundry with an annual production of 12,000 t (5-day working week, 14 hours' production per working day).

It is found from Table 1.3 that the production of wastewater in laundries is 20-60 m³/(tonnes produced). Here 50 m³/(tonnes produced) is estimated

$Q_{yr} = (12,000 \text{ t/yr}) \cdot (50 \text{ m}^3/\text{t}) = 600,000 \text{ m}^3/\text{yr}$

Number of working days per year, $t_{d,yr}$, is estimated at

45 weeks · 5 d/week = 225 d/yr

$Q_{d,av} = Q_{yr}/t_{d,yr} = (600,000 \text{ m}^3/\text{yr})/(225 \text{ d/yr}) = 2,670 \text{ m}^3/\text{d}$

The hourly factor, $t_{h,d}$, is estimated at 12 h/d (the same volume of wastewater is not produced throughout the 14 working hours)

$Q_{h,max} = Q_{d,av}/t_{h,d} = (2,670 \text{ m}^3/\text{yr})/(12 \text{ h/d}) = 220 \text{ m}^3/\text{h}$

Example 1.4

Calculate the maximum hourly flow of water from a camp site with an average of 700 guests during the season May 15 to October 1.

$Q_{yr,pers} = 25\text{-}30 \text{ m}^3/\text{yr}$ is found from Table 1.4, and estimated to

$Q_{yr,pers}$ at 30 m³/(yr · pers).

$Q_{yr} = N \cdot Q_{yr,pers} = (700 \text{ pers}) \cdot (30 \text{ m}^3/(\text{yr} \cdot \text{pers}))$

$Q_{yr} = 21,000 \text{ m}^3/\text{yr}$

The diurnal factor $t_{d,yr}$ equals the number of days where the site is open, that is 135 days

$Q_{d,av} = Q_{yr}/t_{d,yr} = (21,000 \text{ m}^3/\text{yr})/(135 \text{ d/yr}) = 156 \text{ m}^3/\text{d}$

The hourly factor, $t_{h,d}$, is high, perhaps 6 h/d

$Q_{h,max} = Q_{d,av}/t_{h,d} = (156 \text{ m}^3/\text{d})/(6 \text{ h/d}) = 26 \text{ m}^3/\text{h}$

Figure 1.8 shows how infiltration can be assessed. This quantity depends on the length of the sewer system and its condition in general and the groundwater levels in the catchment area. Infiltration is often estimated on the basis of area, measured

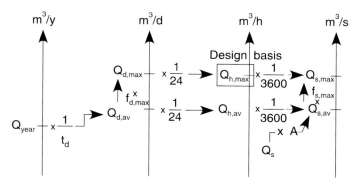

Figure 1.8. Calculation of infiltration.

Infiltration
$Q_{s,A}$ = 0.02 - 0.06 l/(s ha)
$t_{d,yr}$ = 200 - 365
$f_{d,max}$ = 2 - 3
$f_{s,max}$ = 0.1 - 0.2

Q_{yr}	annual infiltration m³/yr	
$Q_{d,av}$	av. infiltration per day in periods with infiltration	m³/d
$Q_{d,av,max}$	av. infiltration per day in maximum month of the year	m³/d
$Q_{h,av}$	av. infiltration per hour in periods with infiltration	m³/h
$Q_{h,max}$	av. infiltration per hour in maximum month of the year with infiltration	m³/h
$Q_{s,A}$	av. infiltration per second per ha in periods with infiltration	l/(s ×ha)
$Q_{s,av}$	av. volume of wastewater per second in periods with infiltration	l/s
$Q_{s,max}$	av. infiltration per sec. in maximum month of the year	m³/s
A	catchment area	ha
$t_{d,yr}$	diurnal factor	d/yr
$f_{d,max}$	diurnal constant in maximum month	-
$f_{s,max}$	seconds constant in maximum month	-

in l/(s· ha), but there are also formulae including the infiltration per length of sewer /13/. The most simple (and unreliable) procedure is to set the infiltration at a certain percentage (for example 50-100 per cent) of the volume of wastewater which has been estimated.

The infiltration may also be negative (= exfiltration). Especially in a warm and dry climate, exfiltration can be up to 50 per cent of the water discharged into the sewers. Exfiltration may be significant in places where the sewers are above the groundwater table and are badly maintained.

Example 1.5

Calculate maximum daily infiltration in a catchment area of 20 ha.
Figure 1.8 states $Q_{s,A} = 0.02 - 0.06$ l/(s · ha), and it is assumed that
$Q_{s,A} = 0.05$ l/(s · ha)
$Q_{d,av} = Q_{s,A} \cdot 3,600$ s/h · 24 t/d · A = 0.05 l/(s · ha) · 3,600 s/h · 24 t/d · 20 ha = 80,400 l/d
= 86.4 m³/d
The diurnal factor, $f_{d,max}$, is estimated at 2.5 on the basis of figure 1.8
$Q_{d,max} = Q_{d,av} \cdot f_{d,max} = (86.4$ m³/d$) \cdot 2.5 = 216$ m³/d
(The value for $Q_{s,max}$ can be directly calculated from $Q_{d,max}$ by division with (24 · 3,600), therefore
$Q_{s,max} = Q_{d,max}/(24 \cdot 3,600) = (216$ m³/d$)/(24 \cdot 3,600$ s/d$) = 0.0025$ m³/s
$Q_{s,max} = 2.5$ l/s

By computing the maximum hours for domestic and industrial wastewater and infiltration, a $Q_{h,max}$ is obtained which can be used in the future design:

$$Q_{h,max} = Q_{h,max}(\text{domestic}) + Q_{h,max}(\text{industry}) + Q_{h,max}(\text{infiltr.}) \qquad (1.1)$$

1.1.4. Population Equivalent and Person Load

Sometimes the volume of wastewater is expressed in the unit Population Equivalent (PE). PE can be expressed in water volume or BOD. The two definitions are used world wide and are the following:

1 PE = 0.2 m³/d
1 PE = 60 g BOD/d

These two definitions are based on the fixed figures given, which cannot be changed. The actual contribution from a person living in a sewer catchment, the Person Load (PL) can vary considerably, see Table 1.5. The reasons for the variation can be working place outside the catchment, ecological lifestyle, household installations like garbage grinders etc.

BOD g/(person·d)	15-80
COD g/(person·d)	25-200
Nitrogen g/(person·d)	2-15
Phosphorus g/(person·d)	1-3
Wastewater m³/(person·d)	0.05-0.40

Table 1.5 Variations in Person Load

PE and PL are often mixed or misunderstood, so be careful when using them. PE and PL are both based on average contributions, and used to give an impression of the loading of wastewater treatment processes. They should not be calculated from data based on short time intervals.

The Person Load varies from country to country, as seen in the yearly values given in Table 1.6.

		Country									
	Pollutant	Denmark	Brazil	Egypt	India	Italy	Sweden	Turkey	Uganda	USA	Germany
kg/(person · yr)	BOD	20-25	20-25	10-15	10-15	18-22	25-30	10-15	20-25	30-35	20-25
	SS	30-35	20-25	15-25		20-30	30-35	15-25	15-20	30-35	30-35
	N-total	5-7	3-5	3-5		3-5	4-6	3-5	3-5	5-7	4-6
	P-total	1.5-2	0.6-1	0.4-0.6		0.6-1	0.8-1.2	0.4-0.6	0.4-0.6	1.5-2	1.2-1.6
	Detergents	0.8-1.2	0.5-1	0.3-0.5		0.5-1	0.7-1.0	0.3-0.5		0.8-1.2	0.7-1.0
g/(person · yr)	Phenols	10-20		3-10		3-10		3-10			
	Hg	0.1-0.2		0.01-0.2		0.02-0.04	0.1-0.2	0.01-0.02			
	Pb	5-10		5-10		5-10	5-10	5-10			
	Cr	2-4		2-4		2-4	0.5-1.5	2-4			
	Zn	15-30		15-30		15-30	10-20	15-30			
	Cd	0.2-0.4					0.5-0.7				
	Ni	2-4					0.5-1.0				
COD = (2-2.5) · BOD			VSS = (0.7-0.8) · SS			NH_4-N = (0.6-0.7) · N-total					

Table 1.6. Person Load in general. For example based on /2/,/3/,/5/,/16/,/17/,/18/

Example 1.6

A catchment area is inhabited by 1000 people. The water consumption is 150 m³/d. The volume of wastewater reaching the treatment plant is 250 m³/d.

Find the infiltration and the Population Equivalent on the basis of the volume of wastewater.

The infiltration can be approximated by the difference between the volume of wastewater and water consumption:

Q (infiltration) = 250 – 150 = 100 m³/d

Population Equivalent is

PE = Q (wastewater)/0.2 = (250 m³/d)/(0.2 m³/(d · PE)) = 1,250 PE

1.1.5. Prognoses

When expanding and rebuilding treatment plants, it is necessary to estimate the volumes of wastewater for the next 10-20 years. In the '50s and '60s, the increases in the specific wastewater volumes were comparatively steady in many parts of the world, but since the beginning of the '70s the development has changed so that the specific volumes of wastewater are either stagnant (households) or actually decrease (industry). Water rates and charges on industry for wastewater service have significantly reduced the industrial consumption of water as well as pollution (in areas where payment is made in proportion to polluting load). Renovation of the sewer system also contributes to reducing the volume of wastewater.

Prognoses can be based on population growth and increased industrial production in the catchment area. The respective figures can be found in urban area development plans.

Example 1.7 shows the calculation, incl. prognoses, of the volumes of wastewater in Nykøbing Falster (Denmark) /14/.

Example 1.7

Area		Note no.	Catchment for North-plant			Catchment for South-plant		
			1980	1987	2005	1980	1987	2005
Annual dry-weather volumes								
Households	mill m³/yr	1)	0.69	1.00	1.24	0.22	0.29	0.35
Slaughterhouse	- " -	2)	0.25	0.37	0.37	-	-	-
Vegetables cannery	- " -	3)	0.05	0.06	0.07	-	-	-
Other industries	- " -	4)	0.55	0.79	1.02	0.11	0.25	0.39
Infiltration	- " -	5)	0.60	0.80	0.80	0.20	0.25	0.25
Total	mill m³/yr		2.14	3.02	3.50	0.53	0.79	0.99
Average dry-weather volume (weekdays)								
Households	m³/d	6)	1,890	2,740	3,400	600	790	960
Slaughterhouse	- " -	7)	1,000	1,480	1,480	-	-	-
Vegetables cannary	- " -	8)	250	300	350	-	-	-
Other industries	- " -	7)	2,200	3,160	4,080	440	1,000	1,560
Infiltration	- " -	6)	1,640	2,190	2,190	550	680	680
Total	m³/d	9)	6,980	9,870	11,500	1,590	2,470	3,200
Average max. dry-weather hour (weekdays)								
Households	m³/h	10)	160	230	285	50	65	80
Slaughterhouse	- " -	11)	165	245	245	-	-	-
Vegetables cannery	- " -	11)	40	50	60	-	-	-
Other industries	- " -	12)	275	395	510	55	125	195
Infiltration	- " -		70	90	90	25	30	30
Total	m³/h		710	1,010	1,190	130	220	305

1) Assumed water consumption: 1980: 55 m³/yr, 1987: 65 m³/yr and 2005: 80 m³/yr per capita.
2) Estimate made on the basis of water consumption during the years 1971-76. The slaughterhouse has planned an expansion of the production to max. 8,000 slaughterings per week (in 1979 approx. 4,800 pigs were slaughtered per week).
3) Estimate made on the basis of water consumption during the years 1971-76.
4) Total volume (based on water consumption and reduced by large-scale consumers) is divided according to industrial area into catchment for North- and South-plant, respectively. Newly developed industrial areas have for example been included by 5,000 m³/(ha · yr). The water consumption for existing industries is assumed to be unchanged in the planning period.
5) Infiltration is estimated at 0.15 m/yr (approx. 0.05l/(s·ha)).
6) Consumption period: 365 d/yr.
7) Consumption period: 250 d/yr.
8) Consumption period: 200 d/yr.
9) In the first part of 1979 an average of approx. 6,400 m³/d has been measured at the North-plant on weekdays with dry weather.
10) Hourly factor for the average day : 2.
11) Hourly factor for the average day: 4, based on measurements in 1979.

Calculation of wastewater volumes, Nykøbing F (Denmark). The slaughterhouse is FSA and the vegetables cannery Samodan /14/.

1.2. Wastewater components

Wastewater components can be divided into different main groups as shown in Table 1.5. In the following the composition of different types of wastewater is shown, based on domestic wastewater and municipal wastewater without any essential industrial influence.

Component	Of special interest	Environmental effect
Micro-organisms	Pathogenic bacteria, virus and worms eggs	Risik when bathing and eating shellfish
Biodegradable organic materials	Oxygen depletion in rivers, lakes and fjords	Changes in aquatic life (less diverse)
Other organic materials	Detergents, pesticides, fat, oil and grease, colouring, solvents, phenol, cyanide	Toxic effect, aesthetic inconveniences, bio accumulation
Nutrients	Nitrogen, phosphorus, ammonia	Eutrophication, oxygen depletion, toxic effect
Metals	Hg, Pb, Cd, Cr, Cu, Ni	Toxic effect, bio accumulation
Other inorganic materials	Acids, for example hydrogen sulphide, bases	Corrosion, toxic effect
Thermal effects	Hot water	Changing living conditions for flora and fauna
Odour (and taste)	Hydrogen sulphide	Aesthetic inconveniences, toxic effect
Radioactivity		Toxic effect, accumulation

Table 1.5. Components in wastewater, partly according to /15/

1.2.1. Domestic wastewater/municipal wastewater

The composition of domestic wastewater and municipal wastewater varies significantly both in terms of place and time. This is partly due to variations in the discharged amounts of substances. However, the main reason is variations in water consumption, infiltration and exfiltration. The composition of typical domestic wastewater/municipal wastewater is shown in Tables 1.7, 1.8, 1.9, 1.10, 1.11 and 1.12. Concentrated wastewater represent cases with low water consumption and/or infiltration. Dilute wastewater represents high water consumption and/or infiltration.

Wastewater components

Analysis parameters	Symbol	Unit[1]	Wastewater type			
			Concentrated	Moderate	Diluted	Very diluted
Biochemical oxygen demand, BOD						
– infinite	$C_{BOD}\lozenge$	g O_2/m³	530	380	230	150
– 7 days	C_{BOD7}	g O_2/m³	400	290	170	115
– 5 days	C_{BOD}	g O_2/m³	350	250	150	100
– dissolved	S_{BOD}	g O_2/m³	140	100	60	40
– dissolved, very easily degradable	S_{BOD}	g O_2/m³	70	50	30	20
– after 2h settling	$S_{BOD}(2h)$	g O_2/m³	250	175	110	70
Chemical oxygen demand with dichromate, COD						
– total	C_{COD}	g O_2/m³	740	530	320	210
– dissolved	S_{COD}	g O_2/m³	300	210	130	80
– suspended	X_{COD}	g O_2/m³	440	320	190	130
– after 2h settling	$C_{COD}(2h)$	g O_2/m³	530	370	230	150
– inert, total	C_I	g O_2/m³	180	130	80	50
– dissolved	S_I	g O_2/m³	30	20	15	10
– suspended	X_I	g O_2/m³	150	110	65	40
– degradable, total		g O_2/m³	560	400	240	160
– very easily degradable	$S_{HAC,COD}$	g O_2/m³	90	60	40	25
– easily degradable	$S_{S,COD}$	g O_2/m³	180	130	75	50
– slowly degradable	X_{COD}	g O_2/m³	290	210	125	85
– heterotrophic biomass	X_H	g O_2/m³	120	90	55	35
– denitrifying biomass	$X_{H,D}$	g O_2/m³	80	60	40	25
– autotrophic biomass	X_A	g O_2/m³	1	1	0,5	0,5
Chemical oxygen demand with permanganate, COD_P						
– total	C_{CODP}	g O_2/m³	210	150	90	60
Total organic carbon	C_{TOC}	g C/m³	250	180	110	70
– carbohydrate		g C/m³	40	25	15	10
– proteins		g C/m³	25	18	11	7
– fatty acids		g C/m³	65	45	25	18
– fats		g C/m³	25	18	11	7
Fats, oil and grease		g/m³	100	70	40	30
Phenol		g/m³	0.1	0.07	0.05	0.02
Phtalates, DEHP		g/m³	0.3	0.2	0.15	0.07
Phtalates, DOP		g/m³	0.6	0.4	0.3	0.15
Nonylphenoles, NPE		g/m³	0.08	0.05	0.03	0.01
Polycyclic aromatic hydrocarbons, PAH		mg/m³	2.5	1.5	0.5	0.2
Detergents, anion[2]		g LAS/m³	15	10	6	4

[1] g/m³ = mg/l = ppm
[2] LAS = Lauryl Alkyl Sulphonate

Table 1.7. Typical average contents of organic matter in domestic wastewater /17/, /25/ ,/26/.

Analysis parameters	Symbol	Unit [3]	Wastewater type							
			Concentrated		Moderate		Diluted		Very diluted	
Total nitrogen	C_{TN}	g N/m^3	80		50		30		20	
Ammonium nitrogen [1]	S_{NH4}	g N/m^3	50		30		18		12	
Nitrite nitrogen	S_{NO2}	g N/m^3	0.1		0.1		0.1		0.1	
Nitrate nitrogen	S_{NO3}	g N/m^3	0.5		0.5		0.5		0.5	
Organic nitrogen	$C_{org.N}$	g N/m^3	30		20		12		8	
Kjeldahl nitrogen [2]	C_{TKN}	g N/m^3	80		50		30		20	
Total phosphorous	C_{TP}	g P/m^3	14	(23) [4]	10	(16)	6	(10)	4	(6)
Orthophosphate	S_{PO4}	g P/m^3	10	(14)	7	(10)	4	(6)	3	(4)
Polyphosphate	S_{p-P}	g P/m^3	0	(5)	0	(3)	0	(2)	0	(1)
Organic phosphate	$C_{org.P}$	g P/m^3	4	(4)	3	(3)	2	(2)	1	(1)

[1] $NH_3 + NH_4^+$
[2] $org-N + NH_3 + NH_4^+$
[3] g/m^3 = mg/l = ppm
[4] figures in parenthesis for catchment areas where detergents with phosphate is used

Table 1.8. Typical content of nutrients in domestic wastewater. Based on /17/, /25/, /26/.

Analysis parameter	Symbol	Unit [1]	Wastewater type			
			Concentrated	Moderate	Diluted	Very diluted
Aluminium	C_{Al}	mg Al/m^3	1000	650	400	250
Arsenic	C_{As}	mg As/m^3	5	3	2	1
Cadmium	C_{Cd}	mg Cd/m^3	4	2	2	1
Chromium	C_{Cr}	mg Cr/m^3	40	25	15	10
Cobalt	C_{Co}	mg Co/m^3	2	1	1	0.5
Copper	C_{Cu}	mg Cu/m^3	100	70	40	30
Iron	C_{Fe}	mg Fe/m^3	1500	1000	600	400
Lead	C_{Pb}	mg Pb/m^3	80	65	30	25
Manganese	C_{Mn}	mg Mn/m^3	150	100	60	40
Mercury	C_{Hg}	mg Hg/m^3	3	2	1	1
Nickel	C_{Ni}	mg Ni/m^3	40	25	15	10
Silver	C_{Ag}	mg Ag/m^3	10	7	4	3
Zinc	C_{Zn}	mg Zn/m^3	300	200	130	80

[1] mg/m^3 = µg/l = ppb

Table 1.9. Typical content of metals in domestic wastewater, data for example from /17/

Analysis /matter	Symbol	Unit	Wastewater type			
			Concentrated	Moderate	Diluted	Very diluted
Suspended solids	X_{SS}	g SS/m^3	450	300	190	120
Suspended solids, volatile	X_{VSS}	g VSS/m^3	320	210	140	80
Precipitate after 2h	X_{SS}	ml/l	10	7	4	3
Precipitate after 2h, suspended solids	X_{SS}	g/m^3	320	210	140	80
Precipitate suspended solids, volatile	X_{VSS}	g/m^3	220	150	90	60
Suspended solids after 2h		g SS/m^3	130	90	50	40
Coliform bacteria		No./100ml	10^8	10^8	10^8	10^8
Absolute viscosity	μ_a	kg/(m·s)	0.001	0.001	0.001	0.001
Surface tension		dyn/cm^2	50	55	60	65
Conductivity		mS/m [1)]	120	100	80	70
pH		-	7-8	7-8	7-8	7-8
Alkalinity	TAL	eqv/m^3 [2)]	3-7	3-7	3-7	3-7
Sulphide [3)]		g S/m^3	0.100	0.100	0.100	0.100
Cyanide	C_{CN}	g/m^3	0.050	0.035	0.020	0.015
Chloride [4)]		g Cl/m^3	500	360	280	200
Boron		g B/m^3	1.0	0.7	0.4	0.3

[1)] mS/m = 10 µS/cm = 1 m mho/m
[2)] 1 eqv/m^3 = 1 m eqv/l = 50 mg CaCO$_3$/l
[3)] H$_2$S + HS$^-$ + S^{--}
[4)] with 100 gCl/m^3 in the water supply

Table 1.10. Different parameters in domestic wastewater. Data for example from /17/.

Table 1.11. gives an idea of the concentration of microorganisms in raw and biologically treated domestic wastewater.

	Raw sewage	Biol. treated
E.Coli	10^7	10^4
Cl.perfringens	10^4	$3 \cdot 10^2$
Fecal streptococcae	10^7	10^4
Salmonella	200	1
Campylobacter	$5 \cdot 10^4$	$5 \cdot 10^2$
Listeria	$5 \cdot 10^3$	50
Staphyllococus aureus	$5 \cdot 10^4$	$5 \cdot 10^2$
Coliphages	10^5	10^3
Giardia	10^3	20
Roundworms	10	0.1
Enterovirus	5000	500
Rotavirus	50	5
Susp. matter (mg/100 ml)	30	2

Table 1.11. Microorganisms in wastewater (No./100 ml)

The ratio between various substances in wastewater influences the selection and function of treatment processes. Table 1.12. shows typical ratios. A high COD/BOD ratio indicates organic matter difficult to degrade. A high COD/TN ratio favours denitrification and a high VSS/SS ration indicates a high fraction of organic matter in the suspended solids.

Ratio	Low	Typical	High
COD/BOD	1.5-2.0	2.0-2.5	2.5-3.5
COD/TN	6-8	8-12	12-16
COD/TP	20-35	35-45	45-60
BOD/TN	3-4	4-6	6-8
BOD/TP	10-15	15-20	20-30
COD/VSS	1.2-1.4	1.4-1.6	1.6-2.0
VSS/SS	0.4-0.6	0.6-0.8	0.8-0.9
COD/TOC	2-2.5	2.5-3	3-3.5

Table 1.12. Ratios in domestic wastewater

It is not only the wastewater that comes from the catchment that a treatment plant has to handle. Often a number of internal wastewater recycles and external deliveries has to be dealt with. These are typically

– digester supernatant
– supernatant from thickeners
– reject water from sludge dewatering
– filter wash water
– septic tank sludge
– landfill leachates

Septic tank wastes are often trucked to the treatment plant, while landfill leachates can be trucked or pumped to the treatment plant. These extra waste streams will in many cases contribute significantly to the total waste load that the plant has to deal with.

Table 1.13 shows the typical composition of leachate and septic sludge.

Compound	Leachate		Septic sludge		Unit
	High	Low	High	Low	
BOD total	12000	300	30000	2000	g/m^3
BOD soluble	11900	290	1000	100	g/m^3
COD total	16000	1200	90000	6000	g/m^3
COD soluble	15800	1150	2000	200	g/m^3
Total nitrogen	500	100	1500	200	gN/m^3
Ammonium nitrogen	475	95	150	50	gN/m^3
Total phosphorus	10	1	300	40	gP/m^3
Ortho-phosphate	10	1	20	5	gP/m^3
Suspended solids	500	20	100000	7000	g/m^3
Volatile suspended solids	300	15	60000	4000	g/m^3
Precipitate after 2h	~0	~0	900	100	ml/l
Chloride	2500	200	300	50	g/m^3
Sulphide	10	1	20	1	g/m^3
pH	7.2	6.5	8.5	6	–
Alkalinity	40	15	40	10	ekv/m^3
Lead	300	20	30	10	mg/m^3
Total iron	600	50	200	20	g/m^3
Cadmium	10	1	4	1	mg/m^3
Mercury	1	0,1	2	1	mg/m^3
Chromium	600	50	40	10	mg/m^3
Faecal Coliforms	200	5	10^8	10^6	No./100 ml

Table 1.13. Composition of leachate and septic sludge

Table 1.14 shows examples of the composition of thickener supernatant and digester supernatant.

Compound	Thickener supernatant		Digester supernatant		Unit
	High	Low	High	Low	
BOD total	1000	300	4000	300	g/m^3
BOD soluble	900	100	1000	100	g/m^3
COD total	2500	700	9000	700	g/m^3
COD soluble	1500	650	2000	200	g/m^3
Total nitrogen	300	50	800	120	gN/m^3
Ammonium-nitrogen	60	30	500	100	gN/m^3
Total phosphorus	25	5	300	15	gP/m^3
Ortho-phosphate	10	4	20	5	gP/m^3
Suspended solids	1000	100	10000	500	g SS/m^3
Volatile suspended solids	650	65	6000	250	g VSS/m^3
Precipitate after 2h	200	10	100	5	ml/l
pH	7.5	6.0	8	6	-
Hydrogen sulphide	5	0,2	20	2	g S/m^3
Alkalinity	7	2	40	3	eqv/m^3

Table 1.14 Examples of supernatant concentrations.

Compound	Reject water from sludge dewatering		Filter wash water		Unit
	High	Low	High	Low	
BOD total	1500	300	400	50	g/m^3
BOD soluble	1000	250	30	10	g/m^3
COD total	4000	800	1500	300	g/m^3
COD soluble	3000	600	200	40	g/m^3
Total nitrogen	500	100	100	25	gN/m^3
Ammonium-nitrogen	450	95	10	1	gN/m^3
Total phosphorus	20	5	50	5	gP/m^3
Ortho-phosphate	5	1	5	1	gP/m^3
Suspended solids	1000	100	1500	300	g/m^3
Volatile suspended solids	600	60	900	150	g/m^3
pH	7.5	6.0	8.0	6.5	-
Alkalinity	10		10	2	eqv/m^3
Total iron	600	50	50	5	g/m^3
Sulphide	20	0,2	0.1	0.01	g S/m^3

Table 1.15 Example of reject water and filter wash water composition.

The special external and internal loadings can contribute significantly to the overall loading of the treatment plant, as shown in Table 1.16

Loads	% of raw wastewater flow	% of raw wastewater BOD load
Leachate	0.1-5	1-40
Septic sludge	0.1-5	1-60
Thickening supernatant	1-2	5-10
Digester supernatant	0.5-2	5-15
Reject water from sludge dewatering	0.2-0.5	1-2
Filter wash water	5-10	10-20

Table 1.16 Internal and external water and wasteload contributions, typical values.

The development of alternative and ecological settlements, makes it interesting from where the various waste fractions in domestic wastewater have their origin, and the magnitude of these contributions. Table 1.17 gives an overview of this.

	Unit	Toilet		Kitchen	Bath & wash	Total
		total	urine			
Water	m^3/yr	19	11	18	18	55
BOD	kg/yr	9.1	1.8	11	1.8	21.9
COD	kg/yr	27.5	5.5	16	3.7	47.2
Nitrogen	kg/yr	4.4	4.0	0.3	0.4	5.1
Phosphorus	kg/yr	0.7	0.5	0.07	0.1	0.87
Potassium	kg/yr	1.3	0.9	0.15	0.15	1.6

Table 1.17. Sources for household wastewater components and their values for non-ecological living /27,28, 30/

Wastewater from houses does not need to have the composition that has been known for many years. The composition is a result of the installations, the habits of the inhabitants and the standard of living, and that composition can be changed.

Separation of toilet wastes from the rest of the wastewater will result in grey and black wastewater generation, the characteristics of which can be seen in Table 1.18.

Compound	Grey wastewater		Black wastewater		Unit
	High	Low	High	Low	
BOD total	400	100	600	300	$g\ O_2/m^3$
COD total	700	200	1500	900	$g\ O_2/m^3$
Total nitrogen	30	8	300	100	$g\ N/m^3$
Total phosphorus	7	2	40	20	$g\ P/m^3$
Potassium*	6	2	90	40	$g\ K/m^3$

* Exclusive of the content in the water supply

Table 1.18 Characteristics of grey and black wastewater. Low values can be due to high water consumption or dislocation of some of the inhabitants part of the day (or night). Low water consumption or high pollution load from kitchens can cause high values. Based on /27,28,29,30/.

1.2.2. Variations

Daily, weekly and monthly fluctuations in the loads of water and pollutants are important for the design, operation and control of the treatment plant. Figure 1.9 shows the diurnal variations of organic matter (measured as BOD_7) for the Lundtofte treatment plant. It is seen that there is a factor 10 between the lowest and the highest hourly influent.

An example of diurnal variation of nitrogen is shown in Figure 1.10. Often the daily influents of municipal wastewater show a clear weekly variation. It may be particularly important for the operation of a biological treatment plants. For example the COD/N-ratio will often be lower on Saturdays and Sundays than during the rest of the week. This may cause operational problems in connection with denitrification.

In Figure 1.11 a fractile diagram for the influent of BOD and COD to a treatment plant is shown.

The temperature of wastewater varies over the year. The result is that the temperature also varies in the tanks. In Figure 1.12 the temperature variation in two activated sludge plants in temperate climate is shown.

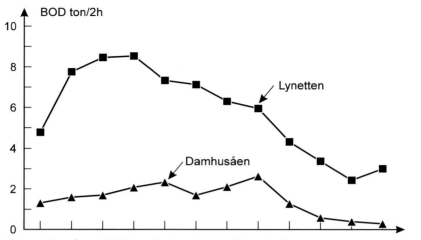

Figure 1.9. Influent BOD load to Lynetten and Damhuså wastewater treatment plants.

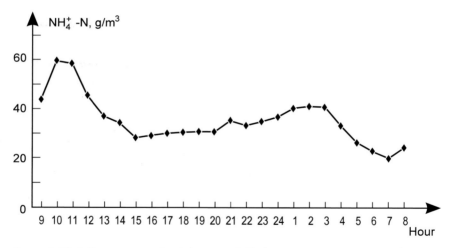

Figure 1.10. Influent ammonia-nitrogen to Galindo wastewater treatment plant, Spain.

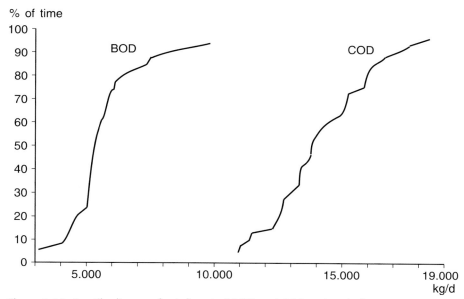

Figure 1.11. Fractile diagram for influent of BOD and COD to Lundtofte wastewater treatment plant (Denmark). On this basis the load to be considered for the design can be determined. Maximum design load can for example be determined corresponding to the 85% fractile. Data from /24/.

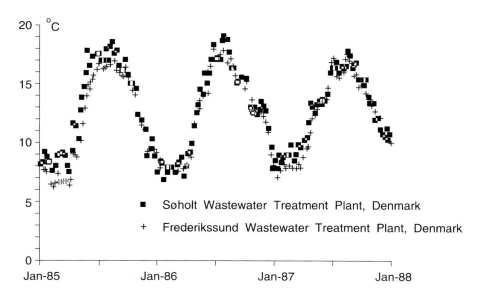

Figure 1.12. Temperature variation in activated sludge tanks, Søholt and Frederikssund wastewater treatment plants (Denmark) /23/.

References

/1/ Gernay, K. and Petersen,B.(2000): Personal communication

/2/ Triebel, W. (red.) (1982): Lehr- und Handbuch der Abwassertechnik. (Wastewater Technique: Textbook and Manual) Bd. II. 3rd edition. Verlag von Wilhelm Ernst & Sohn, Berlin.

/3/ Henze, M. (1977): Approaches and Methods in Estimation of the Polluting Load from Municipal Sources in the Mediterranean Area. Paper for the Meeting of Experts on Pollutants from Land-based Sources, Geneva 19-24 Sept. l977. United Nations Environment Programme (UNEP Project Med X). WHO Europe, Copenhagen, Denmark.

/4/ ASCE & WPCF (1977): Wastewater Treatment Plant Design. American Society of Civil Engineers and The Water Pollution Control Federation, New York, N.Y. (ASCE manuals and reports on engineering practice No. 36) (WPCF manual of practice No. 8).

/5/ Andersson, L. (1978): Föroreningar i avloppsvatten från hushåll. (Pollutions in Dometic Wastewater). Statens Naturvårdsverk, Stockholm, Sweden. (SNV PM 1103).

/6/ Østlandskonsult A/S (1978): Avløpsvannets mengde og sammensætning. (The Volume and Composition of Wastewater). PRA 1.1. Statens Forurensningstilsyn/Østlandskonsult, Oslo, Norway.

/7/ Barnes, D. et al. (1981): Water and Wastewater Engineering Systems. Pitman Books Ltd., London.

/8/ Buky, J.B. (1981): Appropriate Water Supply and Sanitation Technology Activities of the World Bank. Lecture at meeting on October 13 in Hygiejneteknisk Faggruppe, The Danish Society of Chemical, Civil, Electrical and Mechanical Engineers, concerning "United Nations Drinking Water and Sanitation Decade".

/9/ DIF (1981): Dansk Ingeniørforenings norm for vandforureningskontrol (Advice for Water Pollution Supervision). Teknisk Forlag, Copenhagen, Denmark. (Normstyrelsens publikationer NP-150-R).

/10/ SFT (1983): Retningslinjer for dimensionering av avløpsrenseanlegg (Guidelines for the Design of Wastewater Treatment Plants). Revised edition. Statens Forurensningstilsyn, Oslo, Norway. (TA-525).

/11/ Andersen, E. Bahl (1973): Om vandforbrug. I: Kursus i vandforsyningsteknik XXII (About Water Consumption. Course in water supply technique XXII), The University of Århus, pp 127-157. Danish Water Supply Association, Århus, Denmark.

/12/ VAV (1980): Kontroll av industriavlopp (Control of Industrial Effluent). Svenska Vatten- och Avloppsverksföreningen, Stockholm, Sweden (VAV Publikation P36).

/13/ Andersen, E. Bahl et al. (1976): Afløbsteknik (Sanitary Engineering). 2nd edition. Polyteknisk Forlag, Lyngby, Denmark.

/14/ Cowiconsult (1980): Planlægning af spildevandsbehandling i Nykøbing F (Planning of Wastewater Treatment in Nykøbing F). Cowiconsult Consulting Engineers and Planners AS, Virum, Denmark.

/15/ Forureningsrådet (1971): Vandrensning (Water Treatment). Forureningsrådets Sekretariat, Copenhagen, Denmark. (Publication No. 11).

/16/ EPA (1977): Process Design Manual. Wastewater Treatment Facilities for Sewered Small Communities. U.S. Environmental Protection Agency, Cincinnati, Ohio. (EPA 625/1-77-009).

/17/ Henze, M. (1982): Husspildevands sammensætning (The Composition of Domestic Wastewater). *Stads- og Havneing.*, **73**, 386-387.

/18/ Lønholdt, J. (1973): Råspildevands indhold af BI5, N og P (The Content in Raw Wastewater of BOD, N and P). *Stads- og Havneing.*, **64**, 138-144.

/19/ Cowiconsult (1979): Sjælsø rensningsanlæg. Planlægning af udbygning (Sjælsø Treatment Plant. Planning the Expansion). Cowiconsult Consulting Engineers and Planners AS, Virum, Denmark.

/20/ Cowiconsult (1979): Spildevandsplanlægning for Odense Kommune. Bearbejdning af måledata for årene 1975-1978 for rensningsanlæggene Ejby Mølle og Nordvestanlægget (Planning of Wastewater for the Municipality of Odense. Analysis of Measuring Data for the years 1975-1978 for the Treatment Plants Ejby Mølle and the North-West Plant). Cowiconsult Consulting Engineers and Planners AS, Virum, Denmark.

/21/ Jensen, K.H. and Jørgensen, M. (1974): Variationer i spildevands sammensætning. Individuelt kursus. (Variations in the Composition of Wastewater). Department of Environmental Engineering, The Technical University of Denmark, Lyngby, Denmark.

/22/ Villadsen, I. (1982): Statistiske metoder til dimensionering og drift af rensningsanlæg. Eksamensprojekt (Statistical Methods for the Design and Operation of Treatment Plants. Master thesis). The Department of Environmental Engineering and the Institute of Mathematical Statistics og Operations, The Technical University of Denmark, Lyngby, Denmark.

References

/23/ VKI (1988): Temperaturforhold i danske renseanlæg - indledende undersøgelser (Temperatures in Danish Treatment Plants - Initial Examinations). Water Quality Institute, Hørsholm, Denmark.

/24/ Krüger (1989): Dispositionsforslag for N- og P-reduktion. Lyngby-Tårbæk Kommune. Renseanlæg Lundtofte. (Proposal for N and P Reduction. The Municipality of Lyngby-Tårbæk. Lundtofte Treatment Plant) I. Krüger AS, Copenhagen, Denmark.

/25/ Henze, M. (1992): Characterization of Wastewater for Modelling of Activated Sludge Processes. *Water Sci. Technol.*, **25**, (6), 1-15.

/26/ Ødegaard, H. (1992): Norwegian Experiences With Chemical Treatment of Raw Wastewater. *Water Sci. Technol.*, **25**, (12), 255-264.

/27/ Henze, M. (1997): Waste design for households with respect to water, organics and nutrients. *Water Sci. Technol.*, **35**, (9), 113-120.

/28/ Naturvårdsverket (1995): Vad innehåller avlopp från hushåll? (What is the content in household waste?) Swedish EPA, Stockholm. (Rapport nr.4425).

/29/ Almeida, M.C., Butler, D. and Friedler, E. (2000): At-source domestic wastewater quality. *Urban Water*, **1**, 49-55.

/30/ Eilersen A.M., Nielsen S.B., Gabriel S., Hoffmann B., Moshøj C.R., Henze M., Elle M. and Mikkelsen P.S. (1999): Assessing the sustainability of wastewater handling in non-sewered settlements. Department of Env. Science and Engineering, Technical University of Denmark. Submitted to *Ecological Engineering*.

Concentration of wastewater in the sewer system during high tide is seen in many low-lying towns. This may be the reason why fish are sometimes found the the sewer system, and hence it is a wrong conclusion that fish like wastewater. In this picture the street is used as stormwater tank during high tide. (Bangkok, Thailand).

2 Characterization of Wastewater and Biomass

by Mogens Henze

Different analytical methods of a mixed origin are used for the characterization of wastewater and biomass. Many of them have been specially developed for treatment plants and treatment processes. Below some of the analysis and characterization methods are described. Details on the subject can be found under the references /1/,/2/,/3/.

2.1. Suspended solids

The division of pollutants into dissolved and suspended solids is essential as many of the treatment processes are only effective against one of these.

The division between dissolved and suspended solids is not well defined. In most countries, filters of a pore size of 1 µm (GF/C), or possibly 0.45 µm, are used. In Denmark, a filter of a pore size of 1.6 µm (GF/A filter) is used. Of course the fine filters remove more suspended solids, than the coarse ones.

The solids passing a filter are defined as the dissolved solids, S, whereas the solids held by the filter are the suspended solids, X. See figure 2.1.

For the total amount of solids, C,

$$C = S + X \tag{2.1}$$

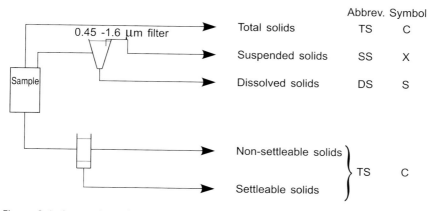

Figure 2.1. Separation of dissolved and suspended solids, from Danish Standard /4/.

Suspended solids

In connection with chemical precipitation, particles up to and including 0.1 µm can be precipitated, and hence the determination of suspended solids by filtration through an 0.1 µm filter may be a good idea.

The distribution of particles in wastewater is essential to separation and precipitation processes. Figure 2.2 shows how particles may be distributed. In spite of the varying concentration of the wastewater at the three dates of sampling, a certain pattern is seen in the appearance of the curve. The wastewater has its fingerprint.

Sometimes the solubility index, is used which is defined as

$$S/C \tag{2.2}$$

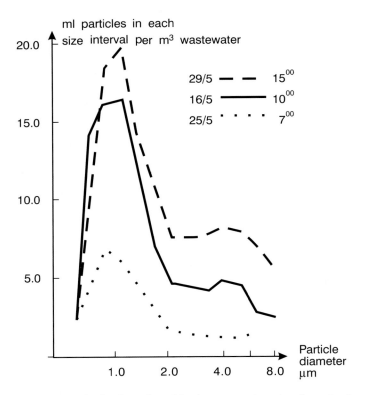

Figure 2.2. Distribution of particles in raw wastewater, from the Lundtofte treatment plant (Denmark)/19/.

2.1.1. Settleable solids

Settleable solids are normally ascertained after 2 hours' settling. It is the difference between the total content of solids in the untreated sample of wastewater and in the upper part of the water column after 2 hours' settling.

2.2. Organic matter

Wastewater normally contains thousands of different organics. A measurement of each individual organic compound would be impossible. Therefore different collective analyses are used which comprise a greater or minor part of the organics. The possibilities of collective analyses can be illustrated by an equation of reaction for the oxidation of organic matter, here symbolized by the average composition $C_{18}H_{19}O_9N$ /5/:

$$C_{18}H_{19}O_9N + 17.5\ O_2 + H^+ \rightarrow 18\ CO_2 + 8\ H_2O + NH_4^+ \qquad (2.3)$$

If organic matter is oxidized, the consumed amount of oxygen (the BOD-, COD-, TOD-analyses) or the produced amount of carbon dioxide (the TOC-analysis) can be measured.

The different measuring methods yield different results and therefore the individual methods cannot replace each other indiscriminately. Table 2.1 shows an example of analysis values for the different measuring methods for two different types of wastewater.

The choice of analysis parameters has been, and still is, the subject of discussion. Generally, chemical analyses may be said to be fast, but they do not always measure what is relevant. Biochemical analyses (C_{BOD}, $S_{S,COD}$, $X_{S,COD}$) are slow, but on the other hand they measure, in many cases, what is relevant in relation to the treatment process design and operation.

Analysis	Name	Typical conc. (g O_2/m^3)	
		Raw wastewater	Biol. treated wastewater without nitrification
C_{CODp}	Chemical oxygen demand (with potassium permanganate in alkaline solution)	180	30
C_{BOD}	5-day biochemical oxygen demand	280	25
C_{BOD7}	7-day biochemical oxygen demand	320	30
$C_{BOD\infty}$	Total biochemical oxygen demand	400	35
C_{COD}	Chemical oxygen demand (with potassium dichromate)	600	100
$S_{S,COD}$	Easily biodegradable matter, measured as COD	60	5
$X_{S,COD}$	Slowly biodegradable matter	200	10
C_{TOD}	Total oxygen demand (at 900 °C, platinum catalyst)	800	230
C_{TOD} (theoretical)	Theoretical total oxygen demand*	850	270
C_{TOC}	Total organic carbon (at 800° C)	200**	35**
C_{TOC}(theoretical)	Theoretical TOC*	200**	35**

* Calculated stoichiometrically, if the content of organic matter is assumed to be known
** Unit: g C/m^3

Table 2.1. Analyses for organics in wastewater and their interrelationship

Organic matter

Biochemical oxygen demand, BOD (BOD$_5$)

BOD$_5$ means 5-day biochemical oxygen demand. The BOD analysis was developed in Britain shortly before the turn of the century. The idea of the analysis was that in polluted water there is an oxygen demand caused by micro-organisms. The required oxygen demand was used to measure the extent of the pollution. As the oxygen demand increases by increasing temperatures and time of reaction - see figure 2.3 - it was originally decided to use the standard values 65° F (approx. 18°C) and 5 days. Now the temperature is standardized at 20°C.

The BOD-analysis is used to measure the oxygen demand of the micro-organisms for oxidation of organic matter and ammonium. In the course of 5 days, the major part of the biodegradable matter will be oxidized which corresponds to the horizontal level of the 20 degree curve in figure 2.3.

As the oxygen demand is normally only measured after 5 days, no information is available about the form of the curve as shown in figure 2.3. The form of the curve may vary considerably as shown in figure 2.4. The analysis does not reveal the amount of degradable organic matter and ammonium oxidized.

Different organics have different oxygen demands per gramme of matter and this means that the BOD-analysis only gives an approximate estimate of the weight of oxidized organic matter.

The measurement of BOD is carried out as shown in figure 2.5. In the classical dilution method polluted water is mixed with clean aerated water /11/. The mixture is poured into a bottle which is sealed so that there is no air above the liquid. The micro-organisms in the polluted water consume oxygen, and it is necessary to ensure that oxygen is present for the process all 5 days. If the oxygen is depleted, or if its concentration becomes low, the measuring result cannot be used. It is therefore necessary to make dilution series so that at least one bottle is obtained in which the oxygen content is still sufficiently high after 5 days.

Based on the oxygen demand in the 5 days and the volume of polluted water in the laboratory flask, the BOD can be calculated. The unit is g/m^3 (or mg/l).

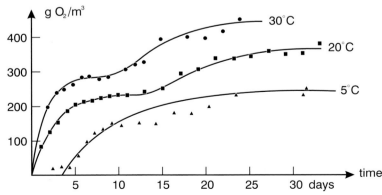

Figure 2.3. Typical BOD-curves. From /9/.

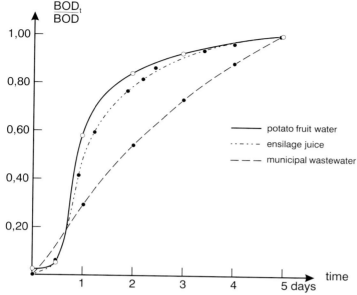

Figure 2.4. Example of different degradation rates for industrial wastewaters and ordinary municipal wastewaters. The curves are standardized as regards the 5-day oxygen demand (BOD). From /10/.

The BOD-value represents only a part of the biologically degradable matter which makes it difficult to use BOD for the calculations of mass balances, for example on treatment plants. A more complete analysis is obtained by extending the

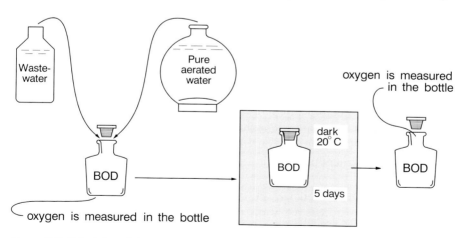

Figure 2.5. Principles of the BOD-analysis /12/. For ordinary municipal wastewaters, the dilution with clean water is approx. 100 times. Originally, the BOD analysis was not developed to be used for the design of treatment plants. Nevertheless, it is still extensively used for this purpose, even if fractionated COD analyses begin to gather momentum in more advanced designs, for example when using computer models.

duration of the analysis from the 5 days to somewhere between 15 and 20 days. For ordinary raw wastewater, the value (BOD∞) thus found will typically be 40-50 per cent higher than the BOD (5 days), that is, BOD/BOD∞ is approx. 0.6-0.7. However, the ratio BOD/BOD∞ may vary greatly and will be reduced by biological treatment.

Autoanalyzers exist which are less time consuming. These methods are based on the fact that a wastewater sample (often without dilution) is placed directly in a container with air above the liquid. The oxygen demand is either measured by means of the low pressure which is created when the oxygen in the sealed container is consumed, or by means of the amount of oxygen to be fed to the container to sustain the pressure. The automated analyses may yield results which deviate considerably from the dilution method. Poisonous substances may for example impede the oxidation process and result in a lower BOD-value in the undiluted analyses. This illustrates the problem of using the BOD-analysis: great uncertainty and variable results both when using the same methodology carried out by different laboratories/laboratory assistants and when using different methodologies.

Example 2.1

BOD (5 days) has been measured at 240 g O_2/m^3 in raw wastewater. Estimate the value for BOD∞.

Normally BOD/BOD∞ = 0.6-0.7

Here it is estimated that BOD/BOD∞ = 0.7

BOD∞ = BOD/0.7 = (240 g/m³)/0.7 = 340 g O_2/m^3

(It is true that the calculation gives BOD = 342.8571429 g O_2/m^3, but as an **estimate** has been asked for, 2 indicating numbers will suffice, that is, BOD = 340 g O_2/m^3).

Modified BOD

If the oxidation of organic matter and ammonium is to be separated, a modified BOD-analysis may be performed. A substance is added to the laboratory flask (for example Thiourea) which impedes the ammonium oxidation but does not influence the oxidation of organic matter. The analysis procedure itself is unchanged. The analytical result then only covers the oxidation of organic matter.

The modified BOD-analysis is sometimes used for measurement on the effluent from biological treatment plants.

Chemical oxygen demand (COD_p and COD)

Different chemical oxidants are used: potassium permanganate and potassium dichromate.

Potassium permanganate is used to determine the chemical oxygen demand (COD_p). It is a very incomplete oxidation of organic matter, and the analysis should therefore only be used to estimate the necessary dilutions in the BOD-analysis.

With potassium dichromate, a more thorough oxidation is obtained, including the oxidation of different inorganic materials (NO_2^-, S_2^-, $S_2O_3^{2-}$, Fe^{2+}, SO_3^{2-}). Ammonium and ammonia, which are released by the oxidation of organic nitrogen, are not oxidized. Certain nitrogen containing compounds like trimethylamine, typical in fish industry wastewater, and cyclic nitrogen compounds like pyridine, are not oxidized in the COD analysis. Overall, the COD analysis gives a fair estimate of the content of organic matter in municipal wastewater, may be in the range of 90-95 % of the theoretical oxygen demand by full oxidation of all organic matter present.

The COD-analysis may be carried out automatically and relatively fast (1-2 hours), and the COD-values are suitable for the calculations of the mass balances in treatment plants.

Both COD_P and COD are stated in the unit oxygen, for example g O_2/m^3. This means that the consumption of potassium permanganate or potassium dichromate is converted into an equivalent oxygen demand (the amount of oxygen which will be consumed if the oxidation had taken place by using oxygen).

COD-fractionation

The COD-value covers a number of organic materials of varying biological qualities. For the purpose of calculations in biological treatment plants, COD may, for example, be divided into the following fractions:

$$C_{COD} = S_S + S_I + X_S + X_I + X_B \tag{2.4}$$

where S_S is (dissolved) easily biodegradable organic matter,
S_I is dissolved biological inert organic matter,
X_S is (suspended) slowly biodegradable organic matter,
X_I is suspended biological inert organic matter.
X_B is biomass

For many purposes, this division is sufficient. The fractions S_S, X_S and X_B can be further divided in sub fractions if needed, as illustrated below. A major topic for discussions in relation to characterisation, has been the distinction between soluble and suspended substrate (S_S and X_S). In relation to the biological processes the distinction is not important, but for separation purposes it can be. The symbols used for the degradable substrates, S_S and X_S, just gives an indication of the main component of the fraction. It must be borne in mind that particles can be easily degradable (and thus belong to S_S) and that soluble organics can be slowly degradable (and thus belong to the fraction X_S). The biomass, X_B, present in various amounts in wastewater will influence the biomass composition in the treatment processes. This inoculation can be beneficial for the process or create problems. The last is the case if filamentous or foam creating organisms are being seeded into the process from the influent wastewater.

The development of wastewater and biomass component fractionation has been significant in the last 20 years/13,21,22/. The biological processes in question and the degree of accuracy, with which the problem has to be solved, determine the degree of details needed in a given case. A detailed fractionation for use in relation to biological phosphorus removal/21/ is given below.

Soluble COD fractions, $S_?$

S_A, Fermentation products, $[M(COD) L^{-3}]$. Compounds that can be used by PAOs under anaerobic conditions. In the modelling assumed to be acetate. Since fermentation is included in the biological processes, the fermentation products must be modelled separately from other soluble organic materials. They are end products of fermentation. For all stoichiometric computations, it is assumed that S_A is equal to acetate, which also dominates this fraction in practice, but in reality a whole range of other compounds can be present.

S_F, Fermentable readily biodegradable organic material, $[M(COD) L^{-3}]$. This fraction of the soluble COD is directly available for biodegradation by heterotrophic organisms. It is in modelling often assumed that S_F may serve as a substrate for fermentation, therefore it does not include fermentation products.

S_I, Inert soluble organic material $[M(COD) L^{-3}]$. The prime characteristic of S_I is that these organics cannot be further degraded in the traditional wastewater treatment plants. This material is assumed to be part of the influent and it is also assumed to be produced in the context of hydrolysis of particulate substrates X_S.

S_S, Readily biodegradable substrate $[M(COD) L^{-3}]$. This component was introduced in the Activated Sludge Model No.1, ASM1/13/. In ASM2d /21/, S_S is replaced by the sum of $S_F + S_A$.

Particulate COD fractions, $X_?$

X_S, Slowly biodegradable substrates $[M(COD) L^{-3}]$. Slowly biodegradable substrates are high molecular weight, colloidal and particulate organic substrates, which must undergo cell external hydrolysis before they are available for further degradation. In some models this compound is hydrolysed to S_S and then further degraded by direct metabolism. In other models X_S is hydrolysed to S_F, which can further be fermented.

X_B or X_H, Heterotrophic organisms $[M(COD) L^{-3}]$. These organisms are assumed to be the 'allrounder' heterotrophic organisms, they may grow aerobically and anoxically (denitrification) and may be active anaerobically (fermentation). They are responsible for hydrolysis of particulate substrates and can use all degradable organic substrates under all relevant environmental conditions. They are assumed not to produce methane under anaerobic conditions.

X_{PAO}, Phosphate-accumulating organisms: PAO $[M(COD) L^{-3}]$. These heterotrophic organisms are assumed to be representative for all types of polyphosphate-accumulating organism. All are assumed to grow aerobically and part of them anoxically, sometimes called $X_{PAO,DEN}$. PAOs are for modelling purposes assumed not to take part in fermentation under anaerobic or anoxic conditions. The concentration of X_{PAO} does not include the cell internal storage products X_{PP} and X_{PHA}, but only the 'true' biomass. X_{PAO} and X_B can in biological processes be converted to each other in a reversible manner. This means that under certain conditions (anaerobic zones) X_B can become X_{PAO} and vice versa.

$X_{B,A}$ or X_{AUT}, Nitrifying organisms [M(COD) L^{-3}]. Nitrifying organisms are responsible for nitrification; they are obligate aerobic, chemo-litho-autotrophic. It is in many models assumed that nitrifiers oxidize ammonium S_{NH4} directly to nitrate S_{NO3}, which means that they in that case include both ammonia oxidisers and nitrite oxidisers (often called Nitrosomonas and Nitrobacter).

X_I, Inert particulate organic material [M(COD) L^{-3}]. This material is not degraded within the systems of interest. It is flocculated onto the activated sludge. X_I is produced in the context of biomass decay and may be a fraction of the influent COD.

X_{PHA}, Cell internal storage products of phosphorus-accumulating organisms, PAO [M(COD) L^{-3}]. In most modelling it includes primarily poly-hydroxyalkanoates (PHA) and glycogen. It occurs mainly associated with X_{PAO}; it is, however, not included in the mass of X_{PAO}. X_{PHA} can occur in other organisms, like filamentous, but in these it does not have the reversible character it has in the PAOs. X_{PHA} cannot be directly compared with analytically measured PHA concentrations; X_{PHA} is a functional component required for modelling but not directly identifiable chemically. X_{PHA} may, however, be recovered in COD analysis, where it must satisfy COD conservation. For stoichiometric considerations, X_{PHA} is assumed to have the chemical composition of poly-hydroxy-butyrate $(C_4H_6O_2)_n$.

Figure 2.6 shows the COD fractions for biological phosphorus removal and indicates the characterisation tools available at present. Characterisation of wastewater and biomass is still under strong development, as standardisation methods have not yet been agreed upon.

Below it is briefly outlined how the individual COD-fractions can be determined.

- Inert dissolved organic matter, S_I, is determined as dissolved COD after 15-20 days' oxidation of the wastewater mixed with activated sludge. Parallel blind tests, solely with activated sludge, should be carried out.
- Easily degradable organic matter, S_S, is determined on the basis of a respiration test /18/. An example of such a test is shown in figure 2.7. Knowing the yield constant, Y_H, S_S can be determined by means of an oxygen mass balance

$$r_{V,O2} \cdot V = Q \cdot S_S \cdot (1 - Y_H)$$

$$S_S = r_{V,O2} \cdot \theta / (1 - Y_H) \qquad (2.5)$$

as $(1 - Y_H)$ is the part of the easily degradable organic matter which is oxidized whereas Y_H is the part which is transformed into biomass. θ is the hydraulic retention time. A BOD-curve may also be used for an estimate on S_S as the first "quick" part of the curve may be assumed to represent S_S.
- Fermentation products, S_A, are part of the easily degradable matter. A considerable constituent is acetic acid which may be determined specifically by gas chromatography, ion chromatography etc. The concentration of materials of a molar weight less than 800 - 1000 constitutes the major part of this fraction (however, there is a small concentration of dissolved inert matter with a low

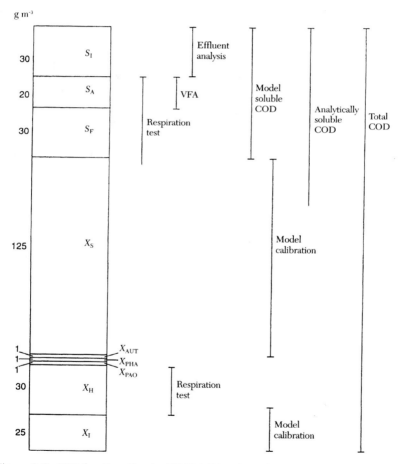

Figure 2.6. COD fractionation in ASM2d /21/. The column shows a typical distribution of COD in primary effluent from municipal wastewater treatment. Various analytical techniques can analyze part of the COD as indicated on the figure. X_I is obtained through modelling using sludge production. X_S is found by modelling using oxygen uptake rate/ nitrate uptake rate (OUR/NUR) test results.

Characterization of Wastewater and Biomass

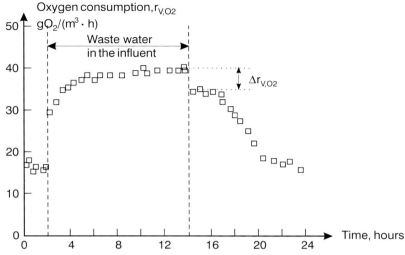

Figure 2.7. Analyses for the determination of easily degradable organic matter. Based on $\Delta r_{V,O2}$, S_S can be calculated using Expression (2.5). From /13/.

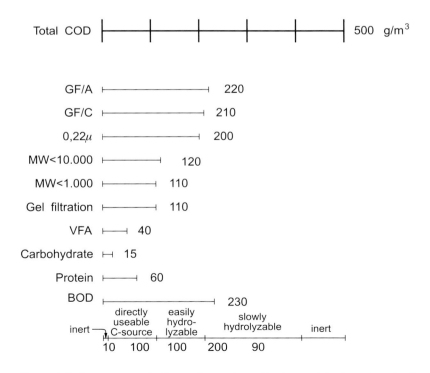

Figure 2.8 shows concentrations of organic matter in municipal wastewater from Lyngby, Denmark. Various methods have been used for the characterisation.

53

molecular weight). Fermentation products can be estimated through phosphate release tests with phosphorus accumulating biomass. The release of phosphate under anaerobic conditions is normally proportional with the amount of S_A present.
- Fermentable organic matter, S_F, is difficult to measure. An estimate can be made by the VFA potential test /24/ where the sample is kept anaerobic for a period of 2-6 days, and the amount of acetate generated is measured.
- Suspended inert organic matter is determined on the basis of the sludge production rate by tests with varying sludge ages. X_I can be estimated by means of curve adaptation with a mathematical model /17/. X_I may also be estimated on the basis of a long-term aerobic sludge stabilization.
- Suspended easily degradable organic matter is found as the difference between the total COD and the other fractions, that is,
$X_S = C_{COD} - X_I - S_A - S_S - S_I$
The concentration of bacteria, X_B, may be calculated on the basis of respiration tests /16/, where the respiration at 20°C per kg bacteria in the wastewater may, for example, be estimated at 200 g O_2/h. The fraction of denitrifying bacteria may be determined by a parallel respiration test with nitrate as the electron acceptor (respiration approx. 70 g NO_3^-/N/(kg bacteria · h)).

The concentration of bacteria may also be estimated by counting or ATP-measurements, dehydrogenase measurements, etc.

Total oxygen demand

By oxidizing at a high temperature and by using a suitable catalyst, a total chemical oxygen demand may be determined, called TOD. By using this analysis, the few organic compounds are oxidized which are not oxidized by the COD-analysis. Besides, ammonium is oxidized. TOD-values are therefore somewhat higher than COD-values, see Table 2.1

Total organic carbon

In this analysis, organic matter is oxidized to carbon dioxide by heating. The difference between the carbon dioxide concentration before and after oxidation is used for the calculation of the TOC.

TOC is not unambiguously tied to other parameters for organic matter as it states the amount of carbon atoms, but does not say anything about their state of oxidation, and hence nothing as to how much oxygen should be used for the oxidation.

In Tables 2.2 and 2.3, the TOD, BOD and BOD_{20}-values are given for a number of compounds.

Substance	Formula	Carbon %	TOD*	BOD	BOD$_{20}$
			g O$_2$/g substance		
Methane	CH_4	75	4	-	-
Ethane	C_2H_6	80	3.74	-	-
Hexane	C_6H_{14}	84	3.54	-	-
Ethylene	C_2H_4	86	3.43	-	-
Acethylene	C_2H_2	92	3.07	-	-
Trichloromethane	$CHCl_3$	10	0.36	-	-
Tetrachloromethane	CCl_4	8	0.21	-	-
Ehtyl ether	$C_4H_{10}O$	65	2.59	-	-
Acetone	C_3H_6O	62	2.21	0.54	-
Formic acid	CH_2O_2	26	0.35	0.09	0.25
Acetic acid	$C_2H_4O_2$	40	1.07	0.70	0.90
Propionic acid	$C_3H_6O_2$	49	1.52	1.30	1.40
Butyric acid	$C_4H_8O_2$	55	1.82	1.15	1.45
Valeric acid	$C_5H_{10}O_2$	59	2.04	1.40	1.90
Palmitic acid	$C_{16}H_{32}O_2$	75	2.88	1.68	1.84
Stearic acid	$C_{18}H_{36}O_2$	76	2.93	1.13	1.59
Oxalic acid	$C_2H_2O_4$	27	0.18	0.10	0.12
Succinic acid	$C_4H_6O_4$	41	0.95	0.64	0.84
Maleic acid	$C_4H_4O_4$	41	0.83	-	-
Lactic acid	$C_3H_6O_3$	40	1.07	0.54	0.96
Tartaric acid	$C_4H_6O_6$	31	0.53	0.35	0.46
Citric acid	$C_6H_8O_7$	37	0.75	0.46	0.67
Glycine	$C_2H_5O_2N$	31	0.96**	0.55	-
Alanine	$C_3H_7O_2N$	40	1.35**	0.94	-
Valine	$C_5H_{11}O_2N$	51	1.84**	-	-
Glutanic acid	$C_5H_9O_4N$	41	1.14**	-	-
Tyrosine	$C_9H_{11}O_3N$	60	1.81**	-	-
Methanol	CH_4O	37	1.50	0.96	1.26
Ethanol	C_2H_6O	52	2.09	1.35	1.80
Isopropanol	C_3H_8O	60	2.40	1.42	-
Amyl alcohol	$C_5H_{12}O$	68	2.73	1.27	1.73

* Theoretically calculated
** Including oxidation of nitrogen to nitrate

Table 2.2. Content of carbon, TOD, BOD and BOD$_{20}$ in organic materials /14/.

Organic matter

Substance	Formula	Carbon %	TOD*	BOD	BOD$_{20}$
			\multicolumn{3}{c}{g O$_2$/g substance}		
Glycol	$C_2H_6O_2$	39	1.29	0.49	-
Glycerine	$C_3H_8O_3$	39	1.22	0.72	0.94
Mannitol	$C_6H_{14}O_6$	40	1.14	0.68	0.94
Glucose	$C_6H_{12}O_6$	40	1.07	0.64	0.95
Lactose	$C_{12}H_{22}O_{11}$	42	1.12	0.61	0.91
Dextrins	$C_6H_{10}O_5$	45	1.19	0.52	0.84
Starch	$C_6H_{10}O_5$	45	1.19	0.68	0.90
Benzene	C_6H_6	92	3.07	-	-
Toluene	C_7H_8	91	3.13	-	-
Naphthalene	$C_{10}H_8$	94	3.00	-	-
Phenol	C_6H_6O	77	2.39	1.70	2.00
o-cresol	C_7H_8O	78	2.52	1.60	1.80
α–Naphthol	$C_{10}H_8O$	83	2.56	0.93	1.60
Pyrocatechol	$C_6H_6O_2$	65	1.89	0.90	0.90
Benzoic acid	$C_7H_6O_2$	69	1.97	1.25	1.45
Salicylic acid	$C_7H_6O_3$	61	1.62	0.95	1.25
Benzyl alcohol	C_7H_8O	78	2.52	1.55	1.95
Aniline	C_6H_7N	77	2.66**	1.49	-
Pyridine	C_5H_5N	76	2.53**	1.15	-
Chinolin	C_9H_7N	84	2.66**	1.71	-

* Theoretically calculated
** Including oxidation of nitrogen to nitrate

Table 2.3. Content of carbon, TOD, BOD and BOD$_{20}$ in organic materials /14/.

Example 2.2

Two wastewater samples show a TOC-value of 12 g/m^3. One sample contains pure methane and the other pure acetic acid.

Calculate the COD-values of the two samples (if everything is oxidized by the COD-analysis)

1. Methane CH$_4$.

1 mole of CH$_4$ contains 1 mole of carbon = 12 g, signifying that the wastewater contains 1 mole of CH$_4$.

Oxidation of methane:

$$CH_4 + 2\ O_2 \rightarrow CO_2 + 2\ H_2O \tag{2.6}$$

Oxygen demand = 2 O$_2$ = 2 · (16 · 2) = 64 g oxygen.

COD = 64 g/m^3.

2. Acetic acid CH_3COOH

½ mole contains 1 mole of carbon = 12 g, signifying that the wastewater contains ½ mole of acetic acid.

Oxidation of ½ mole of acetic acid:

$$\frac{1}{2} CH_3COOH + O_2 \rightarrow CO_2 + H_2O \tag{2.7}$$

Oxygen demand = O_2 = 16 · 2 = 32 g

COD = 32 g/m^3

It appears that in spite of the same TOC-value, the two water types have very different COD-contents.

2.3. Nitrogen

Similar to organic matter, nitrogen in wastewaters may be divided into a number of fractions.

$$C_{TN} = S_{NOX} + S_{NH4} + S_{I,N} + X_{S,N} + X_{I,N} \tag{2.8}$$

where C_{TN} is total nitrogen,
S_{NOX} is nitrite + nitrate nitrogen,
S_{NH4} is ammonium- + ammonia nitrogen,
$S_{I,N}$ is dissolved inert organic nitrogen,
$X_{S,N}$ is suspended easily degradable organic nitrogen,
$X_{I,N}$ is suspended inert (organic) nitrogen.

The nitrogen fractions are determined by conventional analysis techniques (the Kjeldahl nitrogen, nitrite, nitrate, ammonium). Often it may be assumed that the nitrogen content in the different organic fractions is rather constant from wastewater to wastewater. Nitrogen can thus be calculated like this:

$$S_{I,N} = f_{SI,N} \cdot S_{I,COD} \tag{2.9}$$

$$X_{I,N} = f_{XI,N} \cdot X_{I,COD} \tag{2.10}$$

$$X_{S,N} = f_{XS,N} \cdot X_{S,COD} \tag{2.11}$$

where $f_{SI,N}$, $f_{XI,N}$ and $f_{XS,N}$ typically are 0.04-0.08. The concentration of dissolved inert nitrogen, $S_{N,I}$, may vary considerably (1-4 g N/m^3) in ordinary municipal wastewater and may therefore sometimes cause problems in connection with the fulfillment of low effluent requirements for nitrogen.

Table 2.4 gives typical fractions of nitrogen and phosphorus in organic matter in municipal wastewater.

Symbol	Component	Typical range	
		N	P
S_S	Readily (fermentable) biodegradable substrate	2-4	1-1.5
S_A	Volatile acids (acetate)	0	0
S_I	Inert, non-biodegradeable organics	1-2	0.2-0.8
X_I	Inert, non-biodegradeable organics	0.5-1	0.5-1
X_S	Slowly biodegradable substrate	2-4	1-1.5
X_H	Heterotrophic biomass	5-7	1-2
X_{PAO}	Phosphorus-accumulating organisms	5-7	1-2
X_{PHA}	Stored poly-hydroxy-alkanoate	0	0
X_A	Autotrophic, nitrifying biomass	5-7	1-2

Table 2.4. Per cent of nitrogen and phosphorus in organic matter COD in municipal wastewater /20/

2.4. Phosphorus

Phosphorus in wastewaters may be divided into the following fractions:

$$C_{TP} = S_{PO4} + S_{p\text{-}P} + S_{org.P} + X_{org.P} \qquad (2.12)$$

where C_{TP} is total phosphorus,
        S_{PO4} is dissolved inorganic orthophosphate,
        $S_{p\text{-}P}$ dissolved inorganic polyphosphate,
        $S_{org.P}$ is dissolved organic phosphorus,
        $X_{org.P}$ is suspended organic phosphorus.

Normally a more detailed division of phosphorus will not be necessary. An analysis of these fractions is performed by using the well-known laboratory methods.

2.5. Alkalinity (TAL)

The alkalinity of the wastewater is important for its ability to resist acid/base influences. The alkalinity is measured by conventional titration with acid to an end pH of 4.5. The higher value, the greater buffering capacity.

Different processes for the treatment of wastewater change the alkalinity, this applies to nitrification, denitrification and chemical precipitation. Normally municipal wastewaters with an alkalinity over 5 eqv/m^3 will not cause problems in connection with nitrification, denitrification and simultaneous precipitation, whereas lower alkalinity may cause a drop in the pH and low efficiency of the processes. In the event of preprecipitation and post precipitation, low alkalinity will,

CHARACTERIZATION OF WASTEWATER AND BIOMASS

on the other hand, often be an advantage as smaller quantities of chemicals will then be required to obtain the desired pH-value.

2.6. Sludge volume index etc.

The sludge volume tells something about the flocculation and settling characteristics of the activated sludge. The sludge volume index is the reciprocal sludge concentration in the sludge phase after 0.5 hour's settling in a cylinder glass:

$$SVI = \frac{1}{X_{0.5}} \text{ (unit normally ml/g SS)}$$

where $X_{0.5}$ is the sludge concentration in the sludge phase after 0.5 hour's settling.

The lower the sludge volume index is, the better flocculation and settling characteristics. By stirring in the cylinder glass during settling, the stirred sludge volume index may be determined which is smaller than than the "conventional" one.

For high SVI (>200 ml/g SS) diluted SVI (DSVI) is measured, e.g. by a factor 2 dilution. The DSVI is calculated by

$$DSVI = f_D \cdot \frac{1}{X_{0.5}}, \text{ where } f_D \text{ is the dilution factor.}$$

SVI values of 100 ml/g are usually considered to be satisfactory to ensure a good functioning of activated sludge plants. In this way it will be possible to maintain a satisfactory return sludge concentration and have a well-functioning settling.

The sludge density index (SDI) is the sludge concentration in the sludge phase after 30 min. settling (= $X_{0.5}$).

The sludge percentage states the percentage which is taken up by the sludge phase volume (compared to the start volume) after 30 min. settling.

Example 2.3

A sample of activated sludge is taken from an aeration tank and is filled into a 1 l standard cylinder glass. The sludge concentration is 5 g SS/l (5 kg SS/m^3)

After 30 min. settling the sludge phase takes up 400 ml, the water above is clear. What is the sludge volume index, the sludge density index and the sludge percentage?

Sludge volume index: 5 g sludge takes up 400 ml.

SVI = 400 ml/5 g = 80 ml/g

Sludge density: 5 g sludge takes up 400 ml

SDI = 5 g/400 ml = 5 g/0.4 l = 12.5 g/l

Sludge percentage: (400 ml/1000 ml) · 100% = 40%

2.7. Respiration rate of sludge

The oxygen uptake rate for sludge, OUR (or nitrate utilization rate, NUR) may give essential information about the condition of the sludge. Figure 2.9 shows the result of a respiration experiment where raw wastewater is oxidized to a suitably high oxygen concentration (8-12 g/m^3). Subsequently the oxidation is stopped, while the wastewater is still being stirred. The slope of the curve in combination with the measurement of dry sludge solids (COD, SS or VSS) may be used to calculate the respiration of sludge in raw wastewaters. It is here approx. 50 g O$_2$/(kg VSS · h). A respiration rate for activated sludge of 20-40 g O$_2$/(kg VSS · h) signifies that the sludge is activated (many living micro-organisms), and that sufficient substrate (organic matter) is present. A low respiration rate (5-10 g O$_2$/(kg VSS · h)) may signify something else, for example that

- the sludge is poisoned,
- no easily degradable organic matter is present,
- the sludge has been stabilized (for example by aerobic sludge stabilization).

By parallel measurement of the nitrate respiration rate, the fraction of denitrifying bacteria, η_g, can be found as

$$\eta_g = \text{resp. with nitrate (m e-eqv/h)/resp. with oxygen (m e-eqv/h)} \qquad (2.13)$$

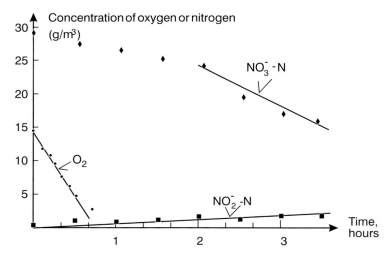

Figure 2.9. Respiration tests with raw wastewater. Both oxygen (OUR) and nitrate respiration (NUR) /15/.

Conversion from oxygen and nitrate, respectively, into electron eqvivalents takes place by means of the conversion factors:

32 g oxygen = 1 mole of O_2 = 4 e eqv

14 g nitrate-nitrogen = 1 mole of $NO_3^- - N$ = 5 e eqv

The respiration rate for ammonium (AUR) may be used to evaluate the amount of nitrifying bacteria (autotrophic bacteria) /16/.

Example 2.4

The respiration rate in an activated sludge sample has been measured at 32 g O_2/(kg VSS · h) and 7 g $NO_3^- - N$/(kg VSS · h), respectively. Find the fraction of denitrifying bacteria, η_g.

32 g O_2/(kg VSS · h) corresponds to (32 g O_2/32 g O_2) · 4 e eqv = 4 e eqv/(kg VSS · h). It is a good respiration rate which indicates that the sludge is active, and that organic matter is available as a substrate.

7 g $NO_3^- - N$/(kg VSS · h) corresponds to (7 g $NO_3^- - N$/(14 g $NO_3^- - N$) · 5 e eqv = 2.5 e eqv/(kg VSS · h).

7 g N/(g VSS · h) is a high denitrification rate. A good, easily degradable carbon source must be available.

η_g = 2.5/4 = 0.63

This is a normal fraction of denitrifying organisms.

References

/1/ APHE (1985): Standard Methods for the Examination of Water and Wastewater. 16th edition. American Public Health Association, Washington D.C.

/2/ VCH (1989): Deutsche Einheitsverfahren zur Wasser-, Abwasser- und Schlamm-Untersuchung Bd. I-III (German Standard Procedures for the Examination of Water, Wastewaters and Sludge, Volumes I-III). VCH Verlagsgesellschaft, Weinheim.

/3/ Danske standarder for analysemetoder for vand (Danish Standards for Methods of Analysis for Water). Danish Standards Association, Copenhagen.

/4/ DS 207 (1985): Vandundersøgelse. Suspenderet stof og gløderest (Total Non Filtrable Residue and Fixed Matter in Non Filtrable Residue). Danish Standards Association, Copenhagen.

/5/ Pöpel, F. (editor) (1975/88): Lehrbuch für Abwassertechnik und Gewässerschutz (Textbook for Wastewater Engineering and Protection of Aquatic Environment). Deutcher Fachschriften-Verlag, Wiesbaden.

/6/ Institut für Wasserversorgung, Abwasserreinigung und Gewässerschutz der Technischen Hochschule Wien (1972): Kläranlage Wien-Inzersdorf, Blumental. Betriebsbericht für den Zeitraum Oktober 1969-Oktober 1972 (Institute for Water Supply, Wastewater Treatment and Protection of Aquatic Environment, Vienna Technical University (1972): Treatment plant Wien-Inzersdorf, Blumental. Report on the operation for the period October 1969 to October 1972).

/7/ Lyngby-Tårbæk Kommune (1977): Analyser fra Lyngby-Tårbæk Kommunes rensningsanlæg i Lundtofte 1970-1977 (The Municipality of Lyngby-Tårbæk (1977): Analyses from the treatment plant in Lundtofte, owned by the Municipality of Lyngby-Tårbæk, 1970-1977), Lyngby, Denmark.

/8/ Metcalf & Eddy, Inc. (1991): Wastewater Engineering. Treatment, Disposal and Reuse. McGraw-Hill Book Company. New York, N.Y.

/9/ Gotaas, H.B. (1948): Effect of Temperature on Biochemical Oxidation of Sewage. **Sewage Works Journal, 20**, 441-477.

/10/ Frederiksen, J. (1963): Om nedbrydningshastigheden af sukkerholdigt spildevand (About the degradation rate for wastewater containing carbohydrate). **Hedeselskabets Tidsskrift, 84**, 136-143.

/11/ DS (1977): 5 døgns biokemisk oxygenforbrug (BOD) (Water analysis - Determination of biochemical oxygen demand (BOD)). Danish Standards Association, Copenhagen. (DS/R 254).

/12/ Henze, M. (1986): BOD (eller BI_5) analysen (The BOD (or BOD_5) analysis). **Vand & Miljø, 3**, 30-31).

/13/ Henze, M., Grady, C.P.L., Gujer, W., Marais, G. v. R. and T. Matsuo (1987): Activated Sludge Model No. 1. IAWPRC, London. (IAWPRC Scientific and Technical Reports No. 1).

/14/ Busse, H.J.(1975): Instrumentelle Bestimmung der organischen Stoffe in Wässern (Instrumental Determination of Organic Matter in Waters). **Z.f. Wasser- und Abwasser-Forschung, 8**, 164-176.

/15/ Henze, M. (1986): Nitrate Versus Oxygen Utilization Rates in Wastewater and Activated Sludge System. **Wat. Sci. Tech., 18**, (6), 115-122.

/16/ Kristensen, G.H., Jørgensen, P.E. and M. Henze (1992): Characterization of Functional Groups and Substrate in Activated Sludge and Wastewater by AUR, NUR and OUR. **Wat.Sci.Tech, 25**, (6), 43-57.

/17/ Gujer, W. and M. Henze (1991): Activated Sludge Modelling and Simulation. **Wat.Sci.Tech, 23**, Kyoto, 1011 - 1023.

/18/ Kappeler, J. and W. Gujer (1992): Estimation of Kinetic Parameters of Heterotrophic Biomass Under Aerobic Conditions and Characterization of Wastewater for Activated Sludge Modelling. **Wat.Sci.Tech, 25**, (6), 125-139.

/19/ Gillberg, L. Eger, L. and S.E. Jepsen (1990): Effect of Coagulants on Particle Distribution and Concentration. I: Chemical Water and Wastewater Treatment, Ed: H.H. Hahn and R. Klute. Springer Verlag, Heidelberg.

/20/ Henze, M., Gujer, W., Mino, T., Matsuo, T., Wentzel, M.C. and G. v. R. Marais (1995): Activated Sludge model no. 2. IAWQ Scientific and Tecnical Report no. 3, IAWQ, London.

/21/ Henze, M., Gujer, W., Mino, T., Matsuo, T., Wentzel, M.C., Marais, G.v.R. & van Loosdrecht, M.C.M. (1999): Activated sludge model No. 2d, ASM2d. **Wat.Sci.Tech., 39**, (1), 165-182.

/22/ Gujer W, Henze M, Takashi M, Van Loosdrecht MCM. (2000): Activated Sludge Model No. 3. In: Activated sludge models ASM1, ASM2, ASM2D and ASM3 (Henze, M., Gujer, W., Mino, T. and Loosdrecht, M.C.M., editors). Scientific and Technical Report No. 9, 99-121, IWA Publishing, London.

/23/ Henze, M., Kristensen, G.H. and Strube, R. (1992): Determination of organic matter and nitrogen in wastewater. Department of Environmental Engineering, Technical University of Denmark, Lyngby, Denmark

/24/ Lie, E. and Welander, T. (1997) Determination of volatile fatty acid potential of wastewater from different municipal treatment plants. **Water Research, 31,** 1269-1274

Biological wastewater treatment.

Top: Kirkeskoven wastewater treatment plant, Søllerød, Denmark, 1989. This plant from 1922 was the first activated sludge plant in continental Europe. The settling tank from 1922 was in use until 1992. How many tanks in new treatments plants will be in operation 70 years after their construction?

Bottom: Morigosaki wastewater treatment plant, Tokyo, Japan. The lower part shows covered primary settling tanks and behind to the left, covered activated sludge tanks are giving room for elevated soccer field and tennis courts.

3 Basic Biological Processes

By Mogens Henze

"There must be some kind of unsatisfactory scientific explanation".
(Freely translated) (Leif Panduro, Danish author)

In this chapter we will discuss the most essential biological processes which are used in biological wastewater treatment. The complexity of biological processes in wastewater treatment is so that the "scientific explanations" might seem unsatisfactory. In order to deal with the processes as engineers, all descriptions have been simplified. However, they are sufficiently detailed to describe the processes. In this chapter only conversion processes are discussed, and for the purpose of our discussion there rates are always considered to be positive. In the subsequent chapters, where mass balances are set up, the individual conversion terms are signed.

3.1. The biology in biological treatment plants

The biological processes in treatment plants are carried out by a very diversified group of organisms. It is only possible roughly to list which species are present as the fauna in a treatment plant is very dependent on the external conditions. This will be further discussed under biological selection in the next section.

3.1.1. The organisms

All organisms in a biological treatment plant must necessarily have there origin from the outside, that is, they come from the wastewaters, from the air, the soil, as additions or from animals which live close to the plant (seagulls, flies, rats, etc.). An essential part of the individual organisms have grown in the plant itself, and it is here the selection described in the next section is important.

An important hygienic aspect in connection with biological treatment plants is the presence of pathogenic organisms. These will not be further discussed in this book, but it is important to know that large amounts of pathogenic organisms will normally be present at all stages of biological treatment, with the exception of the effluent resulting from an efficient disinfection process. In connection with the danger of infection from the pathogenic organisms, the animals living in the vicinity of a treatment plant contribute to the dispersion of these organisms in the environment. It is thus possible to contract a wastewater-borne disease, even without being in close contact with the wastewater itself.

The organisms in a biological treatment plant can be divided into the following groups /1/:

a. bacteria
b. fungi
c. algae
d. protozoa
e. metazoa

The two main types of biological treatment plants: biofilters and activated sludge plants, offer different conditions of living which are reflected in the biology. The most varied animal life is found in biofilters. Due to the design and operation of these filters, they have very changing environments where different animals can flourish. In activated sludge plants the animals are evenly distributed and are not nearly as varied in terms of species.

Bacteria occur in large numbers both in biofilters and in activated sludge plants. There are relatively more bacteria in activated sludge plants. The main task of the bacteria is the primary transformation and degradation of dissolved organic matter. Furthermore they contribute to the degradation of suspended organic matter through the production of extracellular enzymes (exoenzymes). The content of bacteria in activated sludge is 10^{10}-10^{12} per litre /3/.

Fungi must compete for food with the bacteria, and this competition normally turns out to the advantage of the bacteria, and hence many more bacteria than fungi are found. If the pH in the plant is low, this will, however, favour the fungi. Relatively more fungi are present in biofilters than in activated sludge plants.

Algae are found in the surface of biofilters where the conditions are good (light and food) and in algae ponds and polishing ponds.

Protozoa are common in biofilters. In activated sludge plants they occur in varying numbers depending on the loading of the plant. A low-loaded plant will have many protozoa in the activated sludge. The protozoa graze on the bacteria, eat fungi, algae and suspended organic matter and fulfil an important role for the secondary settling of the wastewater.

Metazoa are higher animals. They have a distribution pattern corresponding to that of the protozoa, that is, they occur in biofilters and in low-loaded activated sludge plants. Many species may occur, such as Rotifers, Crustaceans, animals and insects.

Table 3.1 alphabetically lists the organisms which are frequently present during the biological treatment.

Achromobacter	a type of bacteria, typically occurring in both biofilters and activated sludge plants.
Acinetobacter	one of the types of bacteria, responsible for the biological phosphorous removal /6/.
Alcaligenes	a type of bacteria, typically occurring in both biofilters and activated sludge plants as well as in sludge digesters.
Bloodworm	red, 1-2 cm long, very movable larvae. Can be found in biofilters and in sludge from treatment plants. Is also known from very polluted streams /4/.
Chironomus	aquatic larvae of various species.
Crustacea	animals, occurring in very low-loaded activated sludge plants and ponds.
Daphnia	Crustacea, may occur in very low-loaded activated sludge plants and in ponds.
Desulfovibrio	bacteria capable of sulphate reduction, typically occurring in sludge digesters. By excessive dominance, the produced hydrogen sulphide may cause odours and toxic problems /2/.
Filter flies	Psychodidae, a two-winged insect, 2-5 mm long, may occur in large numbers in low-loaded biofilters. The flies are of little importance for the functioning of the filter, but they may be a nuisance to the environment.
Flavobacterium	a type of bacteria, typically occurring in biofilters and activated sludge plants as well as in sludge digesters.
GAO	Glycogen accumulating organisms, not phosphate accumulating
Geotrichum	a genus of fungi, occurring in both biofilters and activated sludge plants.
Gordonia	bacteria, producing foam in activated sludge plants (known earlier as Nocardia).
Micrococcus	a type of bacteria, occurring in biofilters and activated sludge plants.
Microtrix	a bacteria that causes bulking (high SVI) in activated sludge plants. Thriving on long chain fatty acids.
Nitrobacter	bacteria, which oxidize nitrite to nitrate. Occur in low-loaded biological treatment plants.
Nitrosomonas	bacteria, which oxidize ammonium to nitrite. Occur in low-loaded biological treatment plants.
PAO	Phosphate accumulating organisms
Pseudomonas	a type of bacteria, occurring in biofilters, activated sludge plants, sludge digesters and denitrifying plants.
Psychodidae	filter flies.
Rotifera	wheel animalcules, typical for low-loaded activated sludge plants. Indicator of an extremely good biological treatment.
Sphaerotilus natans	filamentous bacteria, which can influence settling negatively in activated sludge plants. This unpleasant phenomenon is called bulking. The bacteria can also be found in highly polluted streams /5/.
Tubifex	3-4 cm long red worm. Can be found in biofilters. Also known from highly polluted streams /4/.
Vorticella	are bell-shaped, attached Ciliatea, typically indicate a good biological treatment. Free-swimming ciliates prey on free-swimming bacteria and thereby clean the water. When the attached ciliates dominate, this is due to a shortage of free-swimming food (bacteria) and hence is the indicator of a good treatment /2/.
Zoogloea ramigera	bacteria, occurring in activated sludge plants as well as in biofilters. The bacterium easily flocculates, thanks to the production of a gelatinous slime, and thereby it is important for slime formation as well as for the flocculation of the activated sludge.

Table 3.1. Organisms in biological treatment plants involved in the processes.

Experiences with infection at treatment plants show that it is the direct contact which is dangerous as bacteria, viruses, and parasites in particular, can penetrate sores in the skin or be assimilated in the body if we actually drink the wastewater or inhale aerosols /37/. In earlier days it was popular for mayors to drink the treated wastewater at the inauguration of a treatment plant. These people do not exist any longer. Even treated wastewater contains large amounts of pathogenic micro-organisms.

Diseases caused by airborne infection from treatment plants are rare.

Figure 3.1 shows an example of the biological diversity in activated sludge plants.

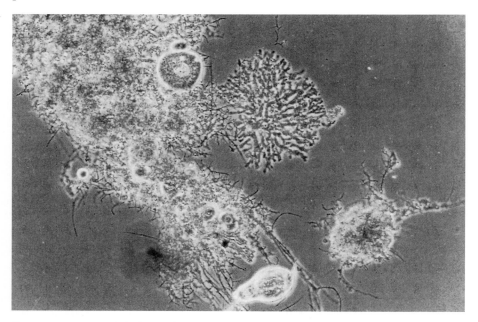

Figure 3.1. Activated sludge with zoogloea (large clusters), Gordonia filaments and protozoa. (Photo: Consorcio des Aguas, Bilbao)

3.1.2. Selection

Selection is a mechanism which favours the growth of a particular group of species of micro-organisms to the exclusion of others. The basis for a selection is a sufficient diversity of species. There must be some species to choose among, and the more there are, the greater the chance is that there is a type of organisms which can flourish in the respective environment.

Apart from selection, an adaptation of specific micro-organisms will also take place.

Biological wastewater treatment is largely based on selection, and it is so fortunate that the diversity of species in raw wastewater ensures a wide selection among very different micro-organisms. This large diversity of incoming species means that without giving much consideration to the environment which the micro-organisms have to put up with, we can still be rather sure that the wastewater is biologically treated. The treatment quality will depend on the environments which are created, but it does not change the fundamental fact that a certain biological treatment is obtained no matter - just about - how silly one behaves.

The two main types of biological treatment are rather different in respect of selection mechanisms. The toughest selection takes place in activated sludge plants, and this is probably the reason why a pure culture was added to the water in the first tests with this treatment method, perhaps on the grounds that it would then be the only type of bacteria which could survive. At that time it was not fully realized that a very varied microbial population is present in raw wastewater.

Selection in biofilters

The selection of organisms which takes place in biofilters can be divided into the following parts as the selection is based on adhesion and growth rate (substrate, temperature, pH, oxygen, etc.)

Selection by growth rate is not constant as both substrate and temperature normally are variables. The substrate may vary significantly if the municipal wastewater contains industrial wastewater from industries which do not operate round the clock or throughout the week, while the temperature will vary as a function of the season. Of the variables, the temperature in particular will be important for the selection to the effect that throughout the year there will be a temperature-dependent variation of species.

Selection in activated sludge plants

The selection taking place in activated sludge plants can be divided into the factors a-f discussed below.

a. Electron acceptor	As the principle in an activated sludge plant is to aerate a mixture of sludge and wastewater, anaerobic bacteria would hardly be able to survive for a longer period of time. Anaerobic conditions may occur in the middle of sludge flocs, but as the flocs are frequently broken up and re-formed, an anaerobic bacterium may not expect to survive and grow. Specifically, facultative bacteria will be able to survive under the varying - however predominantly aerobic - conditions.
	In the denitrifying plant, nitrate will in certain periods be the electron acceptor. In that case bacteria which can utilize both oxygen and nitrate as electron acceptors will have favourable conditions.

b. Substrate	The organisms must be able to utilize the primary and/or secondary substrate. In respect of biological phosphorus removal, phosphate accumulating organisms, PAOs, are selected by their ability to take up small organic molecules (acetic acid, alcohols etc.) under anaerobic conditions. Other bacteria which are normally found in activated sludge do not have this ability /6/.
c. Settling or flocculation characteristics	Settling forms an integral part of an activated sludge process. If an organism is sufficiently large and dense, it will settle by itself and hence be retained in the plant. If the organism, however, is small and light, it must join with other similar organisms to be able to return to the reaction tank through a floc formation.
d. Temperature	If the temperature in the plant is below the temperature at which an organism can grow, the organism concerned will die.
e. Growth rate	In an activated sludge plant, the conditions are kept constant by drawing off surplus sludge. This is one of the mechanisms to remove organisms from the process. Other mechanisms are grazing (predation) and suspended solids in the effluent. If an organism is to survive in the treatment plant, it must be able to reproduce at a faster rate than the rate at which it is removed from the plant. Therefore the growth rate of the organism is important for its survival in the treatment plant. Organisms which degrade complicated, organic compounds, for example toxic waste, will often have a slow growth rate as a result of a heavy work load and small yield. The same applies to nitrifying bacteria in respect of their conversion of ammonium. In respect of the treatment efficiency for special substances in activated sludge plants, the minimum growth rate at which organisms can survive in the plant concerned is of vital importance.
f. Freely suspended life forms	Organisms which require a surface to grow on cannot survive in an activated sludge plant in the long run.

Figure 3.2 shows a chart of the selection mechanisms in activated sludge plants.

Basic Biological Processes

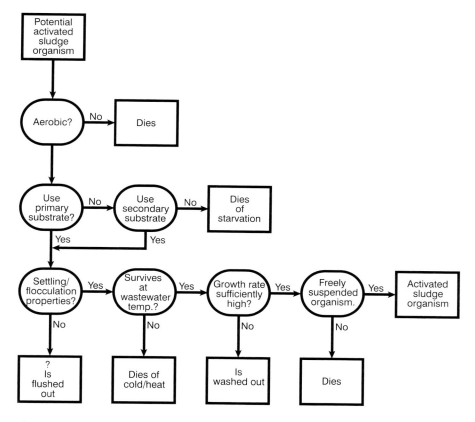

Figure 3.2. Selection in an activated-sludge plant.

Similar to the biofilters, the two selection mechanisms, substrate and temperature, are variables. Two other mechanisms depend to a certain extent on the mode of operation of the plant, that is, the oxygen conditions and the growth rate. The oxygen conditions vary with the varying content of organic matter in the wastewater, or through the regulation of aeration. The necessary minimum growth rate varies with the sludge age in the plant.

3.2. Conversions in biological treatment plants

A number of biologically dependent conversions are of vital importance in biological treatment plants. It concerns

- biological growth
- hydrolysis
- decay

To this, adsorption should be added, which (even if this does not concern an actual conversion) may also influence the process.

Figure 3.3 shows two model representations which combine the conversions concerned. What really happens in a biological treatment plant is far more complicated than what is shown in the figure. The figure represent a degree of detailing which is suitable in many process descriptions.

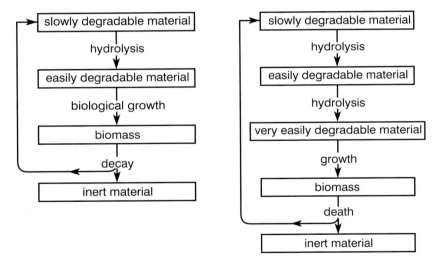

Figure 3.3. Bisubstrate model, biological conversion in treatment plants. Can be used in the process description of conventional activated sludge plants, nitrifying and denitrifying plants and anaerobic plants. The trisubstrate model is particularly applicable for the description of biological phosphorous removal.

The crucial problem in most biological conversions is that only a certain part of the material to be removed/treated in the process is immediately accessible for microbial removal.

Below, the three processes, biological growth, hydrolysis and decay, are discussed, and in chapters 4-9 they are used to describe the biological processes.

3.2.1. Biological growth

Bacteria in wastewater treatment processes are assumed only to be able to utilize very small and simply built molecules for growth. It may be substances such as acetic acid, ethanol, methanol, propionic acid, glucose, ammonium, nitrite etc.

The process can be described using the following expression:

$$r_{V,XB} = \mu_{max} \cdot f(S) \cdot X_B \qquad (3.1)$$

$r_{V,XB}$ is the volumetric biological growth rate
(dimension $M \cdot L^{-3} \cdot T^{-1}$, unit for example kg COD(B)/($m^3 \cdot$ d))

μ_{max} is the maximum specific growth rate
(dimension T^{-1}, unit for example h^{-1} or d^{-1})

$f(S)$ describes the growth kinetics
(for example zero, first order or Monod kinetics)

X_B is the concentration of biomass
(dimension $M_X \cdot L^{-3}$, unit for example kg COD(B)/m^3 or kg SS(B)/m^3)

The substrate consumption corresponding to the biological growth can be found from

$$r_{V,S} = (r_{V,XB})/Y_{max}$$

Y_{max} is the maximum yield constant
(dimension $M_{XB} \cdot M_S^{-1}$, unit for example kg COD(B)/COD(S) or kg VSS(B)/kg COD(S))

Monod kinetics are frequently used, and hence the substrate consumption for growth attains the following description:

$$r_{V,S} = \frac{\mu_{max}}{Y_{max}} \cdot \frac{S}{S + K_S} \cdot X_B \qquad (3.2)$$

It should be noted that the expression applies in a situation where only the substrate, S, is a limiting factor for the growth. Alternatively, μ_{max} can be seen as the maximum specific growth rate under given environmental conditions (temperature, pH, oxygen, nutrients, toxic substances). Variations in the environmental factors may be adjusted as shown in the section about aerobic conversions.

The biomass, X_B, is only a fraction of the organics in the sludge. Figure 3.4 shows the influence of the sludge load (see chapter 4) on the sludge fractions.

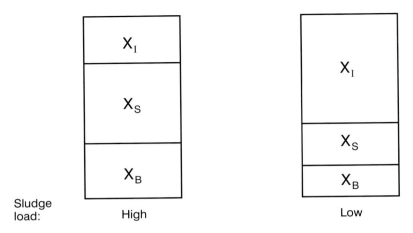

Figure 3.4. Biomass distribution versus sludge load.

3.2.2. Hydrolysis

The hydrolysis process converts larger molecules into small, directly degradable molecules. It may be a degradation of both particulate and dissolved solids. The hydrolysis processes are normally slow compared to the biological growth processes. Hence, in terms of rate of reaction, the hydrolysis will often be the rate limiting step in biological wastewater processes.

Not much is known about the hydrolysis processes /9/, but they are often described as a simple first order process in respect of hydrolyzable materials, for example suspended solids, X_S:

$$r_{V,XS} = k_h \cdot X_S \qquad (3.3a)$$

A similar expression for the hydrolysis of dissolved organic matter is:

$$r_{V,S} = k_h \cdot S_S \qquad (3.3b)$$

It should be noted that the hydrolysis constants, k_h, are not identical in Expressions (3.3a) and (3.3b).

Several models use more complicated expressions of the saturation type where a given biomass, X_B, has a maximum hydrolysis capacity:

$$r_{V,XS} = k_{hX} \cdot \frac{X_S / X_B}{K_X + (X_S / X_B)} \cdot X_B \qquad (3.4)$$

where k_{hX} is the hydrolysis constant (dimension $M_{XS} \cdot M_{XB}^{-1} \cdot T^{-1}$),
        K_X is the hydrolysis saturation constant (dimension $M_{XS} \cdot M_{XB}^{-1}$).

When the ratio between hydrolyzable matter, X_S, and biomass, X_B, corresponds to K_X, the hydrolysis rate is half of the maximum rate.

As regards the release of dissolved substrate:

$$r_{V,S} = r_{V,XS} \cdot v_{X,S}$$

where $v_{X,S}$ is the stoichiometric coefficient which converts suspended solids (X) into dissolved solids (S). If both substances are calculated in the same unit, $v = 1$, for example 1 kg COD(S)/kg COD(B).

The variation of the hydrolysis constant for heterotrophic bacteria under different conditions is illustrated in Table 3.2.

Electron acceptor	Hydrolysis constant, dissolved solids, k_h d^{-1}	Hydrolysis constant, suspended solids, k_h d^{-1}	Hydrolysis constant, k_{hX} kg COD(X)/ (kg COD(B) · d)	Hydrolysis saturation constant, K_X kg COD(X)/ kg COD(B)
Oxygen	3-20	0.6 -1.4	0.6 -1.4	0.02-0.05
Nitrate	1-15	0.15-0.4	0.15-0.4	0.02-0.05
Without oxygen and nitrate	2-20	0.3 -0.7	0.3 -0.7	0.02-0.05

Table 3.2. Hydrolysis constants for organic matter under varying electron acceptor conditions /9/,/11/.

3.2.3. Decay

Microorganisms can be subject to various phenomena that reduces their mass and number. The mechanisms is referred to as decay, lysis, endogenous respiration and maintenace. Often the term decay is used as a bulk description of the phenomena, which also includes the effect of predation or grazing/40/. In relation to biological wastewater treatment, decay is a degradation of the biomass coupled to oxidation of part of the organic matter in the biomass. It can be oxidation of internal structures in the single cell or degradation of the cell mass. Decay is an essential element in the description of the conversions of substances in biological treatment plants. The fact that the organism dies does not change the amounts of the substances, but it means that slowly degradable material is added to the system. This material is hydrolyzed and subsequently causes new growth and for instance oxygen consumption or nitrate consumption, see figure 3.3.

Normally decay is described as a first order process as regards the biomass, X_B

$$r_{V,XB} = b \cdot X_B \tag{3.5}$$

where b is the constant for decay (dimension T^{-1}, unit for example d^{-1}).

In connection with decay, some biologically non-degradable (or **very** slowly degradable) material is released. It is, at any rate, non-degradable (inert) at the retention times we find in biological treatment plants.

3.2.4 Storage

Microorganisms can under certain circumstances store organic and inorganic material internally, see Table 3.3. Internal storage products are stored as polymers inside the cell. Microorganisms can also change organic substrate into exopolymeric substances, EPS. Whether this is a storage product for the microorganism that produces it can be discussed, as the organisms will normally not metabolise it later. In biological wastewater treatment 3 types of storage compounds are found.

PHA, Poly-hydroxy-alkanoates are polymeric substances (lipids) of various fatty acids. These polymers are mainly composed of PHB and PHV, see Table 3.3. They are produced based on the content of fatty acids in the wastewater or as the result of fermentation processes. PHB and PHV can analytically be measured in biomass. For PAOs, Phosphorus Accumulating Organisms, their metabolism is also coupled with the storage of glycogen. PHB can for calculations be assumed to have the chemical composition $(C_4H_6O_2)_n$.

Stored PHA is quickly metabolised, often within 4-6 hours at 20 °C.

Storage compound	Abbreviation	Used by
Exo Polymeric substances	EPS	Heterotrophs, aerobic with carbohydrate load
Poly-β-hydroxybuturate	PHB	Heterotrophs under high acetate load PAOs Some filamentous organisms
Poly-β-hydroxyvalerate	PHV	Heterotrophs under acetate and propionate load PAOs Some filamentous organisms?
Polyhydroxyalkanoates	PHA	As for PHB
Glycogen	GLY	Heterotrophs under glucose load PAOs GAOs
Polyphosphate	PP	PAOs

Table 3.3 Storage products found in biomass in biological wastewater treatment plants.

Glycogen, GLY, a polymer carbohydrate is stored in heterotrophs exposed to carbohydrate rich (industrial) wastewaters/43/ or in PAOs in combination with storage of PHA. Glycogen and PHA are in counter-phase in metabolically active PAOs. As one is being built up, the other is degraded, see figure3.15. Glycogen storage has a long-term effect on the biomass as the storage in wastewater treatment plants can supply energy for 1-2 days.

Polyphosphate, PP, is found in PAOs, where it is used as energy storage, taken into use only under anaerobic conditions when acetate is present and the storage of PHB is possible. The polyphosphate storage can last for many days aerobically, where it is degraded only as a result of decay processes /40/.

In aerobic wastewater treatment, the storage can be modelled as shown in figure 3.4a.

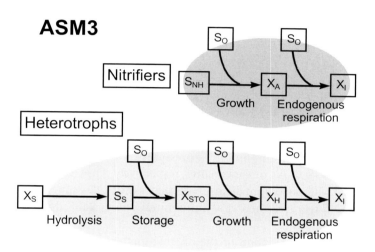

Figure 3.4a. Flow of COD for heterotrophs in ASM3/39/. The model is based on the assumption that all substrate passes storage before being metabolised in the heterotrophic microorganisms. This can be almost true under certain high load conditions, and completely wrong under low load conditions. But as a general model simplification it can be useful.

3.3. Aerobic heterotrophic conversion of organic matter

3.3.1. Reactions, aerobic conversions

Organic matter which is fed to an aerobic biological treatment plant may be exposed to:

a. oxidation to carbon dioxide and different nutrients (specifically in the form of N, P and S-compounds),
b. assimilation in biomass (sludge),

c. unchanged passage, which means that the matter is biologically nondegradable = inert (at least in the plant concerned),
d. conversion into other organic matter.

Normally for municipal wastewater it is the total amount of organic matter to be removed which is interesting. Only in special cases, for example in respect of toxic waste such as cyanide, phenols and chlorinated hydrocarbons, it is the specific substances which are interesting to observe. Many more details of biological conversions in wastewater can for example be found in /36/.

Organic matter in wastewater has the following approximate chemical composition: $C_{18}H_{19}O_9N$ /7/. If the matter is oxidized by micro-organisms to carbon-dioxide, it may result in the following two expressions:

$$C_{18}H_{19}O_9N + 17.5\ O_2 + H^+ \rightarrow 18\ CO_2 + 8\ H_2O + NH_4^+ \quad (3.6)$$
(without nitrification)

$$C_{18}H_{19}O_9N + 19.5\ O_2 \rightarrow 18\ CO_2 + 9\ H_2O + H^+ + NO_3^- \quad (3.7)$$
(with nitrification)

The microbiological oxygen consumption in Expressions (3.6) and (3.7) can be calculated at 1.42 and 1.59 kg O_2/kg organic matter, respectively. The chemical oxygen demand measured with potassium dichromate, that is the COD-value, will in both cases be 1.42 kg O_2/kg organic matter as ammonium is not oxidized in connection with the COD-analysis.

The energy resulting from oxidation of the organic matter can be found by combining the two nominal half-expressions /8/:

$$\frac{1}{70}C_{18}H_{19}O_9N + \frac{28}{70}H_2O \rightarrow \frac{17}{70}CO_2 + \frac{1}{70}HCO_3^- + \frac{1}{70}NH_4^+ + H^+ + e^-$$
$$\Delta G^\circ (W) = -32\ kJ/e\text{-eqv} \quad (3.8)$$

$$\frac{1}{2}H_2O \rightarrow \frac{1}{4}O_2 + H^+ + e^- \qquad \Delta G^\circ (W) = 78\ kJ/e\text{-eqv} \quad (3.9)$$

By combining Expressions (3.8) and (3.9), the total energy yield by aerobic oxidation of organic matter is found:

$$\frac{1}{70}C_{18}H_{19}O_9N + \frac{1}{4}O_2 \rightarrow \frac{17}{70}CO_2 + \frac{1}{70}HCO_3^- + \frac{1}{70}NH_4^+ + \frac{1}{10}H_2O$$
$$\Delta G^\circ (W) = -110\ kJ/e\text{-eqv} \quad (3.10)$$

The organic matter in ordinary domestic wastewater can also be divided into carbohydrates, fats and proteins. In terms of weight, these substances are present in almost equal amounts. In Table 3.4, formulas, oxygen consumption and the percentage contents of carbon and nitrogen in the three groups of substances are listed. It appears from the table that the oxygen consumption varies considerably from one group of substances to another.

Substance	Av. formula	Microbiological oxygen consumption, kg O_2/kg substance	Carbon %	Nitrogen %
Carbohydrate	$C_{10}H_{18}O_9$	1.13	43	0
Fats, oils	$C_8H_6O_2$	2.03	72	0
Protein	$C_{14}H_{12}O_7N_2$	1.20 (1.60)[*]	53	8.8
Av. organic	$C_{18}H_{19}O_9N$	1.42 (1.59)[*]	55	3.6

[*] with nitrification

Table 3.4. Organic substances in wastewater.

Example 3.1

The wastewater from an industry contains glutamic acid, $C_5H_9O_4N$. The concentration is 1.5 kg/m³, the volume of wastewater 300 m³/d.

What is the COD-concentration in the wastewater?

How much oxygen should be added per day to a biological treatment plant which is assumed to oxidize 70 per cent of the organic matter in the wastewater? The plant does not nitrify.

The COD-value is found by an equation of reaction

$$C_5H_9O_4N + 4.5\ O_2 + H^+ \rightarrow 5\ CO_2 + 3\ H_2O + NH_4^+$$

(as the organically bound nitrogen is not oxidized in connection with the COD-analysis). It is found from the equation that 1 mole of glutamic acid (molar weight $5 \cdot 12 + 9 \cdot 1 + 4 \cdot 16 + 1 \cdot 14 = 147$ g/mol) consumes 4.5 moles of oxygen (weight $4.5 \cdot 16 \cdot 2 = 144$ g/mol). Hence the oxygen consumption is 144 g O_2 /147 g glutamic acid = 0.98 g oxygen/g glutamic acid and subsequently the COD-concentration is: 1.5 kg glutamic acid/m³ · 0.98 kg oxygen/kg glutamic acid = 1.47 kg O_2 /m³).

The daily oxygen consumption from the wastewater is

1.47 kg O_2 /m³ · 300 m³/d · 0.70 = 309 kg O_2 /d.

Yield constant, aerobic heterotrophic conversions

The yield in a microbial heterotrophic process is defined as the mass of biomass growth per mass of substrate metabolised

$$Y = \Delta X/\Delta S \qquad (3.11)$$

Heterotrophic microorganisms have typically an overall energy efficiency, β, of around 55-60 per cent /8/. For aerobic growth of microbial biomass, the maximum amount of organic material that is converted to new biomass is around 50 per cent on a COD-basis, which gives a growth yield of 0.5 g COD/g COD(S) /45/. The difference between the 50% biomass yield and the 55-60% energy efficiency is contributed to storage phenomena.

In wastewater treatment systems the yield concept is normally expanded to be the overall sludge mass increase per mass of substrate removed from the wastewater.

Organic matter that is metabolised aerobically (or anoxically) can be used for 4 purposes:

– energy production by oxidation with oxygen (or nitrate) to carbon dioxide
– growth (new biomass)
– storage (various kinds of intracellular polymers)
– extracellular polymeric substances (EPS)

As oxidation is always part of all aerobic activity, it means that the substrate organic material can never be changed 100 per cent into biomass organic material after being metabolised. The carbon can of course be recovered 100 percent, including produced carbon dioxide. For a process that can store substrate, yields as high as 0.95 g COD/g COD(S) can be found. The other extreme is a situation where all supplied substrate is used for maintenance/endogenous respiration with a resulting yield of zero or even negative. For a given amount of substrate, the time available influences the resulting yield. The yield of 0.95 for the bacteria with storage will be reduced down towards a yield of zero, if enough time is considered /47/.

Table 3.5 illustrates how yield can vary depending upon the conditions under which substrate is metabolised.

Organism	Yield g COD/g COD
Bacteria with substrate for growth(1 d)	0.60
Bacteria with much substrate and extensive storage (1 h)	0.95
Bacteria with very little substrate	0.00
Fish(1 year, to 0.5 kg)	0.45
Hen	0.32
Pig (to 65 kg)	0.23
Cow(no milk)	0.18
Human (0-16 years)	0.010
Human (0-70 years)	0.002
Human (16-70 years)	0.000

Table 3.5. Yield for different biological systems, kg body weight produced per kg food consumed.

The maximum yield constant, Y_{max}, for the aerobic growth processes in wastewater (with minor storage included) is around 0.6-0.65 g COD/g COD(S). Values of this magnitude are often used in biological wastewater models /41/. In practise the observed yield, Y_{obs}, will often be lower (0.3-0.5 g COD/g COD) due to maintenance/endogenous respiration. But as mentioned above, an aerobic biological wastewater treatment process with minor or no external carbon source being added (low loading) will result in a negative observed yield. To obtain low yields is often an objective in wastewater treatment. Low loaded processes is one method to lower the yield, another is processes with a high FCF (Food Chain Factor). If the organic material passes more than one organism, more organic matter will be used for energy through oxidation, resulting in a lower overall yield. If the observed yield is 0.5 g COD/g COD for two biological processes that the substrate passes, then the overall yield will be 0.5·0.5 = 0.25 g COD/g COD. Activated sludge plants have a low FCF (around 1.2), while trickling filters have a FCF of maybe 2.0, due to the presence of high numbers of protozoa and metazoa. This results in lover yields for trickling filters than for activated sludge processes.

If the yield constant is known, an equation of reaction can be written for aerobic growth, the ratio between produced amount of biomass (composition $C_5H_7NO_2$) and removed amount of substrate (composition $C_{18}H_{19}O_9N$) being exactly the yield constant.

$$a\ C_{18}H_{19}O_9N + ... \rightarrow b\ C_5H_7NO_2 + ...$$

where
$$\frac{b \cdot MW_{biom}}{a \cdot MW_{org}} = Y_{obs} \qquad (3.12)$$

MW_{biom} and MW_{org} are the molar weights for biomass and organic matter, respectively, and Y_{obs} is the yield constant on a weight/weight basis.

If, for example, Y_{obs} = 0.5 kg biomass/kg of organic matter, Expression (3.12) is:

$$\frac{b \cdot MW_{biom}}{1 \cdot MW_{org}} = 0.5$$

(either a or b can be randomly chosen when evaluating the expression; here a has been set at 1).

The molar weights for biomass ($C_5H_7NO_2$) and organic matter ($C_{18}H_{19}O_9N$), respectively, are in this example MW_{biom} = 113 g/mole, MW_{org} = 393 g/mole.

b is found: $b = \dfrac{0.5 \cdot 1 \cdot 393}{113} = 1.74$

Hence the equation of reaction is:

$$C_{18}H_{19}O_9N + 0.74\ NH_3 + 8.8\ O_2 \rightarrow$$
$$1.74 \cdot C_5H_7NO_2 + 9.3\ CO_2 + 4.52\ H_2O \qquad (3.13)$$

(it is seen that some nitrogen has to be added in the form of ammonia (or ammonium) as there is not sufficient nitrogen in the organic matter for assimilation in the biomass).

Yield constants can be expressed in many different units. In Expression (3.13), Y_{obs} = 0.5 kg biomass/kg organic matter. If the COD unit is used, the biomass and the organic matter can be converted into COD. As the conversion factor with the assumed compositions of biomass and organic matter in both cases is 1.42 kg O_2/kg matter, Y_{obs} = 0.50 kg COD(X)/kg COD(S). Expressed in terms of mole, the yield constant in Expression (3.13), Y_{obs} = 1.74 moles of biom./mole organic matter.

Example 3.2

The observed yield constant for aerobic oxidation of acetic acid (HAc) is 0.55 kg COD/kg COD.

Evaluate the equation of reaction when the biomass is assumed to have the composition $C_5H_7NO_2$.

COD of acetic acid and biomass is calculated:

$$CH_3COOH + 2\ O_2 \rightarrow 2\ CO_2 + 2\ H_2O$$

$$C_5H_7NO_2 + 5\ O_2 + H^+ \rightarrow 5\ CO_2 + 2\ H_2O + NH_4^+$$

1 mole of acetic acid (= 12 + 3 · 1 + 12 + 16 + 16 + 1 = 60 g) corresponds to 2 moles of oxygen (= 2 · 16 · 2 = 64 g), that is, 64 g COD/60 g HAc = 1.07 g COD/g HAc.

1 mole of biomass (= 5 · 12 + 7 · 1 + 14 + 2 · 16 = 113 g) corresponds to 5 moles of oxygen (= 5 · 16 · 2 = 160 g), that is, 160 g COD/113 g biomass = 1.42 g COD/g biomass.

Expression (3.12) converted into COD-units is:

$$\frac{b \cdot MW_{biom} \cdot 1.42}{a \cdot MW_{org} \cdot 1.07} = Y_{obs} \qquad (3.12)$$

If a = 1 and other known values are assigned, b is found

$$\frac{b \cdot 113\ \text{g biom}/\text{mole} \cdot 1.42\ \text{COD}/\text{g biom}}{1 \cdot 60\ \text{g HAc}/\text{mole} \cdot 1.07\ \text{COD}/\text{g HAc}} = 0.55\ \text{g COD}/\text{g COD}$$

b = 0.22

Hence the equation of reaction is:

$$CH_3COOH + 0.22\ NH_4^+ + 0.22\ OH^- + 0.9\ O_2 \rightarrow$$
$$0.22\ C_5H_7NO_2 + 0.9\ CO_2 + 1.78\ H_2O$$

The expression demonstrates that nitrogen is needed for the growth so that it cannot take place solely on the basis of acetic acid. The same applies to other nutrients (P, S, K etc.).

From the expression it is seen that the yield constant on a molar basis is Y_{obs} = 0.22 mole of biomass/mole of HAc.

In biological treatment plants, the yield constant varies, both with the type of wastewater and the load of the plant. This is illustrated in figure 3.5 with an example from an activated sludge plant.

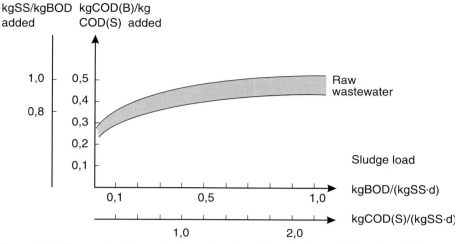

Figure 3.5. Observed yield constant as a function of the sludge load in activated sludge plants.

3.3.3. Nutrients, aerobic heterotrophic conversions

In order that the micro-organisms can grow, nutrients are required. If the chemical composition of the organisms is known, the demand for nutrients can be calculated by using a mass balance. Table 3.6 lists typical concentrations of different materials in micro-organisms taken from aerobic treatment processes. The concentrations can be changed considerably when special organic materials are removed (by means of more specific groups of micro-organisms), or in the case of specific processes, for example biological phosphorus removal.

In domestic and municipal wastewaters without any special industrial load, there are normally sufficient nutrients. There will often be a shortage of nitrogen or phosphorus in industrial wastewaters, see example 3.3.

		g/kg VSS	g/kg COD	g/kg TOC
Carbon	C	400-600	300-400	800-1000
Nitrogen	N	80-120	55-85	150-250
Phosphorus	P	10-25	7-18	25-55
Sulphur	S	5-15	4-11	12-30
Iron	Fe	5-15	4-11	12-30

Table 3.6. Typical concentrations of materials in heterotrophic micro-organisms, aerobic processes.

Example 3.3

Wastewater from breweries with 2.5 kg COD/m³, 15 g N/m³ and 20 g P/m³ must be treated in an aerobic process (activated sludge). The observed yield constant (kg COD (B)/kg COD (S)) is 0.45 kg COD/kg COD added and the contents of nitrogen and phosphorus in the sludge are 7% N/COD and 1.5% P/COD, respectively.

Should nitrogen and phosphorus be added to the treatment process and, if so, how much?

If all the COD is converted, the sludge production, F_{SP}, per m³ wastewater is as follows:

$F_{SP}/Q_1 = C_1 \cdot Y_{obs}$ = 2.5 kg COD/m³ · 0.45 kg COD/kg COD = 1.125 kg COD/m³

Hence the N and P consumptions will be:

$f_{X,N} \cdot F_{SP}/Q_1$ = 0.07 · 1.125 = 79 g N/m³

$f_{X,P} \cdot F_{SP}/Q_1$ = 0.015 · 1.125 = 17 g P/m³

Hence there is sufficient phosphorus in the wastewater whereas 79 − 15 = 64 g N/m³ must be added. If it is not added, the treatment process will be very slow, it will be incomplete and the result may be that the settling and flocculation characteristics of the sludge are bad.

3.3.4. Kinetics, aerobic heterotrophic conversions

Substrate conversions of aerobic micro-organisms can be described by a first order reaction in respect of the biomass

$$r_{V,S} = \frac{\mu_{obs}}{Y_{max}} \cdot X_B$$

where $r_{V,S}$ is the removal rate for substrate ($M \cdot L^{-3} \cdot T^{-1}$)
  X_B is the biomass concentration ($M \cdot L^{-3}$)
  μ_{obs} is the observed specific growth rate ($M \cdot M^{-1} \cdot T^{-1}$)
  Y_{max} is the maximum yield constant ($M_X \cdot M_S^{-1}$)

The growth rate is influenced by a number of environmental factors, such as substrate concentration, S, oxygen concentration, S_{O2}, pH, temperature, T, etc.:

$$\mu_{obs} = \mu_{max} \cdot f(S) \cdot f(S_{O2}) \cdot f(pH) \cdot f(T)$$

Normally, the growth kinetics are described as a Monod expression

$$\mu_{obs} = \mu_{max} \cdot \frac{S_2}{S_2 + K_S} \tag{3.14}$$

where μ_{obs} is the observed specific growth rate,
  μ_{max} is the maximum specific growth rate,
  S_2 is the substrate concentration in the **reactor**,
  K_S is the saturation constant for substrate.

The kinetics according to Expression (3.14) may, in certain cases (for example industrial wastewater with high substrate concentrations), be described by a zero order expression, that is, if $S_2 \gg K_S$:

$$\mu_{obs} = \mu_{max} \quad (3.15)$$

3.3.5. Heterotrophic micro-organisms, aerobic conversions

The aerobic microbiological conversions in treatment plants are carried out by a very large and mixed group of micro-organisms. There is a wide variation in the biomass composition from plant to plant. The reason is the differences in wastewater and differences in plant design and operation. In respect of aerobic conversion of organic matter present in municipal wastewater, the experience is that the process rates obtained are relatively homogenous from place to place in spite of the differences in the biomass.

3.3.6. The influence of the environmental factors, aerobic heterotrophic conversions

The most important factors for aerobic conversions are:

- temperature
- oxygen
- pH
- toxic substances
- nitrogen
- phosphorus

Temperature
The temperature dependency for the biological process can be described by the van't Hoff exponential expression:

$$\mu_{max}(T) = \mu_{max}(20°C) \cdot \exp(\kappa(T-20)) \quad (3.16)$$

For aerobic processes, the expression applies in the temperature range 0-32°C. For the temperature range 32-40°C, the removal rate is constant after which it usually declines drastically to be zero around 45°C.

The processes can also take place in the thermophilic range 50-60°C. Here the process rate will be approx. 50 per cent higher than at 35°C.

Aerobic heterotrophic conversion of organic matter

Oxygen

The oxygen dependency for aerobic processes can be described by a Monod expression

$$\mu_{obs} = \mu_{max} \cdot \frac{S_{O2.2}}{S_{O2.2} + K_{S,O2}}$$

where $S_{O2.2}$ is the oxygen concentration **in the reactor**,
$K_{S,O2}$ is the saturation constant for oxygen.

In combination with (3.14), we have a double Monod expression:

$$\mu_{obs} = \mu_{max} \frac{S_2}{S_2 + K_S} \cdot \frac{S_{O2.2}}{S_{O2.2} + K_{S,O2}} \quad (3.17)$$

The saturation constant, $K_{S,O2}$, depends on floc size/biofilm thickness and on temperature as it reflects diffusional limitations for oxygen into flocs/biofilm.

pH

The aerobic conversions are pH dependent, see figure 3.6. The rather unusual shape of the curve is due to a combination of the pH dependency of the micro-organisms and the selection of individual micro-organisms.

The kinetics of pH can be described as follows:

$$\mu_{max}(pH) = \mu_{max}(opt. pH) \cdot \frac{K_{pH}}{K_{pH} + I} \quad (3.18)$$

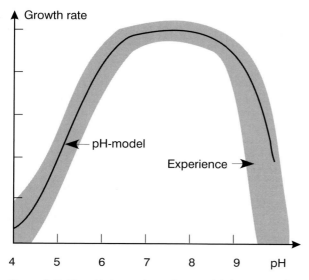

Figure 3.6. The pH dependency for aerobic heterotrophic processes. Model: $K_{pH} = 200$ (see Expression (3.18)).

where K_{pH} is the pH constant

$$I = 10^{|\text{optimum pH-pH}|} - 1$$

Normally it is only the low pH values which cause problems in biological treatment processes. Apart from the fact that low pH values may be caused by the pH in the raw wastewater, low pH values may arise in connection with nitrification, see Section 3.4, and possibly in connection with chemical precipitation of phosphorus.

Toxic substances
Many substances act toxically on aerobic conversions even if aerobic processes in treatment plants are rather robust. Complex formation, chemical precipitations (sulphide precipitates metals) and biodegradation (cyanide, phenols etc.) make it difficult to evaluate the toxic effects. Tests should be performed using the wastewater concerned to obtain a reasonable possibility to evaluate the effects. For more detailed information about the degradation and effect of toxic substances, see for example /36/.

The mechanisms of inhibition include competitive and non-competitive reversible inhibition /8/.

Competitive reversible inhibition influences the kinetics of growth by increasing the saturation constant, K_S,

$$K_S' = K_S \cdot \frac{K_{S,I} + C_I}{K_{S,I}} \qquad (3.19)$$

where K_S' is the saturation constant with inhibition,
K_S is the saturation constant without inhibition,
$K_{S,I}$ is the inhibition constant,
C_I is the concentration of inhibiting substance.

Non-competitive reversible inhibition influences the kinetics of growth by reducing the maximum specific growth rate

$$\mu_{max'} = \mu_{max} \cdot \frac{K_{S,I}}{K_{S,I} + C_I} \qquad (3.20)$$

where $\mu_{max'}$ is max. specific growth rate with inhibition
μ_{max} is max. specific growth rate without inhibition

Nitrogen and phosphorus
Their effect on aerobic processes can be described in a double Monod expression which conceals that the microbial growth is inhibited when the concentrations of nitrogen and phosphorus are low

$$\frac{S_{NH4}}{S_{NH4} + K_{S,NH4}} \cdot \frac{S_{PO4}}{S_{PO4} + K_{S,PO4}} \tag{3.21}$$

where S_{NH4} is the ammonium concentration
S_{PO4} is the phosphorus concentration (orthophosphate)
$K_{S,NH4}$ is the saturation constant, nitrogen
$K_{S,PO4}$ is the saturation constant, phosphorus

Reaction rate constants, aerobic heterotrophic conversions

Table 3.7 lists typical intervals for reaction rate constants and stoichiometric coefficients for micro-organisms in aerobic wastewater treatment processes for municipal wastewater. It should be noted that there is a certain correlation between the individual constants, and therefore sets of constants should be used rather than selecting isolated constants from different sources. Table 3.8 shows such correlating sets of constants.

The determination of constants through tests normally requires an extensive experimental effort in the form of batch tests or continuous flow tests combined with a wide range of chemical analyses. For further details, see /9, 33, 34, 35/.

	Symbol	Unit	Quantity
Maximum specific growth rate	μ_{max}	d^{-1}	4-8
Decay constant	b	d^{-1}	0.1-0.2
Saturation constant for substrate	$K_{S,COD}$	g COD/m³	5-30
Saturation constant for oxygen	$K_{S,O2}$	g O_2/m³	0.5-1
Maximum yield constant	$Y_{max,H}$	g COD/g COD	0.5-0.7
Temperature constant for μ_{max}, k_h, b	κ	°C^{-1}	0.06-0.10
pH constant	K_{pH}	-	150-250
Hydrolysis constant, suspended solids	k_h	d^{-1}	0.6-1.4
Hydrolysis constant, dissolved solids	k_h	d^{-1}	3-20
Hydrolysis constant	k_{hX}	kg COD(X)/kg COD(B) · d	0.6-1.4
Hydrolysis saturation constant	K_X	kg COD(X)/kg COD(B)	0.02-0.05
Saturation constant for nitrogen	$K_{S,NH4}$	g N/m³	0.1-0.5
Saturation constant for phosphorus	$K_{S,PO4}$	g P/m³	0.1-0.2

Table 3.7. Reaction rate constants for heterotrophic aerobic conversions, municipal wastewater (20°C).

Symbol	Unit	Value at 20°C	Value at 10°C
Stoichiometric parameters			
Y_A	g cell COD formed (g N oxidized)$^{-1}$	0.24	0.24
Y_H	g cell COD formed (g COD oxidized)$^{-1}$	0.67	0.67
f_p	dimensionless	0.08	0.08
i_{XB}	g N (COD)$^{-1}$ in biomass	0.086	0.086
i_{XE}	g N (COD)$^{-1}$ in endogenous mass	0.06	0.06
Kinetic parameters			
$\hat{\mu}_H$	day^{-1}	6.0	3.0
K_S	g COD m^{-3}	20.0	20.0
$K_{O,H}$	g O2 m^{-3}	0.20	0.20
K_{NO}	g NO3–N m^{-3}	0.50	0.50
b_H	day^{-1}	0.62	0.20
η_g	dimensionless	0.8	0.8
η_h	dimensionless	0.4	0.4
k_h	g slowly biodegradable COD (g cell COD · day)$^{-1}$	3.0	1.0
K_X	g slowly biodegradable COD (g cell COD)$^{-1}$	0.03	0.01
$\hat{\mu}_A$	day^{-1}	0.80	0.3
K_{NH}	g NH3–N m^{-3}	1.0	1.0
$K_{O,A}$	g O2-N m^{-3}	0.4	0.4
k_s	m^3 · COD (g · day)$^{-1}$	0.08	0.04

Table 3.8. Set of reaction rate constants and stoichiometric coefficients and their temperature dependency /9/.
($f_p = f_I$, $i_{XB} = f_{B,N}$, $i_{XE} = f_{I,N}$, $\hat{\mu}_H = \mu_{max}$)

3.4. Nitrification

Nitrification is a microbiological process that converts ammonium into nitrite and eventually nitrite to nitrate.

The process occurs anywhere in the biosphere, provided that the environments are such that the nitrifying bacteria can exist. The nitrification process is very important for the oxygen conditions in soil, streams, lakes and biological wastewater treatment plants.

3.4.1. Reactions by nitrification

The nitrification process is in practice performed by a limited group of autotrophic micro-organisms. The process takes place in two steps as ammonium is oxidized to nitrite by a group of bacteria, frequently known as Nitrosomonas. Subsequently nitrite is oxidized to nitrate by another group of organisms, frequently known as Nitrobacter. Other bacteria are also involved, like Nitrospira, Nitrococcus and Nitrosocystis. In wastewater treatment the variation in nitrifying species active is considerable. However, the various nitrifying bacteria identified by gene probes, do not seem to have treatment process performance that deviate from the Nitrosomonas and Nitrobacter species. Thus from an engineering conceptual point of view, the process can be thought of as a two step process, with the two bacterial groups, with the well-known stoichiometry and kinetics.

The process for the ammonium oxidizing bacteria is

$$NH_4^+ + \tfrac{3}{2}O_2 \rightarrow NO_2^- + H_2O + 2H^+ \tag{3.22}$$

$\Delta G^\circ(W) = -270 \text{ kJ/mol } NH_4^+ - N$

The process for the nitrite oxidizing bacteria is

$$NO_2^- + \tfrac{1}{2}O_2 \rightarrow NO_3^- \tag{3.23}$$

$\Delta G^\circ(W) = -80 \text{ kJ/mol } NO_2^- - N$

The nitrifying bacteria are characterized by a low growth rate. This is due to the low energy yield which is linked to the oxidation of ammonium and nitrite, respectively. The slow growth is a major problem for the nitrification in biological treatment plants.

Most nitrifying bacteria are autotrophic and thus use carbon dioxide as the carbon source. The carbon dioxide should be reduced before the carbon can form part of the cell mass, and this reduction takes place through the oxidation of the nitrogen source of the organism concerned. For oxidation of ammonium, the expression for growth is:

$$15 \, CO_2 + 13 \, NH_4^+ \rightarrow 10 \, NO_2^- + 3 \, C_5H_7NO_2 + 23 \, H^+ + 4 \, H_2O \tag{3.24}$$

For oxidation of nitrite, the corresponding growth expression is:

$$5 \, CO_2 + NH_4^+ + 10 \, NO_2^- + 2 \, H_2O \rightarrow 10 \, NO_3^- + C_5H_7NO_2 + H^+ \tag{3.25}$$

Once the observed yield constants, $Y_{obs, NH4}$ and $Y_{obs, NO2}$, are known, the equations of reaction can be calculated. The maximum yield constants can be calculated as stated in /8/; however, the metabolic efficiency is lower than normal (β approx. 0.20).

By combining Expressions (3.22) and (3.24) and by using the carbonate equilibrium system (note that nitrogen, oxygen as well as carbon change their oxidation levels), we find for ammonium-oxidizers with $Y_{obs,\ NH4} = 0.1$ g VSS/g $NH_4^+ - N$ (= 0.14 g COD/g $NH_4^+ - N$):

$$80.7\ NH_4^+ + 114.55\ O_2 + 160.4\ HCO_3^- \rightarrow C_5H_7NO_2 +$$
$$79.7\ NO_2^- + 82.7\ H_2O + 155.4\ H_2CO_3 \qquad (3.26)$$

The expression shows that 113 g bacteria ($C_5H_7NO_2$) are produced by the conversion of $80.7 \cdot 14 = 1129.8$ g $NH_4^+ - N$. Hence the yield constant is $113/1129.8 = 0.10$ g VSS/g $NH_4^+ - N$ removed. The yield constant per amount of ammonium nitrogen oxidized is $113/(79.7 \cdot 14) = 0.10$ g VSS/g $NH_4^+ - N$ oxidized.

Example 3.4

From Expression (3.26), the yield constant can also be calculated in other units, for example as g COD/g $NH_4^+ - N$:

The 113 g bacteria $C_5H_7NO_2$ represent a COD-content of 1.42 g COD/g, see Example 3.2. The yield constant is then

(1.42 g COD/g · 113 g)/1129.8 g $NH_4^+ - N$ = 0.14 g COD/g$NH_4^+ - N$

By combining Expressions (3.23) and (3.25), we find for nitrite-oxidizers with $Y_{obs,NO2} = 0.06$ g VSS/g $NO_2^- - N$:

$$134.5\ NO_2^- + NH_4^+ + 62.25\ O_2 + HCO_3^- + 4\ H_2CO_3 \rightarrow$$
$$C_5H_7NO_2 + 134.5\ NO_3^- + 3\ H_2O \qquad (3.27)$$

The overall equation of reaction for nitrification is found by combining Expressions (3.26) and (3.27).

$$NH_4^+ + 1.86\ O_2 + 1.98\ HCO_3^- \rightarrow$$
$$0.020\ C_5H_7NO_2 + 0.98\ NO_3^- + 1.88\ H_2CO_3 + 1.04\ H_2O \qquad (3.28)$$

Example 3.5

Calculate the oxygen consumption for the total nitrification process.

From Expression (3.28) we find that 1.86 moles of O_2 per mole of NH_4^+ are used, that is, the oxygen consumption per g $NH_4^+ - N$ is:

$$\frac{1.86\ mol\ O_2 \cdot 32\ g\ O_2\ /\ mol}{1\ mol\ NH_4^+ - N \cdot 14\ g\ N\ /\ mol} = 4.25\ g\ O_2\ /\ g\ NH_4^+\text{-N}$$

That is, the stoichiometric coefficient $v_{NH4,O2} = 4.25$.

From Expression (3.28), the oxygen consumption per g $NO_3^- - N$ formed can also be calculated:

$$\frac{1.86 \text{ mol } O_2 \cdot 32 \text{ g } O_2/\text{mol}}{0.98 \text{ mol } NO_3^- - N \cdot 14 \text{ g N}/\text{mol}} = 4.34 \text{ g } O_2/\text{g } NO_3^- - N \approx$$

4.34 g O_2/g NH_4^+ − N oxidized

The difference between 4.34 g O_2/g NH_4^+ − N and what we can calculate by combining Expressions (3.22) and (3.23):

$$NH_4^+ + 2 O_2 \rightarrow NO_3^- + 2 H^+ + H_2O$$

$$\frac{2 \text{ mol } O_2 \cdot 32 \text{ g } O_2/\text{mol}}{1 \text{ mol } NO_3^- - N \cdot 14 \text{ g N}/\text{mol}} =$$

4.57 g O_2/g NO_3^- − N = 4.57 g O_2/g NH_4^+ − N oxidized

is due to the fact that inorganic carbon, which the bacteria assimilate, also acts as an oxidizing agent thus reducing the oxygen consumption somewhat.

The oxidation of ammonium to nitrite takes place in several steps while the oxidation from nitrite to nitrate is a single step. The intermediate between hydroxylamine and nitrite is not known:

$$NH_4^+ \xrightarrow{a} NH_2OH \xrightarrow{b} ? \rightarrow NO_2^- \rightarrow NO_3^-$$

Reactions a and b can be selectively inhibited by thiourea and hydrazine /12/. Thiourea is used to limit the nitrification in connection with respiration tests in activated sludge plants and in connection with the modified BOD analysis. 1 ppm is sufficient for inhibition /13/.

3.4.2. Alkalinity

The nitrification process reduces the alkalinity in water. Expressions (3.26) and (3.27) show that it is only the first step of the process which influences the alkalinity. For every mole of NH_4^+ −N which is oxidized to NO_2^--N, approx. 2 moles of HCO_3^- are consumed, and this corresponds to 2 equivalents of alkalinity. This is essential for the nitrification of relatively soft water where the pH in the water can be so low that the nitrification process is limited or stops completely.

Remember that the way a chemical reaction is written does not affect the change in alkalinity in the process. The contribution to the change in alkalinity is calculated for each component in a chemical reaction based on its chemical composition at pH 4.5. A simplified reaction for nitrification can be written in the following two versions:

$$NH_4^+ + 2O_2 \rightarrow NO_3^- + 2H^+ + H_2O$$

Alkalinity change: 0   + 0   0   -2   0   Total change: -2

$$NH_3 + 2O_2 \rightarrow NO_3^- + H^+ + H_2O$$

Alkalinity change: 1   0   0   -1   0   Total change: $-1 - (1) = -2$

It is seen that in both cases the total alkalinity change is – 2 eqv/mole N oxidised.

Example 3.6

Typical domestic wastewater contains about 50 g TN/m³. Part of this ends up in the sludge. The rest might be 30 g N/m³ which can be nitrified.

The alkalinity consumption, ΔTAL, for this purpose is

$$\frac{30 \text{ g N}/m^3}{14 \text{ g N}/\text{mol N}} \cdot 2 \text{ eqv}/\text{mol N} = 4.29 \text{ eqv}/m^3$$

If the wastewater therefore has an alkalinity which is lower than approx. 5 eqv/m³, which many wastewaters have, see Chapter 1, it may be necessary to increase the alkalinity (for example with lime or through denitrification - see Section 3.5).

3.4.3. Kinetics, nitrification

Kinetics can be described by a Monod expression as shown in Expression (3.14). The saturation constants, $K_{S,NH4,A}$ and $K_{S,NO2,A}$, are small - see Table 3.11. For practical calculations in treatment plants, we can consider the nitrification process as a one-step process and use the reaction constants for the total process which are shown in Table 3.11.

3.4.4. The influence of the environmental factors on nitrification

A number of environmental factors influencing the nitrification are discussed below. These are:

– substrate concentration
– temperature
– oxygen
– pH
– toxic substances

It is important to note that in practice it is the oxidation of ammonium which is the rate limiting step in the overall process. This means that nitrite will only appear in large amounts when the considered process is non-stationary, for example because of varying loads, start-up and washout, or other operational problems in the treatment plant.

Temperature dependency

The same applies as for heterotrophic aerobic micro-organisms - see Expression (3.16). The Expression applies at least in the 10-22°C temperature range. At higher temperatures (30-35°C) the growth rate is constant; between 35 and 40°C it starts to decline towards zero - see figure 3.7. At temperatures between 0 and 10°C, the expression probably applies as well.

Just like any other bacteria, nitrifying bacteria are especially sensitive to sudden variations in the temperature. This is seen in figure 3.8. When the temperature rise is fast (in the course of hours), the increase in the growth rate is lower than expected, whereas a sudden temperature drop gives a much higher decline in the activity than could be expected from figure 3.7. As far as we know, nitrifying processes cannot take place at thermophilic temperatures (50-60°C).

Oxygen concentration

The nitrifying bacteria are more sensitive to low oxygen concentrations than the heterotrophic bacteria. The kinetics can be described using a Monod expression corresponding to (3.17). The saturation constants are shown in Table 3.11. Diffusional limitation is an essential feature in activated sludge plants, and hence

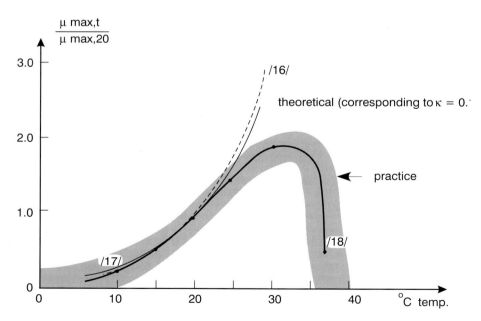

Figure 3.7. Nitrification as a function of temperature. As opposed to the other biological processes in wastewater treatment, thermophilic nitrifying bacteria are unknown.

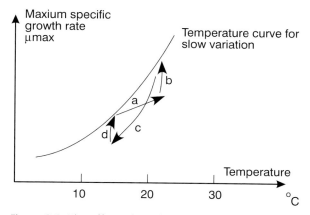

Figure 3.8. The effect of rapid and slow temperature changes on the growth rate. Based on /19/.
a = rapid heating (15 → 25°C)
b = adaptation (slow)
c = rapid cooling (25 → 15°C)
d = adaptation (slow)

the conditions in the plant are determined by floc sizes, substrate load conditions, temperature, etc.

Nitrification can take place at high oxygen concentrations, for example in a pure oxygen plant. 60 g oxygen/m^3 do not inhibit the process /12/.

pH

The nitrification process is pH dependent with an optimum in the 8-9 range - see figure 3.9. Because of the influence of the nitrification process on pH, the pH value

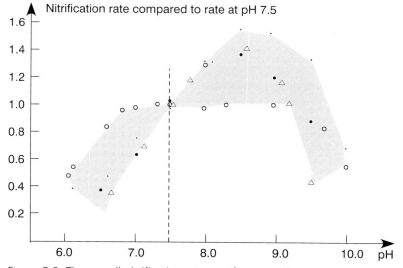

Figure 3.9. The overall nitrification rate as a function of pH /21/.

in flocs and biofilm will be lower than the value which can be observed in the liquid phase.

It is possible that the pH dependency could be linked to the inhibition phenomena from the substrate as we know that free ammonia, NH_3 and free nitrous acid HNO_2, can inhibit both ammonium and nitrite oxidation. At the same time, tests seem to indicate that the uncharged components, NH_3 and HNO_2, are the substrate of the nitrifying bacteria /12/. Thus we have a situation where the same component is a substrate whereas - in a higher concentration - it can also be an inhibi-

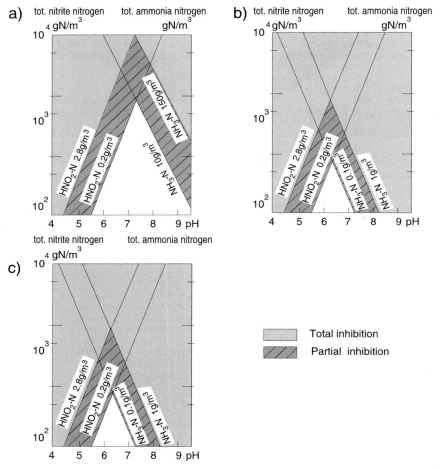

Figure 3.10.
a: Inhibition of ammonium oxidation with NH_3 (0% at 10 g N/m^3, 100% at 150 g N/m^3) and HNO_2 (0% at 0.2 g N/m^3, 100% at 2.8 g N/m^3), based on /20/.
b: Inhibition of nitrite oxidation with NH_3 (0% at 0.1 g N/m^3, 100% at 1 g N/m^3) and HNO_2 (0% at 0.2 g N/m^3, 100% at 2.8 g N/m^3), based on /20/.
c: Inhibition of the overall nitrification process as a function of NH_3, HNO_2 and pH, based on /20/.

tor. As the equilibria NH_3/NH_4^+ and HNO_2/NO_2^- both are pH dependent, a situation can be reached as illustrated in figure 3.10 a-c. Note that a partial inhibition of the processes does not necessarily mean a reduced nitrification degree, but merely that the process is slower, and this should be taken into consideration in the practical design of a treatment plant as it means that the plant must be bigger if the same treatment result is required as without inhibition.

Inhibiting substances

The nitrification process in activated sludge plants can be inhibited by many different substances. As nitrifying treatment plants are designed just to allow for the nitrification process to take place at the design temperature, even a limited inhibition could cause the nitrification to stop completely. However, this will not take place instantaneously, but after a washout period over several weeks. Such a stop in nitrification is therefore not a result of the nitrifying bacteria being 100 per cent inhibited by the toxic substance concerned, but it is a result of the washing out of the nitrifying bacteria /1/. As a general rule, we can say that the nitrifying bacteria are not more sensitive to toxic substances than other bacteria in a treatment plant - see figure 3.11. Metals also inhibit nitrification. Table 3.9 contains some information from the literature. The table shows that just as for almost all other properties, there is a big difference between pure cultures and activated sludge cultures. The nitrifying bacteria cannot be expected to be less susceptible to the influence of metals, just because they are present in activated sludge, but the metal ion activities in the liquid phase and in the sludge phase are very different, and this explains why the bacteria in the activated sludge can "resist" higher metal ion concentrations. Some organic materials, such as sulphur components, aniline components, phenols and cyanide, have, however, an especially strong inhibiting effect. Table 3.10 shows the inhibiting effects on nitrification for a number of substances.

If the micro-organisms are exposed to several inhibiting substances simultaneously, the effects of the individual substances will usually be more powerful (a synergy effect).

Metal	Concentration (g/m³)	Effect
Cu	0.05-0.56	Nitrosomonas activity inhibited (pure culture)
Cu	4	No essential inhibition in activated sludge
Cu	150	75% inhibition of activated sludge
Ni	> 0.25	Nitrosomonas growth inhibited (pure culture)
Cr^{+++}	> 0.25	Nitrosomonas growth inhibited (pure culture)
Cr^{+++}	118	75% inhibition of activated sludge
Zn	0.08-0.5	Inhibition of Nitrosomonas (pure culture)
Co	0.08-0.5	Inhibition of Nitrosomonas (pure culture)

Table 3.9. Inhibition of nitrification with metals/12/, /22/, /31/.

Compound	Formula	Molecular weight	Concentration necessary to inhibit oxidation of ammonia by 75%	
			mg/l	molarity
Thiourea	$(NH_2)_2CS$	76	0.076	$\times 10^{-6}$
Thioacetamide	$CH_3.CS.NH_2$	75	0.53	7×10^{-6}
Thiosemicarbazide	$NH(NH_2).CS.NH_2$	91	0.18	2×10^{-6}
Methyl isothiocyanate	$CH_3.NCS$	73	0.8	1.1×10^{-5}
Allyl isothiocyanate	$CH_2:CH.CH_2.NCS$	99	1.9	1.9×10^{-5}
Dithio-oxamide	$NH_2.CS.CS.NH_2$	120	1.1	9.2×10^{-6}
Potassium thiocyanate	KCNS	97	300*	3.1×10^{-3}*
Sodium methyl dithiocarbamate	$CH_3.NH.CS.S$ Na	129	0.9	7×10^{-6}
Sodium dimethyl dithiocarbamate	$(CH_3)_2.N.CS.S$ Na	143	13.6	9.5×10^{-5}
Dimethyl ammonium dimethyldithio-carbamate	$(CH_3)_2.N.CS.S.NH_2(CH_3)_2$	166	19.3	11.6×10^{-5}
Sodium cyclopenta-methylene-dithiocarbamate	$C_5H_9.NH.CS.S$ Na.$2H_2O$	219	23	10.5×10^{-5}
Piperidinium cyclo-pentamethylene-dithiocarbamate	$C_5H_9.NH.CS.S.NH_2C_5H_{10}$	246	57	2.3×10^{-4}
Methyl thiuronium sulphate	$[NH_2.C(:NH).S.CH_3]_2H_2SO_4$	278	6.5	2.3×10^{-5}
Benzyl thiuronium chloride	$[NH_2.C(:NH).S.CH_2(C_6H_5)]HCl$	203	49	2.4×10^{-4}
Tetramethyl thiuram monosulphide	$(CH_3)_2.N.CS.S.CS.N(CH_3)_2$	208	16	7.5×10^{-5}
Tetramethyl thiuram disulphide	$(CH_3)_2.N.CS.S.S.CS.N(CH_3)_2$	240	30	1.2×10^{-4}
Mercaptobenzothiazole	$C_6H_4.SC(SH):N$	167	3	1.8×10^{-5}
Benzothiazole disulphide	$C_{14}H_8N_2S_4$	332	38	1.2×10^{-4}
Phenol	$C_6H_5.OH$	94	5.6	6×10^{-5}
o-cresol	$CH_3.C_6H_4.OH$	107	12.8	1.2×10^{-4}
m-cresol	$CH_3.C_6H_4.OH$	107	11.4	1.06×10^{-4}
p-cresol	$CH_3.C_6H_4.OH$	107	16.5	1.53×10^{-4}
Aniline	$C_6H_5.NH_2$	93	7.7	8.3×10^{-5}
2-4 dinitrophenol	$C_6H_4(NO_2)_2$	184	460	2.5×10^{-3}
Allyl alcohol	$CH_2:CH.CH_2OH$	58	19.5	3.4×10^{-4}
Allyl chloride	$CH_2:CH.CH_2Cl$	76.5	180	2.4×10^{-3}
Di-allyl ether	$(CH_2:CH.CH_2)_2O$	98	100	1×10^{-3}
Sodium cyanide	NaCN	27	0.65	2.4×10^{-5}
Dimethyl p-nitroso-aniline	$(CH_3)_2N.C_6H_4.NO$	150	19	1.3×10^{-5}
Guanidine carbonate	$[(NH_2)_2.C:NH]H_2CO_3$	180	16.5	9.2×10^{-5}
Diphenyl guanidine	$(NH.C_6H_5)_2.C:NH$	211	50*	2.5×10^{-4}*
Diguanide	$NH_2.C(:NH)NH.C(:NH)NH_2$	101	50	5×10^{-4}
Dicyandiamide	$NH_2.C(:NH)NH.CN$	84	250	3×10^{-3}
Skatole	$C_6H_4NHCH:CCH_3$	131	7.0	5.3×10^{-5}
Strychnine hydrochloride	$C_{21}H_{22}O_2N_2.HCl.2H_2O$	407	175	4.3×10^{-4}
2-chloro-6-trichloromethyl-pyridine	$C_5H_3NCl(CCl_3)$	231	100	0.43×10^{-3}
Ethyl urethane	$NH_2.CO.OC_2H_5$	89	1,780	2×10^{-2}
EDTA	$[(COOH.CH_2)_2.N.CH_2]_2$	292	350*	1.2×10^{-3}*
Hydrazine	$NH_2.NH_2$	32	58	1.8×10^{-3}
Methylamine hydrochloride	$CH_3.NH_2HCl$	67.5	1,550	2.3×10^{-2}
Trimethylamine	$N(CH_3)_3$	59	118	2×10^{-3}
Sodium azide	NaN_3	65	23	3.6×10^{-4}
Methylene blue	$C_{16}H_{18}N_3SCl.3H_2O$	373.5	100*	3×10^{-4}*
Carbon disulphide	CS_2	76	35	0.46×10^{-3}
Ethanol	C_2H_5OH	46	2,400	5×10^{-2}
Acetone	$CH_3.CO.CH_3$	58	2,000	3.5×10^{-2}
Chloroform	$CHCl_3$	119.4	18	1.5×10^{-4}
8-hydroxyquinoline	$C_9H_6N.OH$	145	72.5	5×10^{-4}
Streptomycin	$C_{21}H_{39}N_7O_{12}$	581.6	400*	6.9×10^{-4}*

* Highest concentration tested, but not effective.

Table 3.10. Inhibition of nitrification. From /22/.

Basic Biological Processes

Figure 3.11. The myth that nitrifying bacteria are more sensitive to toxic substances than heterotrophs is just as stubborn and tenacious as an elephant. The myth is not true, just as the elephant in this photo is not true, for it is a sculpture made by Giovanni Lorenzo Bernini, the greatest sculptor of the baroque period. His works are found in Rome, such as the high alter in St Peter's, the Fountain in the middle of Piazza Navona and the elephant with the obelisk on its back at Piazza Minerva.

Reaction rate constants

In Table 3.11, reaction rate constants for nitrifying bacteria are listed.

	Symbol	Unit	Ammonia oxidation	Nitrite oxidation	Total process
Maximum specific growth rate	$\mu_{max,A}$	d^{-1}	0.6-0.8	0.6-1.0	0.6-0.8
Saturation constant	$K_{S,NH4,A}$	g NH_4–N/m^3	0.3-0.7	0.8-1.2	0.3-0.7
Saturation constant	$K_{S,O2,A}$	g O_2/m^3	0.5-1.0	0.5-1.5	0.5-1.0
Maximum yield constant	$Y_{max,A}$	g VSS/g N[1]	0.10-0.12	0.05-0.07	0.15-0.20
Decay constant	b_A	d^{-1}	0.03-0.06	0.03-0.06	0.03-0.06
Temperature constant for $\mu_{max,A}$ and b_A	κ	$°C^{-1}$	0.08-0.12	0.07-0.10	0.08-0.12

[1] per g NO_3^- – N formed

Table 3.11. Reaction rate constants for nitrification at 20°C /14, 15, 12, /16/.

3.5. Denitrification

By denitrification, micro-organisms convert nitrate into atmospheric nitrogen. The process is anaerobic as nitrate is the oxidizing agent.

$$A_{red} + NO_3^- \rightarrow A_{ox} + 0.5\ N_2$$

When nitrate is the oxidizing agent, the process is called anoxic. Denitrification is widespread in nature; it occurs anywhere where nitrate is present, provided that no oxygen (or not too much) is present at the same time. Most denitrifying micro-organisms are facultative and they prefer oxygen as an oxidizing agent, provided it is present.

Many of the commonest bacteria have the ability to change their metabolism from using oxygen as the final electron acceptor, to using nitrate instead. The electron transport system in a denitrifying organism is identical with the electron transport system under aerobic conditions, with the exception of the last step, the nitrate (or nitrite) reductase. The "choice" made by the bacteria of the definitive terminal acceptor depends on the redox potential between the last cytochrome in the electron transport system and oxygen or nitrate. This choice favours oxygen, so that in a situation, where both oxygen and nitrate are present, the bacterium will not denitrify but instead respire with oxygen.

The denitrification process, also known as dissimilatory nitrate reduction, occurs stepwise as shown in figure 3.12. The figure also shows how the assimilatory nitrate reduction occurs.

The intermediate products by denitrification (and nitrification) are all toxic or undesirable. This applies to nitrite, NO_2^-, nitric oxide, NO, as well as to dinitrogen oxide, N_2O. Nitrite is inhibiting for micro-organisms and is used as a preservative. Nitric oxide is converted into nitrogen dioxide in the atmosphere. It was used as poison gas during World War I and can be found in the exhaust from cars. Dinitrogen oxide is used for anaesthetization and is a greenhouse gas. All the intermediate products are formed anywhere in nature in the upper stratum of the soil.

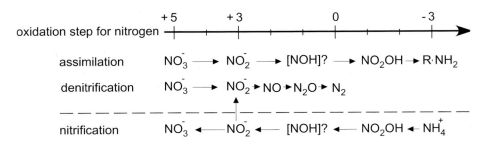

Figure 3.12. Reaction sequences for microbiological nitrogen conversions.

In connection with denitrification, the amount of intermediate products will normally be on a per thousand level. If the process is stressed (for example is short of organic matter or other nutrients) or is subject to strong dynamic influences, the release of intermediate products can be increased.

3.5.1. Reactions by denitrification

The energy yielding process for denitrifying bacteria which uses organic matter in wastewater as an energy and carbon source can be written as a combination of the two half-expressions

$$\tfrac{1}{70} C_{18}H_{19}O_9N + \tfrac{28}{70} H_2O \rightarrow \tfrac{17}{70} CO_2 + \tfrac{1}{70} HCO_3^- + \tfrac{1}{70} NH_4^+ + H^+ + e^- \quad \Delta G°(W) = -32 \text{ kJ/e-eqv}$$

and

$$\tfrac{1}{10} N_2 + \tfrac{3}{5} H_2O \rightarrow \tfrac{1}{5} NO_3^- + \tfrac{6}{5} H^+ + e^-$$

$\Delta G°(W) = + 71$ kJ/e-eqv

By combination we find:

$$\tfrac{1}{70} C_{18}H_{19}O_9N + \tfrac{1}{5} NO_3^- + \tfrac{1}{5} H^+ \rightarrow \tfrac{1}{10} N_2 + \tfrac{17}{70} CO_2 +$$

$$\tfrac{1}{70} HCO_3^- + \tfrac{1}{70} NH_4^+ + \tfrac{1}{5} H_2O \tag{3.29}$$

$\Delta G°(W) = -103$ kJ/e-eqv

The energy developed, –103 kJ/e-eqv, ends up as heat or in the form of biological growth.

3.5.2. Yield constant by denitrification

Provided that all the energy is used for growth, the maximum yield constant of the process is around 0.40 kg biomass/kg organic matter which is 15% lower than that of aerobic heterotrophic conversion. The corresponding equation of reaction, provided that the organism assimilates ammonium, is

$$0.52 \ C_{18}H_{19}O_9N + 3.28 \ NO_3^- + 0.48 \ NH_4^+ + 2.80 \ H^+ \rightarrow$$

$$C_5H_7NO_2 + 1.64 \ N_2 + 4.36 \ CO_2 + 3.8 \ H_2O \tag{3.30}$$

The equation of reaction, provided that nitrate is assimilated, does not differ much from (3.30)

$$0.57\ C_{18}H_{19}O_9N + 3.73\ NO_3^- + 3.73\ H^+ \rightarrow$$

$$C_5H_7NO_2 + 1.65\ N_2 + 5.26\ CO_2 + 3.80\ H_2O \qquad (3.31)$$

but it should be remembered that the maximum yield constant is changed, in this case, to 0.36 kg biomass/kg organic matter. If ammonium is present, the bacteria will always use it as the nitrogen source. This will be the case in (almost) all types of ordinary wastewater.

Example 3.7

A denitrifying biomass degrades phenol, C_6H_6O, without access to ammonium. Evaluate the equation of reaction for the process inclusive of growth (observed yield constant = 1.2 kg COD/kg phenol). The composition of the biomass is assumed to be $C_6H_8NO_2$.

COD of the biomass is:

$$C_6H_8NO_2 + 6.25\ O_2 + H^+ \rightarrow 6\ CO_2 + NH_4^+ + 2.5\ H_2O$$

that is, 6.25 mole oxygen = 6.25 mole COD/mole biomass.

Y_{obs} = 1.2 kg COD/kg phenol =

$$\frac{1{,}200\ g\ COD\ /\ (32\ g\ COD\ /\ mole)}{1{,}000\ g\ phenol\ /\ (94\ g\ phenol\ /\ mole)} = 3.53\ \text{mole COD / mole phenol}$$

Hence the yield constant on a molar basis (mole biomass/mole phenol) is:

$$\frac{3.53\ \text{mol COD / mole phenol}}{6.25\ \text{mol COD / mole biomass}} = 0.56\ \text{mole biomass / mole phenol}$$

The equation of reaction is then:

$$C_6H_6O + \ldots \rightarrow 0.56\ C_6H_8NO_2 + 2.64\ CO_2 \ldots$$

Carbon changes the oxidation step corresponding to +14.0. Nitrate must then be decreased correspondingly. 0.56 mole of NO_3^- is reduced to ammonium and is assimilated in the biomass. This corresponds to a decline in the oxidation step of 0.56 · 8 = 4.48. The remaining decline (14.0 − 4.48 = 9.52) is due to denitrification, that is, $\frac{9.52}{5}$ = 1.90 moles of nitrate each of which is reduced by 5 in oxidation step.

$$C_6H_6O + (1.90 + 0.56)\ NO_3^- + \ldots \rightarrow$$
$$0.56\ C_6H_8NO_2 + 2.64\ CO_2 + 0.95\ N_2 + \ldots$$

A harmonization of charges with H^+ gives the following final appearance:

$$C_6H_6O + 2.46\ NO_3^- + 2.46\ H^+ \rightarrow$$
$$0.56\ C_6H_8NO_2 + 2.64\ CO_2 + 0.95\ N_2 + 1.98\ H_2O$$

In this example we see that even by denitrification a significant part of the nitrogen can be assimilated. Here it is $\frac{0.56}{2.46}$ · 100% = 23%.

3.5.3. Nutrients, denitrification

The nutrient requirement of the denitrifying bacteria corresponds to that of the aerobic, heterotrophic organisms. The organisms prefer ammonium to nitrate as a nitrogen source in assimilation. In municipal wastewater there are usually sufficient nutrients whereas industrial wastewater may be short of phosphorus.

3.5.4. Alkalinity

The denitrification process increases the alkalinity of the water. For every mole of nitrate converted, one equivalent alkalinity is produced which appears from Expression (3.31). If ammonium is used as a nitrogen source, this will reduce the alkalinity production by one equivalent per mole of ammonium assimilated. It appears from Expression (3.30) that the alkalinity production is 2.80 eqv/3.28 mole $NO_3^- - N$ corresponding to 0.85 eqv/mole $NO_3^- - N$.

Example 3.8

By denitrification of wastewater, 25 g $NO_3^- - N/m^3$ are transformed. The process is assumed to follow Expression (3.30) which means that ammonium is used for assimilation.

The alkalinity of the water before the denitrification, TAL_1, is 4.1 eqv/m³. What is the alkalinity, TAL_3, after the denitrification process?

From Expression (3.30) it is found that 0.85 eqv alkalinity/mole $NO_3^- - N$ converted is produced.

$$25 \text{ g } NO_3^- - N/m^3 = \frac{25 \text{ g } NO_3^- - N \big/ m^3}{14 \text{ g } NO_3^- - N \big/ \text{mole}} = 1.79 \text{ mole } NO_3^- - N/m^3$$

The change in alkalinity, ΔTAL, is then

$\Delta TAL = 0.85$ eqv/mole $NO_3^- - N \cdot 1.79$ mole $NO_3^- - N/m^3 = 1.52$ eqv/m³

$TAL_3 = TAL_1 + \Delta TAL = 4.1 + 1.52 = 5.6$ eqv/m³

3.5.5. Kinetics, denitrification

Normally, the kinetics are described by a Monod expression, see Expression (3.14). As it is nitrate we want to convert, we aim at a substrate with no limiting effects. In such cases, a zero order expression, such as (3.15), can be used.

If an external substrate is added, for example acetic acid or methanol, it is possible to prevent substrate limitation. If the existing content of organic matter in the wastewater is used, it will, in many cases, result in a limitation of the removal rate. This is not to be confused with an absolute limitation where the process stops due to the shortage of substrate.

The substrate removal rate can be written as follows:

$$r_{V,S} = \frac{\mu_{max}}{Y_{max}} \cdot \frac{S_{NO3}}{S_{NO3} + K_{S,NO3}} \cdot \frac{S}{S + K_S} \cdot X_B \qquad (3.32)$$

where $\dfrac{S_{NO3}}{S_{NO3} + K_{S,NO3}}$ describes the effect of the nitrate concentration.

$K_{S,NO3}$ is the saturation constant for nitrate,
S   is the concentration of energy source (for example organic matter in wastewater),
K_S is the saturation constant for the energy source.

In many cases, Expression (3.32) can be simplified to a zero order expression in respect of nitrate and energy source:

$$r_{V,S} = \frac{\mu_{max}}{Y_{max}} \cdot X_B$$

as $S_{NO3} >> K_{S,NO3}$
$S >> K_S$

Compared with the aerobic processes, the essential difference is that by denitrification it is normally nitrate (the electron acceptor) we want to control, whereas by the aerobic heterotrophic processes it is normally the organic matter (the electron donor), we want to control.

Example 3.9

Calculate the removal rate for substrate, COD, and nitrate, $NO_3^- - N$, for a process where the following constants apply:

$\mu_{max} = 3\ d^{-1}$

$Y_{max,COD} = 0.50$ kg COD(B)/kg COD(S)

$K_{S,NO3} = 0.1$ g $NO_3^- - N/m^3$

$K_S = 20$ g COD/m^3

$X_B = 2$ kg COD(B)/m^3

The process is assumed to correspond to the one discussed in Example 3.7.

The process occurs in an ideally mixed tank, with influent and effluent concentrations of:

	in	out	
S_{NO3}	20	1	g $NO_3^- - N/m^3$
S_{COD}	200	10	g COD/m^3

Expression (3.32) is used:

$$r_{V,S} = \frac{\mu_{max}}{Y_{max}} \cdot \frac{S_{NO3}}{S_{NO3} + K_{S,NO3}} \cdot \frac{S}{S + K_S} \cdot X_B \qquad (3.32)$$

by evaluating the expression it is found

$$r_{V,S} = \frac{3 \text{ d}^{-1}}{0.5 \text{ kg COD(B) / kg COD(S)}} \cdot \frac{1}{1 + 0{,}1} \cdot \frac{10}{10 + 20} \cdot 2 \text{ kg COD(B) / m}^3$$

$$= 3.6 \text{ kg COD(S) / (m}^3 \cdot \text{d)}$$

The corresponding removal of nitrate is obtained by multiplication with the stoichiometric coefficient which can be found from the equation of reaction in Example 3.7:

1 mole phenol ~ 2.46 mole NO_3^- – N (from Example 3.7).

COD of phenol:

$C_6H_6O + 7 O_2 \rightarrow 6 CO_2 + 3 H_2O$

1 mole phenol consumes 7 mole O_2.

COD of phenol = 7 mole COD(S)/mole phenol.

1 mole phenol · 7 mole COD(S) · mole phenol = 7 mole COD(S) converted by means of ~ 2.46 mole NO_3^- – N.

That is, the stoichiometric coefficient, $v_{NO3,COD}$ is

$v_{NO3,COD}$ = (7 mole COD(S) · (32 g COD/mole COD))/(2.46 mole NO_3^- – N · 14 g N/mole N) = 6.50 g COD(S)/g NO_3^- – N or

$v_{NO3,COD}$ = 6.50 kg COD(S)/kg NO_3^- – N.

That means:

$r_{V,S(COD)} = v_{NO3,COD} \cdot r_{V,S(NO3)}$

$r_{V,S(NO3)} = r_{V,S (COD)} / v_{NO3,COD}$

$r_{V,S(NO3)}$ = (3.6 kg COD(S)/(m^3 · d))/((6.50 kg COD/S)/1 kg NO_3^- – N) = 0.55 kg NO_3^- – N/(m^3 · d)
(or 23 g NO_3^- – N/(m^3 · h)).

Using the stoichiometric coefficient, the maximum yield constant in relation to nitrate can also be determined:

$Y_{max,NO3} = Y_{max,COD} \cdot v_{NO3,COD}$ = 0.5 · 6.50 = 3.25 kg COD/kg NO_3^- – N.

By substituting this yield constant, $r_{V,S(NO3)}$ can be found from Expression (3.32).

3.5.6. The influence of the environmental factors, denitrification

Energy sources (substrate)
Denitrifying bacteria can utilize a broad spectrum of energy sources. Inorganic materials can also be used. Table 3.12 lists some of the energy sources used. Among the organic materials, the interest is particularly linked to the organic materials in wastewater and sludge, the so-called internal energy sources.

Acetone	Molasses
Newsprint	Methane
Hydrogen (potable water)	**Methanol**
Wastewater from breweries	Olive oil
Acetic acid	Org. matter in wastewater
Ethanol (potable water)	Raw syrup
Glucose	Sawdust
Cherry juice	Sulphur
Marmalade	

Table 3.12. Reductants used in denitrification experiments.
The substances in bold have been extensively used in practice /19/.

Among the external carbon and energy sources, methanol and acetic acid are the most interesting if we do not happen to have access to industrial wastewater, for example wastewater from breweries.

Example 3.10

In connection with water supply, acetic acid is a possible energy source for denitrification. In such a separate denitrification process, the observed yield constant will be higher than in combined nitrification-denitrification processes. The yield constant is assumed to be 0.5 kg COD/kg COD corresponding to 0.38 kg biomass/kg HAc. (This can be calculated on the basis of the two oxidation equations for the biomass, $C_5H_7NO_2$, and acetic acid, $C_2H_4O_2$, see Example 3.2).

The process looks as follows:

$$4.96\ C_2H_4O_2 + 3.94\ NO_3^- + NH_4^+ + 2.94\ H^+ \rightarrow$$
$$C_5H_7NO_2 + 4.92\ CO_2 + 1.97\ N_2 + 9.9\ H_2O$$

The alkalinity changes in this process are special because acetic acid (= negative alkalinity) is added.

For the sake of the process as such, the alkalinity has been increased by 2.94 equivalents per 3.94 moles of NO_3^- – N, that is, 0.75 eqv/mole NO_3^- – N.

Furthermore, a certain amount of alkalinity has been removed by adding acetic acid *before* the denitrification process.

The 4.96 mole HAc per 3.94 mole NO_3^- – N, are equal to 1.26 mole HAc/mole NO_3^- – N. Of each mole of HAc, 36 per cent is dissociated at a pH of 4.5. This means a removal of alkalinity corresponding to 36 per cent of the acetic acid added, that is, $0.36 \cdot 1.26 = 0.45$ eqv/mole NO_3^- – N. (The 36 per cent are derived from Example 3.13).

The final result is that the alkalinity has been increased by 0.30 eqv/mole NO_3^- – N.

The energy source used has an effect on the denitrification rate. In figure 3.13, this is illustrated for the biomass specific rate, $r_{X,S(NO3)}$ ($= r_{V,S(NO3)}/X_B$). Methanol yields a higher rate because it is easily degradable, and because it is a special type of bacteria which uses methanol. Consequently biomass has to be adopted to methanol. Organic matter in raw wastewater has a slower reaction rate whereas endogenous energy sources give the slowest rate. Here it will be the hydrolysis processes which limit the rate.

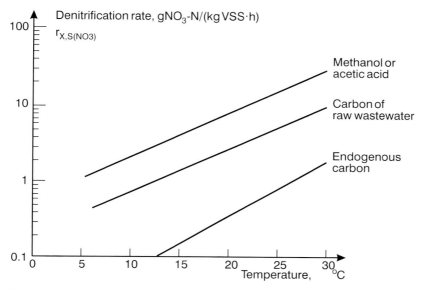

Figure 3.13. The correlation between carbon source, temperature and the biomass specific denitrification rate. From /24/.

Temperature

The temperature dependency of the denitrification process resembles that of the aerobic heterotrophic processes - see Expression (3.16). The denitrification process can also occur thermophilically at 50-60°C, but experiences are few. The removal rate is approximately 50 per cent higher than at 35°C.

Oxygen

Oxygen inhibits the denitrification process. Like for all other factors, it is the oxygen concentration experienced by the micro-organisms themselves, that is, within flocs or biofilms, which is crucial, and not the oxygen concentration which is normally measured in the liquid phase (the bulk phase). However, the effect of oxygen can still be described using the following approximated expression which is multiplied by using the reaction rate Expression (3.32):

$$\frac{K_{S,O2(NO3)}}{K_{S,O2(NO3)} + S_{O2}}$$

where $K_{S,O2(NO3)}$ is the "saturation constant" for oxygen inhibition,
        S_{O2} is the oxygen concentration in the liquid phase.

Denitrification

$K_{S,O2}$ varies with the actual conditions. In an activated sludge plant, $K_{S,O2}$ will be lower than in a biofilm plant. And in an activated sludge plant, $K_{S,O2(NO3)}$ will decrease by decreasing floc sizes (mixing intensity). In model calculations, the same saturation constant value is frequently used for denitrification, $K_{S,O2(NO3)}$, and for aerobic oxidation, $K_{S,O2}$.

pH

The denitrification process pH dependency resembles that of other biological processes. Figure 3.14 illustrates a pH optimum of around 7 to 9, but with variations depending on local conditions. Big differences can be observed between short-term and long-term pH dependency as the microbial system can slowly be adapted to the given pH.

A low pH (< 7) plays an important role for the end product resulting from the denitrification since an increasing amount of nitric oxides, especially N_2O, will be produced when the pH values decline. Nitric oxide, NO, which is a strong toxic gas, will hardly occur in toxic concentrations in practice /24/, /30/.

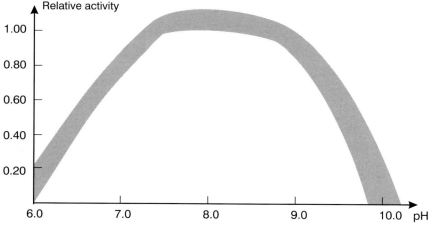

Figure 3.14. Denitrification as a function of pH.

Reaction rate constants for denitrification

Table 3.13 gives a set of reaction rate constants for denitrification.

	Symbol	Unit	Quantity
Maximum specific growth rate*	μ_{max}	d^{-1}	3-6
Maximum specific growth rate, methanol	μ_{max}	d^{-1}	5-10
Decay constant	b	d^{-1}	0.05-0.10
Saturation constant, nitrate	$K_{S,NO3}$	$g\ N/m^3$	0.2-0.5
Inhibition constant, oxygen	$K_{S,O2(NO3)}$	$g\ O_2/m^3$	0.1-0.5
Saturation constant, methanol	$K_{S,MeOH}$	$g\ COD/m^3$	5-10
Saturation constant*	$K_{S,COD}$	$g\ COD/m^3$	10-20
Hydrolysis constant, suspended solids	k_h	d^{-1}	0.15-0.4
Hydrolysis constant, dissolved solids	k_h	d^{-1}	1-15
Hydrolysis constant	k_{hX}	$kg\ COD(X)/kg\ COD(B) \cdot d$	0.15-0.4
Hydrolysis saturation constant	K_X	$kg\ COD(X)/kg\ COD(B)$	0.02-0.05
Maximum yield constant, methanol	Y_{max}	kg COD/kg COD	0.5-0.65
Maximum yield constant*	Y_{max}	kg COD/kg COD	0.5-0.55
Maximum yield constant*	Y_{max}	$kg\ COD/kg\ NO_3^- - N$	1.6-1.8
Temperature constant for μ_{max} and b	k	$°C^{-1}$	0.06-0.12

*organic matter in raw wastewater

Table 3.13. Reaction rate and stoichiometric constants for denitrification, 20°C /19/, /23/, /24/, /39/

3.6. Biological phosphorus removal

In this process bacteria take up large amounts of phosphate. The phosphate is used by the bacteria as an energy reserve which, under anaerobic conditions, can be used to pick up substrate. Regeneration of the phosphate reserve takes place under aerobic as well as anoxic conditions /28/,41/. There are many details of the bio-P process that needs to be investigated, but a broad outline of the metabolism of the bacteria involved is given below.

3.6.1. Microorganisms

Historically it was thought that Acinetobacter was the organism solely responsible for the Bio-P process. Today it is well known that phosphorus accumulating ability is widespread among heterotrophic microorganisms present in wastewater and in biomass in biological treatment processes. All these organisms are called Bio-P bacteria, PAOs /41/ or phosphorus accumulating organisms. The phosphorus accumulation mechanism is not always active, and this makes it difficult to measure for example the concentration of potential Bio-P bacteria in wastewater. In a treatment plant with biological phosphorus removal several groups of heterotrophs are active and competing for the substrate, especially for low molecular fatty acids, that is needed for the phosphorus storage mechanism. Many of the competitors are not PAOs. The result of this competition determines the success or failure of the Bio-P process.

PAOs, non-denitrifying

These organisms take up acetate and propionate under anaerobic and anoxic conditions and store it internally in the cell as PHA (polyhydroxyalkanoates) under simultaneous use of glycogen. The storage is made possible through energy release from polyphosphate degradation, resulting in release as ortho-phosphate. Under certain conditions (pH above 8 – 8.5) part of the released phosphate will precipitate as Calcium or Aluminium or other metal phosphates. Under aerobic conditions these organisms will grow, take up phosphate and store it as polyphosphate and rebuild the glycogen storage. The main energy for this comes from oxidation of PHA. The organisms may also oxidise other organic substrates available under aerobic conditions. Figure 3.15 illustrates the simplified metabolism of PAOs.

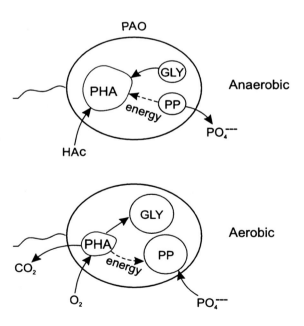

Figure.3.15. Metabolism of PAOs under aerobic and anaerobic conditions. PHA= polyhydroxyalkanoate, GLY=glycogen, PP=polyphosphate and HAc=acetate.

PAOs, denitrifying
These organisms take up acetate and propionate under anaerobic conditions and store it as PHA under simultaneous use of glycogen. The storage is made possible through energy release from polyphosphate degradation, resulting in release as ortho-phosphate. Under certain conditions (pH above 8 – 8.5) part of the released phosphate will precipitate as Calcium or Aluminium or other metal phosphates. Under anoxic and aerobic conditions these organisms will grow, take up phosphate and store it as polyphosphate and rebuild the glycogen storage. The energy for this comes from oxidation of PHA. The growth yield will be bigger for aerobic growth than for anoxic growth.

Non-PAOs, non-denitrifying
These organisms can ferment organic compounds under anaerobic conditions. They will oxidise most carbon sources under aerobic conditions, but without any storage of polyphosphate.

Non-PAOs, denitrifying
These organisms may ferment organic compounds under anaerobic conditions. Under anoxic conditions they will denitrify and grow with organic substrates available. Under aerobic conditions these organisms will grow with a yield coefficient higher that the one for anoxic growth.

GAOs (glycogen accumulating bacteria)
A group of heterotrophic bacteria, which can not store phosphate, but can compete for the substrate, especially for glucose present in the wastewater/43/. In most cases these organisms will not play a role in Bio-P processes.

Microtrix
These organisms take up long chain fatty acids, often present in the wastewater and in the first step anaerobic tank of a Bio-P process. In this tank long chain fatty acids may also be generated from fermentation or hydrolysis processes/42/,/46/. Microtrix results in high SVI of activated sludges, especially under cold conditions (10 – 15 °C). As the anaerobic first step in some case acts as a selector for Microtrix, a different process lay-out should be used (anaerobic tank placed in the sludge recycle).

3.6.2 Reactions, biological phosphorus removal
The reactions in the Bio-P process are complicated/41/. A simplified set of reactions is shown below, taking into account the COD and P transformations only. They are based on the following assumptions for yields in the various sub-processes, and the maximum content of polyphosphate in the PAOs:

BIOLOGICAL PHOSPHORUS REMOVAL

$Y_{PAO,obs}$	Yield coefficient, observed	0.3	g PAO-COD/ g PHA-COD
Y_{PAO}	Yield coefficient, maximum	0.63	g PAO-COD/ g PHA-COD
Y_{PO4}	PP requirement (PO4 – P release) per PHA stored	0.40	g P/ g PHA-COD
Y_{PHA}	PHA requirement for PP storage	0.20	g PHA-COD/ g P
$i_{PP,BM}$	Max PP content in biomass	0.17	g P/g PAO-COD

The PHA in the equations below should be interpreted as the sum of PHA and GLY(glycogen) in the PAOs. The substrate, which is mainly acetate and propionate is expressed as HAc-COD.

Anaerobic process:

$$0.4 \text{ g PP-P} + 1 \text{ g HAc-COD} \rightarrow 0.4 \text{ g PO}_4^{-3} - \text{P} + 1 \text{ g PHA-COD} + 0.04 \text{ g H}^+ \tag{3.33}$$

Equation (3.33) gives the net turnover in the process. The role of glycogen is not specified in the equation, as it acts as a catalyst, being reduced in the anaerobic phase, where some of it is converted to PHA, and being regenerated in the aerobic or anoxic phases. This can confuse if PHA is analytically determined, as approximately 30 % more analytically PHA is being produced (based on C-moles) than the acetate C-moles taken up by the organisms/44/.

Figure 3.16 shows how the concentration of acetate and phosphate varies under anaerobic conditions in a Bio-P process.

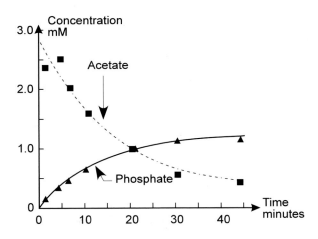

Figure 3.16. Release of phosphate and uptake of acetate under anaerobic conditions. From /25/.

Aerobic storage:

$$0.2 \text{ g PHA-COD} + 1 \text{ g PO}_4^{-3} - P \rightarrow 1 \text{ g PP-P} + 0.1 \text{ g OH}^- \quad (3.34)$$

Aerobic growth (with maximum yield, $Y_{PAO}= 0.63$):

$$1.6 \text{ g PHA-COD} + (-0.6 \text{ g O}_2\text{-COD}) \rightarrow 1 \text{ g PAO-COD} \quad (3.35)$$

Total process, anaerobic plus aerobic (growth and storage with $Y_{PAO,obs}= 0.3$ g COD/g COD):

$$20 \text{ g HAc-COD} + 1 \text{ g PO}_4^{-3} - P \rightarrow 6 \text{ gPAO-COD} + 1 \text{ g PP-P} + 0.1 \text{ g OH}^- \quad (3.36)$$

This gives a total stoichiometric coefficient for polyphosphate uptake of $v_{HAc,P} = 0.05$ g P/g HAc-COD.

Example 3.11

Wastewater contains 50 g acetic acid (HAc)/m³ and 5 g P/m³.

How much phosphorus can be removed by biological phosphorus removal?

From Expression (3.36), the total stoichiometric coefficient is $v_{HAC,P} = 0.05$ g P/g HAc-COD.

Hence 50 g HAc/m³ · 0.05 g P/g HAc-COD = 2.5 g P/m³ (corresponding to 50 per cent) can be removed.

3.6.3. Yield constant, biological phosphorus removal

The maximum yield constant of the phosphorus accumulation bacteria corresponds to that of the aerobic heterotrophs, that is, 0.5-0.6 kg COD(B)/kg COD(S). If it is instead calculated on a weight basis in the form of suspended solids, SS, it will vary, depending on how much polyphosphate the bacteria contain. As a maximum, the bacteria can contain approx. 50 per cent polyphosphate. This corresponds to a phosphorus content of 15-20 per cent and a yield constant of 1.0-1.2 kg SS/kg COD(S).

3.6.4. Alkalinity

The alkalinity is slightly affected by the Bio-P process/41/:

Anaerobic	-0.29	eqv/g P release
Aerobic	+0.10	eqv/g P uptake
Total	-0.19	eqv/g P removed

Thus for municipal wastewater with a removal of 5 g P/m³, there will be a loss of alkalinity of approximately 1 eqv/m³.

3.6.5. Kinetics, biological phosphorus removal

Anaerobic conditions
The acetate uptake kinetics can be described based on the kinetics in the Activated Sludge Model No.2d /41/:

$$r_{V,HAc} = q_{PHA} \cdot \frac{S_{HAc}}{K_{HAc} + S_{HAc}} \cdot \frac{S_{ALK}}{K_{ALK} + S_{ALK}} \cdot \frac{X_{PP}/X_{B,PAO}}{K_{PP} + X_{PP}/X_{B,PAO}} \cdot X_{B,PAO} \quad (3.37)$$

where q_{PHA} is PHA storage rate
S_{HAc} is acetate concentration
S_{ALK} is the alkalinity
X_{PP} is polyphosphate
$X_{B,PAO}$ is the concentration of PAOs
$K_?$ is half-saturation constant for the Monod expressions

The first Monod term in the equation is the substrate regulation, the second the pH regulation and the third the polyphosphate storage regulation, introducing a maximum amount of PP storage.

Aerobic conditions
The kinetics for the phosphate uptake under aerobic conditions can be written as:

$$r_{V,PO4} = q_{PP} \cdot \frac{S_{O2}}{K_{O2} + S_{O2}} \cdot \frac{S_{PO4}}{K_{PS} + S_{PO4}} \cdot \frac{S_{ALK}}{K_{ALK} + S_{ALK}} \cdot \frac{X_{PHA}/X_{B,PAO}}{K_{PHA} + X_{PHA}/X_{B,PAO}} \cdot$$

$$\frac{K_{MAX} - X_{PP}/X_{B,PAO}}{K_{IPP} + K_{MAX} - X_{PP}/X_{B,PAO}} \cdot X_{B,PAO} \quad (3.38)$$

The last Monod term is the limiting of the PP storage in the PAOs.

Anoxic conditions
The kinetics under anoxic conditions can be described using similar expressions as under aerobic conditions. The rates are reduced compared to aerobic ones. How much depends on the percentage of denitrifying organisms among the PAOs, which seems to be around 50-70% in practice.

3.6.6. Environmental factors, biological phosphorus removal

The response of phosphorus accumulating bacteria to environmental factors by and large corresponds to the reaction shown by the denitrifying and the aerobic heterotrophic bacteria. In order for the accumulation of polyphosphate to take place, two requirements are important in practice:

- alternating anaerobic/aerobic conditions
- no nitrate in the anaerobic period

Anaerobic/aerobic conditions
The anaerobic condition is important for the selection mechanism in the treatment plant. By introducing this condition we increase the selection pressure to the advantage of the phosphorus accumulating bacteria, and the result is that a larger part of the biomass will consist of these bacteria.

Nitrate
During the intended anaerobic period, nitrate has severe impact. Denitrification removes some of the easily degradable organic matter which was supposed to be stored in the phosphorus accumulating bacteria. The result is that the phosphorus removal is reduced as the amount of easily degradable organic matter available has been reduced. The denitrification process with acetic acid takes place according to the expression in Example 3.10, from which it appears that 4.96 mole HAc/3.94 mole NO_3^- = 1.26 mole HAc/mole NO_3^- are consumed. Hence the phosphorus removal process has been depleted of the organic matter consumed by denitrification.

Reaction rate constants, phosphorus accumulating bacteria
Table 3.14 gives examples of the reaction rate constants of phosphorus accumulating bacteria.

	Symbol	Unit	Quantity
Maximum specific growth rate	$\mu_{max,PAO}$	d^{-1}	2-4
Maximum yield constant, acetic acid	$Y_{max,PAO}$	kg COD(B)/kg COD(HAc)	0.5-0.65
Maximum yield constant, acetic acid	$Y_{max,PAO}$	kg SS/kg COD(HAc)	0.6-0.8
Maximum yield constant, acetic acid	$Y_{max,PAO}$	kg P/kg COD(HAc)	0.07-0.10
Saturation constant, acetic acid uptake	$K_{S,HAc}$	g HAc/m^3	2-6
Saturation constant, phosphate uptake	$K_{S,PO4}$	g P/m^3	0.1-0.5
PHA storage rate	q_{PHA}	kg COD(HAc)/(kg COD(X)·d)	2-4
PP storage rate	q_{PP}	kg P/(kg COD(X)·d)	1-2
Temperature constant for $\mu_{max,P}$	κ	°C^{-1}	0.02-0.04

Table 3.14. Reaction rate constants for PAO's, phosphorus accumulating bacteria, 20°C. Based on /41, 44/

3.7. Anaerobic processes

In this context anaerobic processes are defined as processes where neither oxygen nor nitrate is present. These processes are carried out by a large and varied group of micro-organisms which normally live in a symbiotic relationship. The energy conditions are extremely complicated and for many bacteria so bad that it is almost impossible to exist, but still they have been successful in many cases.

Many of the bacteria are strictly anaerobic and thus cannot tolerate oxygen at all; this is particularly true for methane-forming bacteria.

The anaerobic degradation process can roughly be divided into the three steps shown in figure 3.17. The two steps are biological, whereas the hydrolysis step is enzymatic. The biological steps are shown in more detail in Figs 3.18 and 3.19.

The acid-forming and the methane-forming bacteria are normally divided into two subgroups as shown in Table 3.15.

Step	Name	Substrate(s)	End product
Acid production	Acid-forming bacteria	carbohydrates amino acids lipids	butyric acid propionic acid
Methane production	Acetoclastic bacteria	acetic acid	methane carbon dioxide
	Methane bacteria	hydrogen carbon dioxide	methane

Table 3.15. Division of bacteria, anaerobic processes.

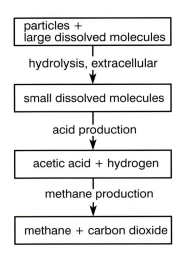

Figure 3.17. Very simplified 3-step anaerobic degradation of organic matter.

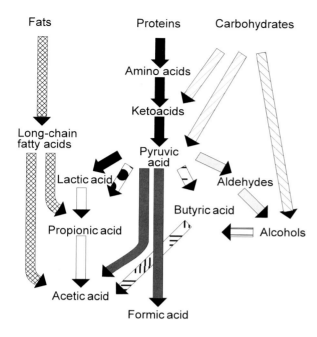

Figure 3.18. The acid step, anaerobic processes, in a simplified representation /27/.

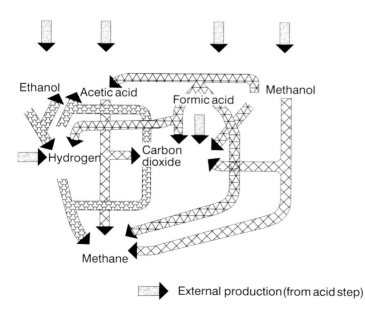

Figure 3.19. Methane step, anaerobic processes, in a simplified representation /27/.

3.7.1. Reactions, anaerobic processes

Compared with methane production, acid production is a quick process. As an example of the complicated removals, the result of a combined propionic acid fermentation and a mixed acid fermentation is given in Expression (3.39). These processes are carried out by a larger collective group of bacteria and the result is as follows:

$$5 \text{ glucose} \rightarrow 2 \text{ lactic acid} + 4 \text{ propionic acid} + 3 \text{ acetic acid} + \text{ethanol} + 4 \text{ carbon dioxide} + 2 \text{ hydrogen} + \text{water} \quad (3.39)$$

or

$$5 \text{ C}_6\text{H}_{12}\text{O}_6 \rightarrow 2 \text{ CH}_3\text{CHOH} \cdot \text{COOH} + 4 \text{ CH}_3\text{CH}_2\text{COOH} + 3 \text{ CH}_3\text{COOH} + \text{CH}_3\text{CH}_2\text{OH} + 4 \text{ CO}_2 + 2 \text{ H}_2 + \text{H}_2\text{O} \quad (3.40)$$

In a simplified form, the conversion can be written:

$$\text{C}_6\text{H}_{12}\text{O}_6 \rightarrow 3 \text{ CH}_3\text{COOH} \quad (3.41)$$

The methane production takes place by one of the two processes:

$$\text{CH}_3\text{COOH} \rightarrow \text{CH}_4 + \text{CO}_2 \quad (3.42)$$

$$4 \text{ H}_2 + \text{CO}_2 \rightarrow \text{CH}_4 + 2 \text{ H}_2\text{O} \quad (3.43)$$

In a simplified form, the conversion from glucose to methane can be written:

$$\text{C}_6\text{H}_{12}\text{O}_6 \rightarrow 3 \text{ CH}_4 + 3 \text{ CO}_2 \quad (3.44)$$

The ratio between methane and carbon dioxide is here 1:1. The ratio will differ for other organic materials. If the oxidation step for carbon is higher than in glucose, relatively more carbon dioxide will be produced, and vice versa if the oxidation step is lower than 0.

In practice, a significant part of the carbon dioxide will be dissolved in the liquid phase so that the methane content of the gas phase will be higher than the equivalent of what is given by the equation of reaction.

3.7.2. Growth and soluble COD yield constants, anaerobic processes

The growth yield constants vary for the different types of bacteria participating in the process. Generally it is low for the acid step and even lower for the methane step.

For the acid step, the maximum yield constant is $Y_{max,S} = 0.2\text{-}0.3$ kg COD(B)/kg COD(S) corresponding to 0.15-0.20 kg VSS/kg COD(S).

With an observed yield constant of for example 0.2 kg COD(B)/kg COD(S) and a biomass composition which can be approximated by $C_5H_7NO_2$, the equation of reaction for the conversion of glucose into acetic acid can be written:

The yield constant can be converted into the unit mole of biomass/mole of glucose in the same manner as shown in Example 3.2, which gives $Y_{obs} = 0.24$ mole bio-mass/mole glucose. Hence the equation of reaction is:

$$C_6H_{12}O_6 + 0.24\,NH_4^+ \rightarrow 0.24\,C_5H_7NO_2 +$$
$$2.40\,CH_3COO^- + 0.72\,H_2O + 2.64\,H^+ \qquad (3.45)$$

The products from the acid step can be used in nutrient removal processes, if methane production is avoided. The hydrolysate is highly reactive and consists of mainly acetic and propionic acid which together accounts for 60 – 80% of the soluble COD after hydrolysis.

The hydrolysis soluble COD yield with respect to total COD varies with temperature, time and quality of the organic matter. The maximum yield can be as high as 10-20% for primary sludge and down to 2-6% for activated sludge.

For the methane step, the maximum growth yield constant is $Y_{max,M} = 0.04$-0.05 kg COD(B)/kg COD(S) corresponding to 0.03-0.04 kg VSS/kg COD(S).

With an observed yield constant of for example 0.03 kg COD(B)/kg COD(S), the equation of reaction can be written (as 0.03 kg COD(B)/kg COD(S) can be converted into 0.012 mole biomass/mole acetic acid, see the principle from Example 3.2):

$$CH_3COOH + 0.012\,NH_4^+ + 0.012\,OH^- \rightarrow$$
$$0.012\,C_5H_7NO_2 + 0.97\,CH_4 + 0.97\,CO_2 + 0.048\,H_2O \qquad (3.46)$$

The total maximum yield constant for an anaerobic process is 0.25-0.35 kg COD(B)/kg COD(S). Note that by calculating in COD-units, the yield constants of the two steps can be added directly.

In practice, the total observed yield constant will be low due to the low load under which the anaerobic processes operate. It will frequently be 0.05-0.10 kg COD(B)/kg COD(S).

3.7.3. Nutrients, anaerobic processes

The contents of N, P and S in anaerobic bacteria are seen in Table 3.16. The most significant difference from aerobic bacteria is the higher content of sulphur.

		g/kg VSS	g/kg COD(B)	g/kg TOC
Nitrogen	N	80-120	55-85	150-250
Phosphorus	P	10-25	7-18	25-55
Sulphur	S	10-25	7-18	25-55
Iron	Fe	5-15	4-11	12-30

Table 3.16. Typical concentrations of nutrients, anaerobic bacteria.

Knowing the yield constant, the demand for the different nutrients can be calculated.

Example 3.12

Calculate the necessary COD(S)/N, COD(S)/P and COD(S)/S ratios for wastewater to be treated in an anaerobic process with a total observed yield constant of 0.23 kg COD(B)/kg COD(S), and N, P and S-contents of 7, 1.4 and 1.4 per cent of COD, respectively.

1 kg COD converted yields 0.23 kg COD (B). The N, P and S-contents are:

N = 0.07 · 0.23   = 0.016 kg N
P = 0.014 · 0.23   = 0.003 kg P
S = 0.014 · 0.23   = 0.003 kg S

Hence the ratios are:

COD(S)/N = 1.0/0.016 = 63/1
COD(S)/P = 1.0/0.003 = 333/1
COD(S)/S = 1,0/0.003 = 333/1

If, in practice, all three ratios are lower, there will be no limitation in nutrients.

3.7.4. Alkalinity, anaerobic processes

The anaerobic processes influence the alkalinity. The acid step reduces the alkalinity and the methane step increases it. The overall result is a small reduction in alkalinity. In Example 3.13, the separate change in alkalinity for the acid step is calculated.

Example 3.13

Calculate the change in alkalinity by conversion of glucose to acetic acid, see Expression (3.45).

2.64 mole H^+ per mole of glucose converted are produced corresponding to a reduction in alkalinity of 2.64 eqv. The produced acetate increases the alkalinity as pKa for HAc is 4.75, which means that about half of the acetate, Ac^-, is converted into acetic acid, HAc, by the titration to pH = 4.5 (by measuring the alkalinity). From /8/ we have:

$$pH = pK_a + \log \frac{[Ac^-]}{[HAc]}$$

By substitution of $pK_a = 4.75$ and $pH = 4.50$ we find:

$\log \dfrac{[Ac^-]}{[HAc]} = -0.25$ or

$\dfrac{[Ac^-]}{[HAc]} = 0.56$

or $[Ac^-] = 0.56 \, [HAc]$

Assuming that $[HAc] + [Ac^-] = 1$, we find:

$[HAc] + 0.56 \, [HAc] = 1 \rightarrow [HAc] = 0.64$ and

$[Ac^-] = 0.36$

that is, at $pH = 4.5$,

$[Ac^-] = 0.36 \cdot ([Ac^-] + [HAc])$ and

$[HAc] = 0.64 \cdot ([Ac^-] + [HAc])$

Hence the contribution to the alkalinity from Ac^- is $0.64 \cdot 2.40 = 1.5$ eqv/mole glucose.

The overall change of the alkalinity, ΔTAL, is:

$\Delta TAL = 1.5 - 2.64 = -1.14$ eqv/mole glucose.

3.7.5. Kinetics, anaerobic processes

Hydrolysis, anaerobic processes

The process can be described using the same expression of kinetics as for aerobic and anoxic processes, that is, either in a very simplified form such as:

$$r_{V,XS} = k_h \cdot X_S \tag{3.3a}$$

$$r_{V,SS} = k_h \cdot S_S \tag{3.3b}$$

or as a saturation type of expression:

$$r_{V,XS} = k_{hX} \cdot \dfrac{X_S / X_{B,S}}{K_X + (X_S / X_{B,S})} \cdot X_{B,S} \tag{3.4}$$

The hydrolysis constants are lower under anaerobic conditions than under aerobic conditions.

Example 3.14

The hydrolysis rate, $r_{V,XS}$, in an anaerobic process is set at 0.38 kg COD(X)/(m³ · d). Furthermore we know:

$X_S = 0.5$ kg COD/m³

$X_{B,S} = 1$ kg COD/m³

$k_{hX} = 0.4$ kg COD(X)/(kg COD(B) · d)

Find the hydrolysis saturation constant, K_X.

From Expression (3.4), K_X can be found:

$$0.38 \text{ kg COD(X)}/(m^3 \cdot d) = 0.4 \text{ d}^{-1} \cdot \frac{0.5/1.0}{K_X + 0.5/1.0} \cdot 1.0 \text{ kg COD}/m^3$$

$$0.38 = \frac{0.2}{K_X + 0.5}$$

$K_X = 0.026$ kg COD(X)/kg COD(B)

Acid step, anaerobic processes

The acid production can be described by Monod kinetics:

$$r_{V,S} = \frac{\mu_{max,S}}{Y_{max,S}} \cdot \frac{S}{S + K_{S,S}} \cdot X_{B,S} \quad (3.47)$$

where S is the substrate in the form of small dissolved molecules of organic matter,
          $X_{B,S}$ is the biomass of acid-forming bacteria,
          $K_{S,S}$ is the saturation constant.

Methane step

The methane production can also be described by Monod kinetics:

$$r_{V,S} = \frac{\mu_{max,M}}{Y_{max,M}} \cdot \frac{S}{S + K_{S,M}} \cdot X_{B,M} \quad (3.48)$$

where S is the concentration of acetic acid,
          $X_{B,M}$ is the biomass of methane producers.

Acetic acid can act as an inhibitor which can be described by using the following kinetics /29/:

$$r_{V,S} = \frac{\mu_{max,M}}{Y_{max,M}} \cdot \left[1 + \frac{K_{S,M}}{S} + \frac{S}{K_I}\right]^{-1} \cdot X_{B,M} \quad (3.49)$$

where K_I is the inhibition constant.

3.7.6. Gas production

The gas production in an anaerobic process consists of methane, carbon dioxide and hydrogen. The gas may further contain free nitrogen and hydrogen sulphide.

If the observed yield constant and the substrate for the anaerobic process are known, an expression can be set up showing how big the production of gas is.

If, for example, a total observed yield constant of 0.08 mole of biomass/mole of glucose is assumed, the following expression is found:

$$C_6H_{12}O_6 + 0.08\,NH_4^+ \rightarrow$$

$$0.08\,C_5H_7NO_2 + 2.8\,CH_4 + 2.8\,CO_2 + 0.24\,H_2O + 0.08\,H^+ \tag{3.50}$$

The gas production is here 2.8 mole of CH_4 and 2.8 mole of CO_2. The composition of the gas depends on the amount of carbon dioxide dissolved in the liquid phase.

The main part (90-95 per cent) of the energy, for example measured as COD found in the substrate for an anaerobic process, can be retrieved in the methane produced.

> Example 3.15
>
> Calculate the percentage of the COD, COD(S), of the substrate which can be retrieved in methane in the process described in Expression (3.50).
>
> COD in 1 mole glucose is found from the expression:
>
> $C_6H_{12}O_6 + 6\,O_2 \rightarrow 6\,CO_2 + 6\,H_2O$
>
> 1 mole $C_6H_{12}O_6 \sim 6 \cdot 32\,g = 192\,g\,O_2 = 192\,g\,COD(S)$.
>
> COD in methane is found from the expression:
>
> $CH_4 + 2\,O_2 \rightarrow CO_2 + 2\,H_2O$
>
> 1 mole $CH_4 \sim 2 \cdot 32 = 64\,g\,O_2 = 64\,g\,COD(M)$.
>
> From Expression (3.50) we see that 2.8 mole CH_4 are produced per mole of glucose removed. It corresponds to a COD-recovery as methane of:
>
> $$\frac{2.8\ \text{mole}\ CH_4 \cdot 64\ g\ COD(M)\,/\,\text{mole}\ CH_4}{1\ \text{mole glucose} \cdot 192\ g\ COD(S)\,/\,\text{mole glucose}} \cdot 100\% = 93\%$$

3.7.7. The influence of the environmental factors, anaerobic processes

Temperature

The temperature dependency of the anaerobic processes corresponds to what is known from the aerobic processes, hence Expression (3.16) can be used. The generally slow reaction rate for anaerobic processes calls for a very long hydraulic retention time and a very long sludge retention time. In many cases, this makes it economically profitable to allow the processes to occur at high temperatures as large tank volumes can be saved. Figure 3.20 illustrates the temperature dependency for anaerobic processes.

pH

The overall anaerobic process occurs at maximum rate in the pH range 6-8. For a pH below 6, the activity of the methane-forming bacteria drops rapidly so that at a pH of 5.5 they have by and large stopped their activity.

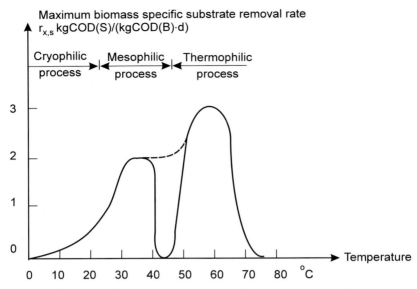

Figure 3.20. Anaerobic removal rate as a funtion of temperature. The process might be unstable or fail between 45°C and 50°C.

As discussed in Section 3.7.4, the overall anaerobic process only influences the alkalinity to a small extent.

Inhibition

The anaerobic processes are influenced by toxic substances following a pattern similar to what is known from aerobic organisms. Therefore they are not specially sensitive. As the processes are designed on the basis of the slow-growing methane bacteria, inhibition will normally be observed for these bacteria at first. Table 3.17 gives some examples of concentrations of toxic substances which influence the processes. Just like other biological processes, the anaerobic processes can develop resistance to toxic substances which can also be seen in the table.

Parameter	Inhibition by single dose	Inhibition by continued supply
pH	< 6	< 5
pH	> 8	> 8,5
Ammonium, NH_4^+	> 100 g/m^3	> 200 g/m^3
Hydrogen sulphide, H_2S	> 250 g/m^3	> 1000 g/m^3
Cyanide, CN^-	> 5 g/m^3	> 100 g/m^3
Trichloromethane	> 1 g/m^3	> 50 g/m^3
Formaldehyde	> 100 g/m^3	> 400 g/m^3
Nickel	> 200 g/m^3	> 50 g/m^3

Table 3.17. Toxic substances, anaerobic processes.

Reaction rate constants, anaerobic processes

Table 3.18 gives examples of reaction rate constants.

	Symbol	Unit	Quantity
Maximum specific growth rate, acid-forming bacteria	$\mu_{max,S}$	d^{-1}	1-3
Maximum specific growth rate, methane-forming bacteria	$\mu_{max,M}$	d^{-1}	0.3-0.5
Maximum yield constant, acid-forming bacteria	$Y_{max,S}$	$\frac{\text{kg COD (B)}}{\text{kg COD (S)}}$	0.2-0.3
		$\frac{\text{kg VSS}}{\text{kg COD (S)}}$	0.15-0.20
Maximum yield constant, methane-forming bacteria	$Y_{max,M}$	$\frac{\text{kg COD (B)}}{\text{kg COD (S)}}$	0.04-0.05
		$\frac{\text{kg VSS}}{\text{kg COD (S)}}$	0.03-0.04
Hydrolysis constant, suspended solids	k_h	d^{-1}	0.3-0.7
Hydrolysis constant, dissolved solids	k_h	d^{-1}	2-20
	k_{hX}	$\frac{\text{kg COD (X)}}{\text{kg COD (B)}} \cdot d^{-1}$	0.3-0.7
Hydrolysis saturation constant	K_X	$\frac{\text{kg COD (X)}}{\text{kg COD (B)}}$	0.02-0.05
Saturation constant, acid step	$K_{S,S}$	kg COD(S)/m³	0.03-0.15
Saturation constant, methane step	$K_{S,M}$	kg COD(S)/m³	0.03-0.10
Inhibition constant, acetic acid, methane step	K_I	kg COD(S)/m³	0.2-0.8

Table 3.18. Reaction rate constants, anaerobic processes, 35°C.

References

/1/ Fenger, B. (1970): Aktiverede slamanlæg og biologiske filtre (Activated Sludge Plants and Biofilters). *Stads- og Havneing.*, **61**, 47-54.

/2/ McKinney, R.E. (1962): Microbiology for Sanitary Engineers. McGraw-Hill Book Company, Inc., New York, N.Y.

/3/ van Gils, H.W. (1964): Bacteriology of Activated Sludge. Research Institute for Public Health Engineering, TNO, Haag. (Report no. 32).

/4/ Liebmann, H. (1962): Handbuch der Frischwasser- und Abwasserbiologie (Manual of Freshwater and Wastewater Biology). R. Oldenburg, München.

/5/ Wanner, J. and Grau, P. (1989): Identification of Filamentous Microorganisms from Activated Sludge: A Compromise between Wishes, Needs and Possibilities. *Water Res.*, **23**, 883-891.

/6/ Arvin, E. (1984): Biological Phosphorus Removal-Systems, Design and Operation. *Vatten*, **40**, 411-415.

/7/ Pöpel, F. (ed.) (1975/88): Lehrbuch für Abwassertechnik und Gewässerschutz (Textbook for Wastewater Engineering and Protection of the Aquatic Environment). Deutscher Fachschriften-Verlag, Wiesbaden.

/8/ Harremoës, P., Henze, M., Arvin, E. and Dahi, E. (1989): Teoretisk vandhygiejne (Water Chemistry). Third edition. Polyteknisk Forlag, Lyngby, Denmark.

/9/ Henze, M., Grady, C.P.L., Gujer, W., Marais, G. v. R., and Matsuo, T. (1987): Activated Sludge Model No. 1. IAWPRC, London. (IAWPRC Scientific and Technical Reports No. 1).

/10/ Busse, H.J. (1975): Instrumentelle Bestimmung der organischen Stoffe in Wässern (Instrumental Determination of Organic Matter in Waters). *Z. f. Wasser- und Abwasser-Forsch.*, **8**, 164-176.

/11/ Henze, M. and Mladenovski, C. (1991): Hydrolysis of Particulate Substrate by Activated Sludge under Aerobic, Anoxic and Anaerobic Conditions. *Water Res.*, **25**, 61-64.

/12/ Sharma, B. and Ahlert, R.C. (1977): Nitrification and Nitrogen Removal. *Water Res.*, **11**, 897-925.

/13/ DS (1977): Vandundersøgelse: 5-døgns biokemisk oxygenforbrug (BOD) (Water Analysis: Determination of Biological Oxygen Demand (BOD)). Danish Standards Association, Copenhagen. (DS/R 254).

/14/ Painter, H.A. (1977): Microbial Transformations of Inorganic Nitrogen. *Prog. Wat. Technol.*, **8**, (4/5), 3-29.

/15/ EPA (1975): Process Design Manual for Nitrogen Control. U.S. Environmental Protection Agency,. Washington D.C.

/16/ Knowles, G., Downing, A.L. and Barrett, M.J. (1965): Determination of Kinetic Constants for Nitrifying Bacteria in Mixed Culture, with the Aid of an Electronic Computer. *J. Gen. Microbiol.*, **38**, 263-278.

/17/ Gujer, W. (1977): Design of a Nitrifying Activated Sludge Process with the Aid of Dynamic Simulation. *Prog. Wat. Tech.*, **9**, (2), 323-336.

/18/ Buswell, A.M. et al. (1954): Laboratory Studies on the Kinetics of the Growth of Nitrosomonas with Relation to the Nitrification Phase of the B.O.D. Test. *Appl. Microbiol.*, **2**, 21-25.

/19/ Henze Christensen, M. and Harremoës, P. (1978): Nitrification and Denitrification in Wastewater Treatment. Chap. 15 in: Mitchell, R. (ed.) Water Pollution Microbiology, Vol. 2, pp. 391-414. John Wiley & Sons, New York, N.Y.

/20/ Anthonisen, A.C. et al. (1976): Inhibition of Nitrification by Ammonia and Nitrous Acid. *J. WPCF*, **48**, 835-852.

/21/ Sutton, P.M. and Jank, B.E. (1976): Design Considerations for Biological Carbon Removal-Nitrification Systems. In: Conference Proceedings - Research Program for the Abatement of Municipal Pollution under Provisions of the Canada-Ontario Agreement on Great Lakes Water Quality, 1975, pp. 206-249. Training and Technology Transfer Division (Water) Environmental protection Service, Fisheries and Environment Canada, Ottawa, Ontario. (Publ. No. 3).

/22/ Tomlinson, T.G., Boon, A.G. and Trotman, G.N.A. (1966): Inhibition of Nitrification in the Activated Sludge Process of Sewage Disposal. *J. Appl. Bact.*, **29**, 266-291.

/23/ Stensel, H.D., Loehr, R.C. and Lawrence, A.W. (1973): Biological Kinetics of Suspended-Growth Denitrification. *J. WPCF*, **45**, 249-261.

/24/ Henze Christensen, M. and Harremoës, P. (1977): Biological Denitrification of Sewage. A literature review. *Prog. Water Techn.*, **8**, (4/5), 509-555.

/25/ Tracy, K.D. and Flammino, A. (1987): Biochemistry and Energetics of Biological Phosphorus Removal. In: Ramadori, R. (ed.): Biological Phosphate Removal from Wastewaters. Proceedings of an IAWPRC specialized conference held in Rome, Italy, 28-30 September 1987, pp.15-26. Pergamon Press, Oxford. (Advances in Water Pollution Control).

/26/ Mino, T., Arun, V., Tsuzuki, Y and Matsuo, T. (1987): Effect of Phosphorus Accumulation on Acetate Metabolism in the Biological Phosphorus Removal Process. In: Ramadori, R. (ed.): Biological Phosphate Removal from Wastewaters. Proceedings of an IAWPRC Specialized Conference held in Rome, Italy, 28-30 September 1987, pp. 27-38. Pergamon Press, Oxford. (Advances in Water Pollution Control)

/27/ Henze, M. and Harremoës, P. (1983): Anaerobic Treatment of Wastewater in Fixed Film Reactors - A Literature Review. *Water Sci. Technol.*, **15**, (8/9), 1-101.

/28/ Arvin, E. (1985): Biological Removal of Phosphorus From Wastewater. *CRC Crit. Rev. Environ. Contr.*, **15**, (1), 25-65.

/29/ Andrews, J.F. (1969): Dynamic Model of the Anaerobic Digestion Process. *J. Sanit. Engng. Div., Proc. ASCE*, **95**, 95-116.

/30/ Pichinoty, F. et al. (1977): Etude de 14 bacteries denitrifiantes appartenant du groupe Pseudomonas stutzeri isolées du sol par culture d'enrichissement en presence d'oxyde nitreux. (Study of 14 denitrifying bacteria from the Pseudomonas stutzeri group isolated from soil and enriched in the presence of nitrous oxide). *Ann. Microbiol. (Inst. Pasteur)*, **128A**, 75-87.

/31/ Loveless, J.E. and Painter, H.A. (1968): The Influence of Metal Ion Concentrations and pH Value on the Growth of a Nitrosomonas Strain Isolated from Activated Sludge. *J. Gen. Microbiol.*, **52**, 1-14.

/32/ Hiraishi, A., Masamune, K. and Kitamura, H. (1989): Characterization of the Bacterial Population Structure in an Anaerobic-Aerobic Activited Sludge System on the Basis of Respiratory Quinone Profiles. *Appl. Env. Microb.*, **55**, 897-901.

/33/ Henze, M. (1992): Characterization of Wastewater for Modelling of Activated Sludge Processes. *Water Sci. Technol.*, **25**, (6), 1-15.

/34/ Ekama, G.A., Dold, P.L. and Marais, G. v. R. (1986): Procedures for Determining COD and the Maximum Specific Gowth Rate of Heterotrophs in Activated Sludge Systems. *Water Sci. Technol.*, **18**, (6), 91-114.

/35/ Kappeler, J. and Gujer, W. (1992): Estimation of Kinetic Parameters of Heterotrophic Biomass under Aerobic Conditions and Characterization of Wastewater for Activated Sludge Modelling. *Water Sci. Technol.*, **25**, (6), 125-139.

/36/ Grady, C.P.L., Daigger, G. and Lim, H.C. (1998): Biological Wastewater Treatment. Theory and Applications. 2nd ed. Marcel Dekker, Inc. New York, N.Y.

/37/ Theerman, J. (1991): Biological Hazards at Wastewater Treatment Facilities. Water Pollution Control Federation, Alexandria, VA.

/38/ Kerrn-Jespersen, J.P. and Henze, M. (1993): Biological Phosphorus uptake under anoxic and aerobic conditions. *Water Res.*, **27**, 617-624.

/39/ Gujer, W, Henze, M, Mina, T., Van Loosdrecht, M.C.M. (2000): Activated Sludge Model No. 3. In: Henze, M., Gujer, W., Mino, T. and Loosdrecht, M.C.M., (eds.) Activated sludge models ASM1, ASM2, ASM2D and ASM3, IWA Publishing, London. (IWA Scientific and Technical Report No. 9).

/40/ van Loosdrecht, M.C.M. and Henze, M. (1999): Maintenance, endogeneous respiration, lysis, decay and predation. *Water Sci. Technol.*, **39**, (1), 107-117.

/41/ Henze, M., Gujer, W., Mino, T., Matsuo, T., Wentzel, M.C., Marais, G.v.R. and van Loosdrecht, M.C.M. (1999): Activated sludge model No. 2d, ASM2d. *Water Sci. Technol.*, **39**, (1), 165-182.

/42/ Andreasen, K. and Nielsen, P.H.(1998): In situ characterization of substrate uptake by Microthrix Parvicella using microautoradiography. *Water Sci. Technol.*, **37**,No. 4-5, 19-26.

/43/ Mino,T., Satoh,H. and Matsuo,T. (1994): Metabolisms of different bacterial populations in enhanced biological phosphate removal processes. *Water Sci. Technol.*, **29**, (7), 67-70.

/44/ Kuba, T., Van Loosdrecht, M.C.M., Murnleitner, E. and Heijnen, J.J (1997): Kinetics and stoichiometry in the biological phosphorus removal process with short cycle times. *Water Res.*, **31**, 918-928.

/45/ Van Loosdrecht, M.C.M., Pot, M.A., Heijnen, J.J. (1997): Importance of bacterial storage polymers in bioprocesses. *Water Sci. Technol.*, **35** (1), 41-47.

/46/ Andreasen, K. and Nielsen, P.H. (2000): Growth of *Microtrix parvicella* in nutrient removal activated sludge plants: Studies of *in situ* physiology. *Water Res.*, **34**, 1559-1569.

/47/ Dircks, K., Pind, P.F., Mosbæk, H. and Henze, M. (1999): Yield determination by respirometry - The possible influence of storage under aerobic conditions in activated sludge. *Water S.A.*, **25**, (1), 69-74.

Activated sludge tank. Metamorfosis wastewater treatment plant (Athens, Greece). Ideally mixed tank with surface aeration. Treatment plants in Greece receive large amounts of septage which influence design as well as operation of the plants.

4 Activated Sludge Treatment Plants

By Mogens Henze

Activated sludge treatment plants are built to remove organic matter from wastewater to avoid the situation described by Joseph Heller in *Something Happened*: "You don't find fish in lakes and rivers anymore. You have to catch them in cans. Towns die. Oil spills. Money talks".

The principle in activated sludge plants is that a mass of activated sludge is kept in suspention by stirring or aeration. Apart from the living biomass, the suspended solids contain inorganic as well as organic particles. Some of the organic particles can be degraded by hydrolysis whereas others are non-degradable (inert).

The amount of suspended solids in the treatment plant is regulated through recycle of the suspended solids and by removing excess sludge. Organic matter that enters an activated sludge process has 3 outlets only: carbon dioxide, excess sludge or the effluent. Figure 4.1. shows the elements of an activated sludge process.

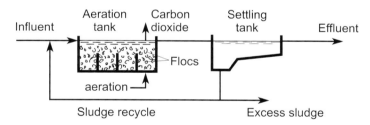

Figure 4.1. Definition of activated sludge process elements.

4.1. Mass balances, activated sludge plants

By setting up mass balances from the expressions of reaction derived in Chapter 3, each term will need a positive or negative sign, depending on which side of the mass balance it is placed. If the substance in question is removed, the sign is negative.

An activated sludge process with symbols is shown in figure 4.2.
Different mass balances can be set up for the plant.

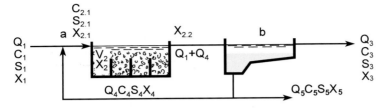

Figure 4.2. Activated sludge plant with recycle.

The water balance looks as follows:

$$Q_1 = Q_3 + Q_5 \tag{4.1}$$

A total mass balance looks as follows:

$$Q_1 \cdot C_1 - r_{V,S} \cdot V_2 = Q_3 \cdot C_3 + Q_5 \cdot C_5 \tag{4.2}$$

The conversion term $r_{V,S} \cdot V_2$ can also be written in such a way that tank volume, V_2, as well as activated sludge concentration, $X_{B,2}$, are included separately, that is as $r_{X,S} \cdot V_2 \cdot X_{B,2}$.

The mass balance can be used for three interesting purposes as shown in Table 4.1, that is, for the design, for the investigation of function, and for the investigation of kinetics.

By writing the mass balance in Expression (4.2), no definitions are implied in respect of the meaning of the individual symbols. C is a concentration (COD, BOD, N, TOC, etc.), $r_{V,S}$ or $r_{X,S}$ is a reaction rate whose units must be defined together with the unit of the volume, V_2, and for $r_{X,S}$ the activated sludge concentration, X_2.

The activated sludge concentration, X_2, can for example be measured in kg of SS/m³, kg of VSS/m³ or kg of COD(B)/m³. What the individual units stand for must be defined in each individual case. VSS may, for example, be the total VSS-content in the sludge, the content of nitrifying bacteria (measured as VSS), the content of denitrifying bacteria, etc. **But** if X_2 is used with a unit which is active biomass (living bacteria), the equivalent reaction rate, $r_{X,S}$, must have the same unit in the denominator.

Purpose	Unknown	Known
Design	Volume, V_2	$Q_1, C_1, X_2, C_3, Q_3, r_{X,S}$
Prediction/analysis of function	Effluent concentration, C_3	$Q_1, C_1, V_2, X_2, Q_3, r_{X,S}$
Kinetics studies	Reaction rate, $r_{X,S}$	$Q_1, C_1, V_2, X_2, Q_3, C_3$

Table 4.1. Calculations which may be carried out by means of a mass balance for an activated sludge plant.
(For symbols see figure 4.2)

The mass balance in Expression (4.2) has been set up in the simplest way as all conversions of the substance are concentrated in one term, that is the term $r_{V,S} \cdot V_2$. The other terms are transport terms describing the input and output of solids.

In practice several processes will frequently influence the substance considered, making the mass balance more complicated.

A simplified process matrix for activated sludge plants without nitrification is illustrated in Table 4.2. It is **assumed** that three processes contribute to the conversion, that is, biological growth, decay and hydrolysis. The reaction rate is shown to the far right, and the coefficients in the Table are stoichiometric coefficients. By means of the Table, a mass balance can be set up, for example for easily degradable organic matter, S_S, in an ideally mixed tank. The transport terms are self-explanatory whereas the two reaction terms are found by multiplying the stoichiometric coefficient from (in this case) the S_S-column by the reaction rate expression in the column to the right in Table 4.2.

$$\overset{\text{input}}{Q_1 \cdot S_{S,1}} + \overset{\text{hydrolysis}}{1 \cdot k_h \cdot X_{S,2} \cdot V_2} + \overset{\text{growth}}{\left(-\frac{1}{Y_{max,H}}\right) \mu_{max,H} \cdot}$$

$$\frac{S_{S,2}}{S_{S,2} + K_S} \cdot \frac{S_{O2,2}}{K_{S,O2,H} + S_{O2,2}} \cdot X_{B,H,2} \cdot V_2 = Q_3 \cdot S_3 \quad (4.3)$$

Table 4.2 also shows that aerobic growth influences the three components S_S, $X_{B,H}$ and S_{O2}. To use the matrix for mass balances, all units must be uniform (in this case COD). For this reason, oxygen has been expressed as negative COD as there is no oxygen demand when aerating a certain amount of oxygen, but this amount of oxygen can remove an equivalent amount of COD!

If the recycle is zero, the activated sludge concentration, $X_{B,2}$, can be found if the observed yield constant, Y_{obs}, is known:

$$X_{B,2} = Y_{obs}(C_1 - C_3) \quad (4.4)$$

The yield constant, Y_{obs}, is here the current value for the plant in question estimated on the basis of the sludge mass leaving the plant. Based on this definition it does not matter whether there is sludge (suspended solids) in the influent or not.

Example 4.1

Find the activated sludge concentration, $X_{B,2}$, in an activated sludge plant without recycle where the influent and effluent concentrations are 500 g COD(S)/m³ and 80 g COD(S)/m³, respectively.

The observed yield constant is known and is 0.35 kg COD(B)/kg COD(S) or 0.25 kg VSS/kg COD(S).

From Expression (4.4) is found:

$X_{B,2} = Y_{obs} (C_1 - C_3) = (0.35$ kg COD(B)/kg COD(S)) $(0.5$ kg COD(S)/m^3 $- 0.08$ kg COD(S)/m$^3) = 0.147$ kg COD(B)/m^3

$X_{B,2}$ can also be calculated in the unit kg of VSS/m^3 if the yield constant with the unit VSS is used, that is, $Y_{obs} = 0.25$ kg VSS/kg COD. Hence $X_{B,2} = 0.105$ kg VSS/m^3.

Example 4.2

Design an activated sludge plant (find the necessary volume) corresponding to the one shown in figure 4.2. The following quantities are known:

Influent: $Q_1 = 1{,}000$ m^3/d
$C_1 = 500$ g COD(S)/m^3

Effluent: $Q_3 = 1{,}000$ m^3/d
$C_3 = 60$ g COD(S)/m^3 (value required by the authorities)

Furthermore the following quantities are known:

$r_{X,S} = 3$ kg COD(S)/(kg COD(B) · d)
$Y_{obs} = 0.3$ kg COD(B)/kg COD(S)

From Expression (4.2) the necessary volume, V_2, of the aeration tank can be found:

$V_2 = (Q_1 \cdot C_1 - Q_3 \cdot C_3)/(r_{X,S} \cdot X_{B,2})$

The only unknown is $X_{B,2}$. It is found in the same way as shown in Example 4.1

$X_{B,2} = Y_{obs} \cdot (C_1 - C_3) = 0.3 \cdot (0.5 - 0.06) = 0.132$ kg COD(B)/m^3.

Substitution of these values gives:

$V_2 = (1{,}000 \cdot 500 - 1{,}000 \cdot 60)/(3 \cdot 0.132) = 1{,}111{,}111$ m^3.

Component	S_S	X_S	X_I	$X_{B,H}$	S_{O2}	Reaction rate, $r_{v,\ldots}$
Process						
Aerobic heterotrophic growth	$-\dfrac{1}{Y_{max,H}}$			1	$\dfrac{1-Y_{max,H}}{Y_{max,H}}$	$\mu_{max,H} \left(\dfrac{S_S}{S_S+K_S}\right)\left(\dfrac{S_{O2}}{K_{S,O2,H}+S_{O2}}\right) X_{B,H}$
Decay of heterotrophs		$1-f_{XB,XI}$	$f_{XB,XI}$	-1		$b_H \cdot X_{B,H}$
Hydrolysis	1	-1				$k_h \cdot X_S$
Unit	kg COD/m^3					
	Easily degradable organic matter	Slowly degradable organic matter	Inert suspended organic matter	Heterotrophic biomass	Oxygen	

Table 4.2. Process matrix for an activated sludge plant without nitrification.

It is a good idea to substitute the specific units and not just the numbers as shown above, thus checking whether commensurable quantities have been substituted and whether the expression as such is correct in terms of design.

Substitution of these values gives:

$V_2 = (1{,}000 \text{ m}^3/\text{d} \cdot 500 \text{ g COD(S)/m}^3 - 1{,}000 \text{ m}^3/\text{d} \cdot 60 \text{ g COD(S)/m}^3)/(3 \text{ kg COD(S)/(kg COD(B)} \cdot \text{d}) \cdot 0.132 \text{ kg COD(B)/m}^3) = 1{,}111{,}111 \text{ m}^3 \cdot \text{g COD(S)/kg COD(S)}$.

However, it is seen that something is wrong. It may be corrected by either substituting C_1 and C_3 into the unit kg of COD(S)/m^3, or by converting the result by means of the conversion factor 1,000 g COD(S)/kg COD(S).

$V_2 = 1{,}111{,}111 \text{ m}^3 \cdot \text{g COD (S)}/((1{,}000 \text{ g COD (S)/ kg COD (S)}))$
$= 1{,}111 \text{ m}^3$,

which is (hopefully) the correct result.

The recycle has the effect that the hydraulic retention time and the sludge age are separated. In this way it is possible to accumulate a biomass consisting of both rapid- and slow-growing microorganisms. This is very important for the settling and flocculation behaviour of the sludge. Figure 4.3 shows the structure of the sludge flocs in activated sludge.

The sludge must be concentrated prior to recycling which normally takes place in a settling tank. Flotation, centrifugation and membrane systems of various kind may also be used.

Mass balances can also be set up at points a and b in figure 4.2.

A mass balance for sludge, set up at point a, may for example be used to calculate the influent concentration of sludge discharged into the aeration tank, $X_{2.1}$

$$Q_1 \cdot X_{B,1} + Q_4 \cdot X_{B,4} = (Q_1 + Q_4) \cdot X_{B,2.1} \tag{4.5}$$

An equivalent mass balance at point b may for example be used to calculate the flow rate for return sludge, Q_4, if the sludge concentrations are known which flow from the top and bottom of the settling tank ($X_{B,3}$ and $X_{B,4}$) as well as the sludge concentration in the effluent from the aeration tank, $X_{B,2.2}$

$$(Q_1 + Q_4) \cdot X_{B,2.2} = Q_3 \cdot X_{B,3} + Q_4 \cdot X_{B,4}$$

$$Q_4 = (Q_1 \cdot X_{B,2.2} - Q_3 \cdot X_{B,3})/(X_{B,4} - X_{B,2.2})$$

The concentration of sludge is increased while passing through the aeration tank. A sludge mass balance around the aeration tank itself may be used to calculate the tank effluent concentration, $X_{B,2.2}$ (see figure 4.2):

$$(Q_1 + Q_4) \cdot X_{B,2.1} + (Q_1 + Q_4) \cdot (C_{2.1} - C_{2.2}) \cdot Y_{obs} = (Q_1 + Q_4) \cdot X_{B,2.2}$$

or

$$X_{B,2.2} = X_{B,2.1} + (C_{2.1} - C_{2.2}) \cdot Y_{obs} \tag{4.6}$$

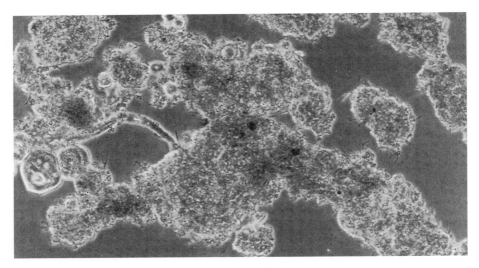

Figure 4.3. Activated sludge flocs magnified 200 times. The flocs include micro-organisms, organic and inorganic particles. They are glued together by polymers produced by the bacteria. Dense regular flocs with protozoa indicate that the biomass is healthy. In the event of inhibition or other mistreatment, the protozoa will die simultaneously with the disaggregation of the flocs. Filamentous bacteria (not visible in this picture) reduce the settling and concentration ability of the flocs.(Photo: Consorcio des Aquas, Bilbao)

Example 4.3 shows how to calculate the effluent concentration from the aeration tank. The amount of the sludge concentration in the aeration tank itself, $X_{B,2}$, depends on the hydraulics. If the tank has an ideal mix

$$X_{B,2} = X_{B,2.2}$$

If the tank has a plug flow system, the concentration varies throughout the tank from $X_{B,2.1}$ in the influent to $X_{B,2.2}$ in the effluent.

For activated sludge plants with recycle it often applies that $X_{2.1} \sim X_{2.2}$. This does **not** apply to plants without recycle.

Example 4.3

Calculate the effluent concentration, $X_{B,2.2}$, of activated sludge from the aeration tank of an activated sludge plant. The following is known (for symbols, see figure 4.2):

$X_{B,2.1} = 4.5$ kg SS/m^3

$C_1 = 0.4$ kg BOD/m^3

$C_3 = 0.015$ kg BOD/m^3 (15 mg BOD/l)

$Y_{obs} = 0.9$ kg SS/kg BOD

$Q_1 = 2{,}500$ m^3/d

$Q_4 = 2{,}000$ m^3/d

$X_{B,2.2}$ can be calculated from Expression (4.6)

$$X_{B,2.2} = X_{B,2.1} + (C_{2.1} - C_{2.2}) \cdot Y_{obs} \tag{4.6}$$

$C_{2.1}$ and $C_{2.2}$ are not known.

$C_{2.1}$ and $C_{2.2}$ are not the total BOD-concentration, but should be assumed to represent the BOD content of the raw wastewater (in the plant it is mixed into the activated sludge with its very high BOD concentration).

It is normally assumed that no reactions take place in the settling tank, that is, $C_{2.2} = C_3$.

$C_{2.1}$ can be found by means of a BOD-balance at point a (figure 4.2):

$$Q_1 \cdot C_1 + Q_4 \cdot C_4 = (Q_1 + Q_4) \cdot C_{2.1}$$

Assuming that removal only takes place in the aeration tank, $C_4 = C_3 = C_{2.2}$; substitution of these values gives:

$$C_{2.1} = (Q_1 \cdot C_1 + Q_4 \cdot C_3)/(Q_1 + Q_4)$$

Substitution of these values by known values gives:

$$C_{2.1} = (2{,}500 \cdot 0.4 + 2{,}000 \cdot 0.015)/(2{,}500 + 2{,}000) = 0.229 \text{ kg BOD/m}^3$$

Substitution of these values by known values into Expression (4.6) gives:

$$X_{B,2.2} = 4.5 + (0.229 - 0.015) \cdot 0.9 = 4.69 \text{ kg SS/m}^3.$$

The example shows that the variation from the influent entering the aeration tank to the effluent leaving the tank is, in this case, only an increase of the sludge concentration of 4-5 per cent. Considering the uncertainties in the determination of the incoming quantities, the variation may often be ignored and the sludge concentration may be assumed to be constant throughout the tank.

NOTE! The variation **cannot** be ignored during the process of calculating the sludge production which takes place in the plant (see next chapter).

4.2. Concepts and definitions of the activated sludge process

Several concepts and definitions are related to activated sludge plants. Some of the concepts and definitions are not very applicable and logical, but it is useful to know what they mean when others use them. Some of the concepts are briefly explained below.

Treatment efficiency is normally used to describe the overall efficiency of the treatment process, and is defined as:

$$E = \frac{C_1 - C_3}{C_1} \tag{4.7}$$

The treatment efficiency can also be used to describe what happens in a single process or tank in the plant.

Furthermore the treatment efficiency can be used as a dimensionless quantity in different mathematical descriptions of the activated sludge process.

Recycle rate is defined as the ratio between return sludge and raw wastewater (see figure 4.2):

$$R = Q_4/Q_1 \qquad (4.8)$$

The typical value is given in Table 4.3.

Volumetric loading in an activated sludge plant is defined as:

$$B_V = Q_1 \cdot C_1/V_2 \qquad (4.9)$$

The volumetric loading can be given in different units, but is usually expressed as kg of BOD/(m³ · d). Volumetric loading is used for the primitive design of activated sludge plants.

Sludge concentration (biomass concentration), X, in wastewater in the aeration tank or in the flow of recycle sludge is often given in the unit suspended solids, X_{SS}, volatile suspended solids, X_{VSS}, or as COD, X_{COD}. Calculations of the activated sludge processes can be made in any required unit. If, for example, SS is used, all quantities must be calculated in this unit, that is, yield constant, sludge mass, sludge production, removal rate, etc. It should be noted that only a part of the sludge concentration measured, for example as COD, is active biomass (living bacteria), that is

$$X_{COD} > X_{B,H} + X_{B,A}$$

where $X_{B,H}$ is the concentration of heterotrophic biomass,
$X_{B,A}$ is the concentration of autotrophic biomass (nitrifying bacteria).

	Symbol	Unit	Quantity
Treatment efficiency for BOD, COD and SS	E	%	80-95
Recycle ratio	R	%	50-100
Sludge index	SVI	ml/g	50-150
Oxygen respiration, 20° C (activated sludge)	$r_{X,O2}$	gO₂/(kgVSS h)	5-40
Sludge percentage	-	%	50-80

Table 4.3. Different parameters in municipal wastewaters and ordinary activated sludge plants.

Sludge mass, M_X, in an activated sludge plant comprises any sludge in the part of the treatment plant handling activated sludge, including the sludge which is not currently being aerated in the activated sludge tank. Sometimes large sludge masses may be hidden in secondary settling tanks. The sludge mass is defined as:

$$M_X = \Sigma V \cdot X \qquad (4.10)$$

Hydrolysis is carried out in the whole sludge mass, irrespective of the location of the sludge, however, at varying speed depending on the electron acceptors at hand in the individual tanks.

Sludge loading states the amount of organic matter applied to the activated sludge mass per day:

$$B_X = Q_1 \cdot C_1/(V_2 \cdot X_2) \qquad (4.11)$$

The sludge concentration is often expressed as kg of SS/m³ and the concentration of organic matter, C_1, as kg of BOD/m³ so that the unit is kg of BOD/(kg SS · d). The sludge loading is often used as design parameter. When using the unit SS, attention should be paid to the wastewater conditions (for example much inorganic suspended solids) or to the operation of the plant (for example the addition of precipitants) which may modify the ratio between VSS and SS in the activated sludge. A more reliable unit for sludge loading for design purposes is kg of BOD/(kg VSS · d).

Sludge production, F_{SP}, states the quantity of sludge leaving the plant per time unit. Figure 4.2 shows that

$$F_{SP} = Q_3 \cdot X_3 + Q_5 \cdot X_5 \qquad (4.12)$$

The sludge in the influent ($Q_1 \cdot X_1$) adds to the sludge production and should therefore not be deducted in Expression (4.12).

The sludge production can be measured in many ways and hence be given in different units, for example kg SS/d, kg VSS/d and kg COD/d.

Table 4.4 shows which components are included in the different measuring methods for sludge and sludge production.

Sludge production component→ Unit for sludge production↓	Inorganic suspended solids, raw wastewater	Organic suspended solids, raw wastewater	Biological growth, aeration tank	Chemical precipitants
kg SS/d	+	+	+	+
kg VSS/d		+	+	(+)
kg COD/d		+	+	

Table 4.4. Measuring methods for sludge and sludge production and the sludge components included in the methods.

Different units for the sludge production are needed, depending on the purpose of application. For sludge dewatering, the unit kg SS/d is relevant; in the case of an anaerobic or an aerobic sludge stabilization, the unit kg of VSS/d or kg of COD(B)/d

Concepts and definitions of the activated sludge process

is relevant. The volume of the sludge production thus depends both on the influent of wastewater to the plant, and on the process in the plant. The sludge production can be **measured** in an existing plant and is calculated from Expression (4.12).

The sludge production can be **estimated** in a design situation by means of Expression (4.13):

$$F_{SP} = Y_{obs}(C_1 - C_3) \cdot Q_1 \tag{4.13}$$

In order to be able to estimate the sludge production by means of Expression (4.13) the yield constant, Y_{obs}, must be estimated. This constant varies, for example according to the loading of the activated sludge plant, see figure 3.5.

Example 4.4

Calculate sludge loading and sludge production for an activated sludge plant for the treatment of municipal wastewaters. The following is known:

$Q_1 = 11{,}000 \text{ m}^3/\text{d}$

$C_1 = 0.4 \text{ kg BOD/m}^3$

$C_3 = 0.04 \text{ kg BOD/m}^3$

The concentration of activated sludge in the aeration tank is 5.2 kg SS/m^3. The volume, V_2, of the aeration tank is 5,600 m^3.

The sludge loading is calculated from Expression (4.11):

$B_X = Q_1 \cdot C_1/(V_2 \cdot X_2)$

Substitution of these values gives:

$B_X = 11{,}000 \cdot 0.4/5{,}600 \cdot 5.2 = 0.15 \text{ kg BOD/(kg SS} \cdot \text{d)}$.

The sludge production is estimated from Expression (4.13):

$F_{SP} = Y_{obs}(C_1 - C_3) \cdot Q_1$ (4.13)

Y_{obs} is estimated (for example from figure 3.5) to 0.8 kg SS/kg BOD. Substitution of these values gives:

$F_{SP} = 0.8 (0.4 - 0.04) \cdot 11{,}000 = 3{,}170 \text{ kg SS/d}$.

(Note that in the denominator the yield constant, Y_{obs}, **must** have the same unit as the concentration, C, in this case BOD. The numerator may have any unit (SS, VSS, COD, TN, TP, etc.)

It is not all of the sludge production which is further treated in the sludge treatment section of the activated sludge plant. Some of the sludge production leaves the treatment plant together with the treated water. It corresponds to the quantity $Q_3 \cdot X_3$ in figure 4.2. Sometimes this amount is significant, with high sludge discharges during storm events.

Excess sludge production, F_{ESP}, is that part of the sludge production which is actively withdrawn and further treated at the treatment plant. Using the symbols from figure 4.2, the quantity is $Q_5 \cdot X_5$. The excess sludge production can be with-

Activated Sludge Treatment Plants

drawn from the recycle flow as shown in figure 4.2. It can be handled, together with the primary sludge by leading it to the influent of the treatment plant if there is mechanical treatment.

If sludge age control is required in the plant, direct excess sludge withdrawal from the aeration tank is the best solution, see figure 4.4.

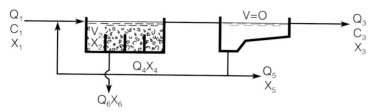

Figure 4.4. Activated sludge with sludge age control by excess sludge withdrawal from the aeration tank.

Sludge age is the mean cell residence time of sludge (biomass) in the plant. It can be estimated on the basis of measurements on an existing plant by means of Expression (4.14):

$$\theta_X = M_X/F_{SP} \qquad (4.14)$$

For the plant shown in figure 4.2:

$M_X = V_2 \cdot X_2$

$F_{SP} = Q_3 \cdot X_3$

and $X_2 = X_3$ (based on ideal mix)

$$\theta_X = V_2 \cdot X_2/Q_3 \cdot X_3 = V_2/Q_3 = \theta \qquad (4.15)$$

where θ_X is sludge age,
θ is hydraulic retention time.

For the plant in figure 4.2., the sludge mass is

$M_X = V_2 \cdot X_2$
(the sludge mass in the settling tank being set at 0 which is not quite correct)

$F_{SP} = Q_3 \cdot X_3 + Q_5 \cdot X_5 = Y_{obs} \cdot (C_1 - C_3) \cdot Q_1$

Substitution into Expression (4.14) gives:

$$\theta_X = V_2 \cdot X_2/(Y_{obs} \cdot (C_1 - C_3) \cdot Q_1) = \theta \cdot X_2/(Y_{obs} \cdot (C_1 - C_3)) \qquad (4.16)$$

Expression (4.16) is general and also applies for an activated sludge plant without recycle where $X_2 = Y_{obs} \cdot (C_1 - C_3)$ (Expression (4.4)), which, if substituted into Expression (4.16), gives $\theta_X = \theta$ (= Expression (4.15)).

Example 4.5

Calculate the hydraulic retention time in the aeration tank (ideal mix) and the sludge age in the activated sludge plant shown in figure 4.4.

The following is known:

$V_2 = 15,000 \text{ m}^3$

$X_2 = 4.0 \text{ kg COD/m}^3$

$Q_1 = 1,500 \text{ m}^3/\text{h}$

$Q_4 = 750 \text{ m}^3/\text{h}$

$Q_5 = 100 \text{ m}^3/\text{d}$

$Q_6 = 1,000 \text{ m}^3/\text{d}$

$X_3 = 0.05 \text{ kg COD/m}^3$

$X_5 = 11.0 \text{ kg COD/m}^3$

$Y_{obs} = 0.4 \text{ kg COD(B)/kg COD(S)}$

The hydraulic retention time is:

$\theta = V_2/Q_1 = 15,000/1,500 = 10 \text{ h}$

The hydraulic retention time is **not**:

$\theta = V_2/(Q_1 + Q_4) = 15,000/(1,500 + 750) = 6.7 \text{ h},$

as it is the passage time (= the mean time for **one** flow through of a water particle).

The sludge age, θ_X, is calculated from the expression:

$\theta_X = M_X/F_{SP}$ (4.14)

$M_X = V_2 \cdot X_2$ (the volume of the settling tank being = 0, so that no sludge is present)

$F_{SP} = Q_3 \cdot X_3 + Q_5 \cdot X_5 + Q_6 \cdot X_6$

Substitution of these values gives:

$\theta_X = V_2 \cdot X_2/(Q_3 \cdot X_3 + Q_5 \cdot X_5 + Q_6 \cdot X_6)$

Here all quantities are known with the exception of X_6 and Q_3. As the Q_6-water flow is withdrawn from the aeration tank, $X_6 = X_2$. Q_3 is found from a water balance for the whole plant.

$Q_1 = Q_3 + Q_5 + Q_6$

$Q_3 = Q_1 - Q_5 - Q_6 = 1,500 \cdot 24 - 100 - 1,000 = 34,900 \text{ m}^3/\text{d}.$

Substituion of Q_3 and X_6 and other known quantities gives:

$\theta_X = 15,000 \cdot 4.0/(34,900 \cdot 0.05 + 100 \cdot 11.0 + 1,000 \cdot 4.0)$

$\theta_X = 8.8 \text{ d}$

Activated Sludge Treatment Plants

NOTE: This example gives more information than needed to come up with an adequate solution (redundant information). Very often engineers have problems with insufficient information, and such systems are therefore likely to be subject to engineering estimates.

Aerobic sludge age is important for nitrification processes and for processes where slowly degradable organic matter must be removed biologically. The aerobic sludge age states the time during which a sludge particle (for example a nitrifying bacteria) stays in the plant under aerobic conditions. The aerobic sludge age is usually somewhat shorter than the sludge age (the total sludge age). The definition is parallel to the sludge age:

$$\theta_{X,aerobic} = M_{X,aerobic} / F_{SP} \qquad (4.17)$$

In a tank, with aerobic conditions half of the time (and in the other half for example anoxic conditions), the aerobic sludge mass will be 50 per cent of the total sludge mass, that is $0.5 \cdot V \cdot X$.

4.3. Types of plants, activated sludge plants

The physical design of activated sludge plants may vary considerably. Many designs and tank depths can be used. The design is normally determined on the basis of economy. Irrespective of the design, a plant must consist of two elements:

– aeration tank, with a certain volume, V_2, and a certain sludge concentration, X_2,
– settling tank or some other separation unit (membranes, flotation etc.) from where treated wastewater is discharged at the top (concentrations C_3 and X_3) and concentrated return sludge is withdrawn from the bottom (concentrations C_4 and X_4).

In certain plant designs the two elements have physically been constructed as **one** element with two functions, see for example Section 4.3.2 concerning oxidation ditches.

4.3.1. Activated sludge with recycle

Figure 4.5 shows examples of activated sludge plants with separate settling tanks. The aeration tank (type a) may be designed as a rectangular or a square tank, or it may be an oxidation ditch or a pond (lined with plastics or placed directly in the ground/in clay).

In type b the tank is long, narrow and with plug flow whereas the aeration tanks in type c (figure 4.5) are ideally mixed tanks and they are usually rectangular.

The plant type with step feeding (type d) normally has an oblong aeration tank with a type of plug flow (resembling a contact stabilization plant, see under contact

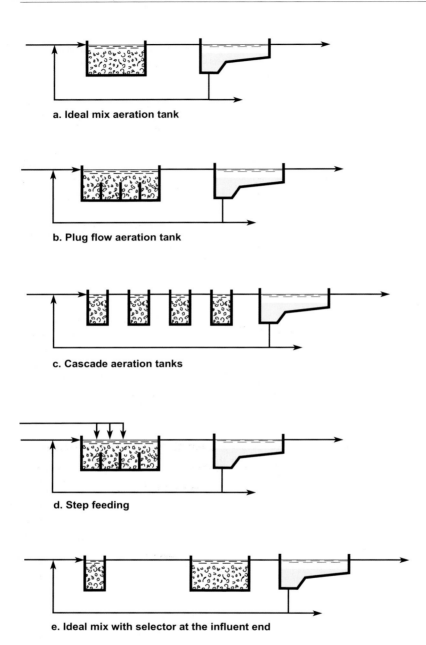

Figure 4.5. Types of activated sludge plants with separate settling tank.

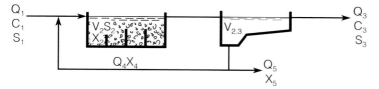

Figure 4.6. Basic layout of the plants in figure 4.5.

stabilization plants) whereas the selector system (type e) normally consists of a small ideally mixed tank followed by a large ideally mixed tank which may be rectangular, square, an oxidation ditch, or a pond.

Irrespective of the physical design, sufficient stirring/turbulence is required to ensure that the sludge is kept suspended, and that it gets into contact with the influent wastewater. This is usually achieved through aeration, but the demand for air (oxygen demand) can be so insignificant that the air supply cannot support the mixing, or the aeration system may be designed in such a way that a good mixing is not ensured.

The basic layout of the plants shown in figure 4.5 is the one shown in figure 4.6 with two separate tanks and return sludge Q_4.

Irrespective of the design, air is continuously supplied to the aeration tank. This is to facilitate an oxygen concentration which ensures a sufficiently high removal rate.

The substrate removal rate, $r_{V,S}$, can for all these plant designs be written as:

$$r_{V,S} = \frac{\mu_{max}}{Y_{max}} \cdot \frac{S_2}{S_2 + K_S} \cdot \frac{S_{O2,2}}{S_{O2,2} + K_{S,O2}} \cdot X_{B,2} \qquad (4.18)$$

where μ_{max} is the maximum specific growth rate for the removal of organic matter (S),
Y_{max} is the maximum yield constant per unit of organic matter (S),
S_2 is the concentration of organic matter in the aeration tank (**not** in the influent),
$S_{O2,2}$ is the concentration of oxygen in the aeration tank,
$X_{B,2}$ is the concentration of sludge (biomass) in the aeration tank.

Like all other expressions describing the biological removals, Expression (4.18) has been simplified. For example it will be more correct to use the concentration of activated (living) heterotrophic biomass, $X_{B,H}$, in Expression (4.18). It implies also that other growth constants, μ_{max} and Y_{max}, must be used.

A mass balance for the whole activated sludge plant for dissolved organic matter is:

input + hydrolysed − removed = output
$Q_1 \cdot S_1$ + $r_{V,XS} \cdot V_2 \cdot v_{XS}$ − $r_{V,S} \cdot V_2$ = $Q_3 \cdot S_3$ (4,19)

$v_{X,S}$ is the stoichiometric coefficient for hydrolysis.

The variation of the activated sludge concentration (biomass) through the plug-flow aeration tanks is usually so insignificant that in terms of calculations it may be assumed that it is constant. This is not possible for the concentration of dissolved organic matter which will vary from a value corresponding to a mix of raw wastewater and return sludge (water) to a value corresponding to that of the treated water.

4.3.2. Single tank activated sludge plants

Activated sludge plants can be built as **one** combined tank functioning both as aeration tank and settling/separation tank. In certain plant types the two tanks are integrated to such an extent that they may in principle be considered as one tank with several functions. Figure 4.7 gives examples of these plant designs.

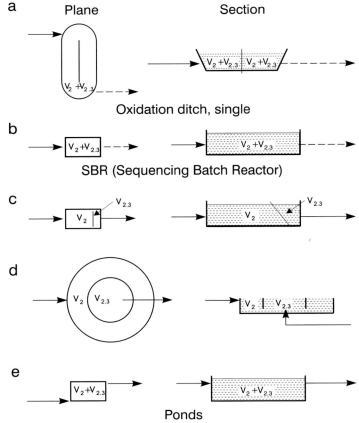

Figure 4.7. Single tank activated sludge plants. Basically the processes in the tanks are not influenced by the geometry of the plant.

When making calculations on the processes in these plants, the problem is that an attempt must be made to "divide" the tank into two parts, the aeration part, V_2, and the settling part, $V_{2.3}$, to achieve a basic layout useful for calculations as the one shown in figure 4.6, however without recycle flow, Q_4. In this type of plant it is usually impossible to define this and the return sludge concentration, X_4, .

The plant types with a certain physical separation between the two functions will allow for this separation of the volume, that is, types c, d and f and partly type e in figure 4.7. In the plant types a and b there is no physical, but a temporal shift between aeration tank and settling tank. If, for example, the oxidation ditch, type a, half of the time is used as a settling tank we may assume:

$V_{TOT} = V_2 + V_{2.3}$

$V_2 = V_{TOT}/2$

$V_{2.3} = V_{TOT}/2$

The same applies for the SBR-plant where the tank part of the time is without aeration and functions as a settling tank.

For all process calculations, the calculated aeration tank volume, V_2, must be used. For example for the calculation of sludge loading, removal of organic matter, etc. For the a and b types of plants in figure 4.7 it applies that the water volume in the tank varies with time which complicates possible calculations. Figure 4.8 shows a plant with integrated aeration tank and settling tank.

Figure 4.8. The settling tank is not always separate. Here it is integrated in the aeration tank (BIOLAK-Wax treatment plant). This is a low-loaded activated sludge plant with nitrification and denitrification. Air is introduced through floating tubes with suspended diffusers.

Example 4.6

An oxidation ditch plant as shown in figure 4.7a holds a total volume of 1,200 m³ just after withdrawal of settled wastewater. The plant has the following operational cycle:

2.5 h aeration

0.5 h settling

1.0 h withdrawal of settled wastewater

The flow of wastewater is 1,200 m³/d = 50 m³/h, the concentration of organic matter in wastewater is 0.3 kg BOD/m³. The sludge concentration (when there is aeration) is 5.8 kg SS/m³.

The design volume of the "aeration tank" is:

The water volume varies from 1,200 m³ to

1,200 + 2.5 h · 50 m³/h = 1,325 m³.

The average value is

$$V \sim \frac{1,200 + 1,325}{2} = 1,263 \text{ m}^3$$

The water volume varies more than estimated here as it is also increased during the half-hour before the withdrawal of settled water, but it has nothing to do with the volume of the aeration tank.

Aeration takes place for 2.5 h for every 4 hours, that is, the tank functions as an aeration tank (2.5/4) · 100% = 63% of the 24-hour period. The design volume, V_2, of the aeration tank is:

$V_2 = 1,263 \text{ m}^3 \cdot 0.63 = 796 \text{ m}^3$.

The sludge loading is:

$$B_x = \frac{Q_1 \cdot C_1}{V_2 \cdot X_2} \tag{4.11}$$

Substitution of these values gives:

$$B_x = \frac{1,200 \text{ m}^3/\text{d} \cdot 0.3 \text{ kg BOD}/\text{m}^3}{796 \text{ m}^3 \cdot 5.8 \text{ kg SS}/\text{m}^3} = 0.078 \text{ kg BOD}/(\text{kg SS} \cdot \text{d})$$

4.3.3. Contact stabilization plants

The principle in contact stabilisation is to save aeration tank volume while maintaining the same sludge mass. For this purpose an aeration tank is placed in the return sludge pipe as shown in the examples in figure 4.9. Plants a, b and c in figure 4.9 are basically identical. Plant d is a mixture between contact stabilization and an "ordinary" activated sludge plant.

The effect of this plant design is that the sludge is stabilized to the same degree as in an activated sludge plant with a corresponding sludge mass in a bigger tank volume. The retention time for wastewater in the aeration tank in the main flow is 0.5-1 h. The treatment efficiently for organic matter is somewhat lower while the nitri-

ACTIVATED SLUDGE TREATMENT PLANTS

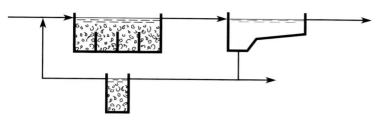

a. Traditional contact stabilization

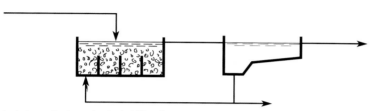

b. Integrated contact stabilization

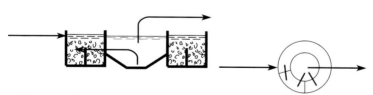

c. Integrated contact stabilization in circular tank

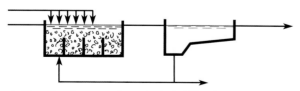

d. Step feeding, semi-contact stabilization

Figure 4.9. Activated sludge, contact stabilization.

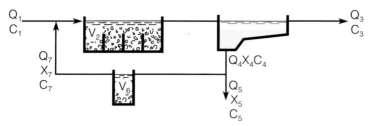

Figure 4.10. Contact stabilization, schematic layout.

fication efficiency is reduced significantly compared to an ordinary activated sludge plant with identical sludge mass (see Chapter 6).

In figure 4.10 a schematic layout of the contact stabilization process is shown. The sludge mass in the plant is:

$$M_X = V_2 \cdot X_2 + V_6 \cdot X_6$$

The mass balances for the contact stabilization processes become more complicated because two tanks have to be included for each process. A contribution from hydrolysis to a mass balance is for example:

$$k_h \cdot X_{S,2} \cdot V_2 + k_h \cdot X_{S,6} \cdot V_6$$

Detailed calculations for the contact stabilization plant can only be carried out by means of mathematical computer models.

4.3.4. Biosorption plants

These plants are very high loaded activated sludge plants where the hydraulic retention time of the wastewater in the aeration tank is 0.2-0.5 h. In such a highly loaded plant most of the suspended organic matter in the wastewater is removed whereas only a minor treatment of the dissolved organic matter takes place. The sludge in a biosorption plant is very active, often much more than in the second step low-loaded activated sludge plant /3/, see figure 4.11a.

The particles of the wastewater are very quickly adsorbed (1-2 minutes) to the activated sludge flocs so that a model description of this adsorption will usually not be necessary. The efficiency of the process is controlled by the settling. It will usually be designed in such a way that a larger amount of suspended solids will be discharged to the effluent than from an ordinary activated sludge settler. The reason is that biosorption is only used as a pretreatment and is normally followed by an activated sludge plant. A plant with a biofilter may also be used as the 2nd stage.

Figure 4.11 shows examples of biosorption plants. The sludge production in a biosorption plant, as the one shown in figure 4.11a, is bigger than in a similar single

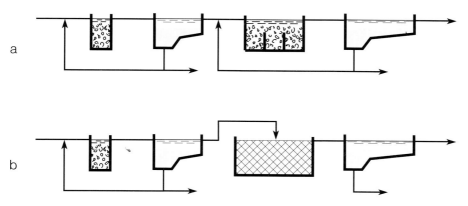

Figure 4.11. Biosorption plant followed by activated sludge or trickling filter.

activated sludge plant. Biosorption saves the energy which would otherwise have been used for oxidation of the extra sludge production. The bigger sludge production by biosorption can be used for increased biogas production. The sludge production from biosorption can be used for a hydrolysis process which can supply a subsequent biological nitrogen/phosphorus removal with easily degradable organic matter.

4.3.5. Design of activated sludge processes

The design can be more or less detailed. The level to be chosen depends on the purpose of the design. Is the sole purpose roughly to estimate the price for the plant, or is the purpose to use the design for a detail planning?

For the activated sludge plants, 3 levels (methods) for the design are discussed below.

1. Volumetric loading
2. Sludge loading
3. Computer-aided process design, dealt with in chapter 11.
   For further details about the design, see for example /4/, /5/, /6/.

4.3.6. Design by means of volumetric loading

This design is based on the BOD-volumetric loading:

$$B_{V,BOD} = Q_1 \cdot C_1/V_2$$

This design method corresponds historically to the first one to be used. If the treatment plants receive wastewater of a uniform composition, and if they are operated in such a way that the sludge concentration in the aeration tank is the same from plant to plant, reasonable results can be obtained with this simple procedure. However, it is seldom that the conditions are so uniform from plant to plant.

Table 4.5 states simple design quantities for activated sludge plants for ordinary municipal wastewaters at 10°C. At lower temperatures, the operation of the plants will be less efficient.

The necessary aeration tank volume, V_2, can be found from the expression:

$$V_2 = Q_1 \cdot C_1 / B_{V,BOD} \tag{4.20}$$

Parameter	Symbol	Unit	Sludge loading kg BOD/(kg SS · d)		
			0.05-0.10	0.20-0.30	0.50-0.60
Sludge concentration	X_{SS}	kg SS/m³	4.0-7.0 [1]	3.0-5.0 [1]	3.0-5.0 [1]
Sludge concentration	X_{VSS}	kg VSS/m³	2.5-4.5 [1]	2.0-3.5 [1]	2.0-3.5 [1]
Sludge concentration	X_{COD}	kg COD/m³	3.5-6.5 [1]	3.0-5.0 [1]	3.0-5.0 [1]
VSS of SS in sludge	-	%	65-70	70-75	70-75
Volumetric loading	$B_{V,BOD}$	kg BOD/(m³ · d)	0.2-0.6 [1]	0.6-1.5 [1]	1.5-3.0 [1]
Sludge loading	$B_{X,BOD}$	kg BOD/(kg VSS · d)	0.08-0.15	0.3-0.45	0.7-0.85
Yield constant	Y_{BOD}	kg SS/kg BOD	0.6-0.9	0.9-1.1	0.9-1.2
Yield constant	Y_{COD}	kg COD/kg COD	0.3-0.5	0.4-0.6	0.4-0.6
Treatment efficiency	E_{BOD}	%	90-95	85-90	80-90
Treatment efficiency	E_{COD}	%	75-85	70-80	65-80
Sludge age	θ_X	d	15-20	3-6	1-3

[1] North American practice corresponds to approx. 50 per cent of the values stated.

Table 4.5. Quantities for different parameters in ordinary activated sludge plants in Europe treating domestic wastewater (concentrations C_1 = 0.2-0.4 kg BOD/m³ and 0.4-0.8 kg COD/m³). T = 10°C.

The volumetric loading is a secondary design parameter and should therefore be used with caution. It cannot be used for the design of more complicated processes.

Example 4.7

On the basis of the volumetric loading, design an activated sludge plant to treat the wastewaters in Table 4.6 to an effluent concentration of 0.020 kg BOD/m³.

The dry-weather and wet-weather treatment efficiency for BOD, respectively, should be:

E_{BOD} = (240 − 20)/240 = 0.92 dry-weather days

E_{BOD} = (160 − 20)/160 = 0.88 wet-weather days

Table 4.5 shows that the volumetric loading, $B_{V,BOD}$, must be 0.2-0.6 kg BOD/(m³ · d) in dry weather to obtain a 90-95 per cent BOD treatment. In wet weather the volumetric loading can be 0.6-1.5.

ACTIVATED SLUDGE TREATMENT PLANTS

Parameter	Unit	Dry-weather	Wet-weather
Q_1	m³/d	5,000	9,000
	m³/h (max.)	420	750
C_1	kg BOD/m³	0.240	0.160
	kg COD/m³	0.510	0.360
	kg TN/m³	0.040	0.028
	kg TP/m³	0.012	0.007
$Q_1 \cdot C_1$	kg BOD/d	1,200	1,440
	kg COD/d	2,550	3,240
	kg TN/d	200	252
	kg TP/d	60	63

Table 4.6. Wastewater data for a municipality for dry-weather and wet-weather days.

In other words, design must in this case be made on the basis of the dry-weather situation. It is estimated that $B_{V,BOD} = 0.4$ kg BOD/(m³ · d).

From Expression (4.20) is found:

$V_2 = Q_1 \cdot C_1/B_{V,BOD} = (1{,}200 \text{ kg BOD/d})/0.4 \text{ kg BOD/(m}^3 \cdot \text{d)} = 3{,}000 \text{ m}^3$

On the basis of the figures in Table 4.6 the population equivalent represented by the wastewater can be calculated:

$PE_{water} = Q_1/0.2 = (5{,}000 \text{ m}^3/\text{d})/(0.2 \text{ m}^3/(\text{PE} \cdot \text{d})) = 25{,}000 \text{ PE}$

$PE_{BOD} = Q_1 \cdot C_1/0.06 =$
$(1{,}200 \text{ kg BOD/d})/(0.06 \text{ kg BOD}/(\text{PE} \cdot \text{d})) = 20{,}000 \text{ PE}$

The two calculations of the population equivalent will usually always deviate somewhat from each other. The deviation can reveal industrial influence, infiltration, exfiltration, high or low water consumption per capita, etc.

4.3.7. Design by means of sludge loading or sludge age

A somewhat more advanced design can be carried out by using sludge loading or sludge age. Sludge loading can be used for BOD/COD removal only whereas the sludge age should be used in connection with nitrification processes and other processes utilizing slowly growing bacteria for the removal of special pollutants in wastewaters, such as phenol and cyanide. When designing on the basis of the BOD-sludge loading

$B_{X,BOD} = Q_1 \cdot C_{BOD,1}/(X_2 \cdot V_2)$

the volume, V_2, of the aeration tank can be found from the expression:

$$V_2 = Q_1 \cdot C_{BOD,1}/(X_2 \cdot B_{X,BOD}) \tag{4.21}$$

Typical treatment degrees as a function of the sludge loading are stated in Table 4.5.

For many years, sludge loading has been preferred as the design parameter for biological treatment plants. With the introduction of processes such as biological phosphorus removal, simultaneous precipitation, nitrification and denitrification it will in many cases be dangerous, difficult or quite impossible to use the sludge loading as the design basis.

If the sludge age is used as the design basis, the aerobic sludge age will in most cases be used as the design basis. Expression (4.14) can be transformed

$$\theta_X = M_X/F_{SP} = V_2 \cdot X_2/F_{SP}$$

$$V_2 = \theta_X \cdot F_{SP}/X_2 \tag{4.22}$$

(it is assumed that $M_X = V_2 \cdot X_2$, and this is usually a good assumption if the aerobic sludge age is used as the design basis and hence is included in the expression).

Example 4.8

Using the sludge loading, design an activated sludge plant to treat the wastewater in Table 4.6 to an effluent concentration of 0.020 kg BOD/m^3.

The sludge concentration in the activated sludge plant is in dry weather 4.5 kg SS/m^3 and in wet weather 4.0 kg SS/m^3.

From Example 4.7:

E_{BOD} = 0.92 dry weather

E_{BOD} = 0.88 wet weather

From Expression (4.21) is found:

$V_{2,dry\ weather}$ = (1,200 kg BOD/d)/(4.5 kg SS/m^3) · $B_{X,BOD}$

Table 4.5 shows that a sludge loading of 0.05-0.1 kg BOD/(kg SS · d) gives a treatment efficiency for BOD of 90-95 per cent. As a treatment efficiency of 92 per cent is needed, the necessary $B_{X,BOD}$ ~ 0.07 (dry weather) is estimated. For wet weather $B_{X,BOD}$ ~ 0,2 is assumed.

$V_{2,dry\ weather}$ = 1,200/(4.5 · 0.07) = 3,810 m^3

$V_{2,wet\ weather}$ = 1,440/(4.0 · 0.20) = 1,800 m^3

The aeration tank must be approx. 3,800 m^3, as the dry-weather loading here proves to be the design basis. The volume found here is greater than the 3,000 m^3 which was found in Example 4.7. The reason is that the sludge concentration in the aeration tank is somewhat low. As design by means of volumetric loading does not take this into consideration, the necessary volume is underestimated. (The result may at worst be that the plant does not meet the effluent requirements, and that those who designed the plant may be held liable in damages).

References

/1/ Harremoës, P., Henze, M., Arvin, E. and Dahi, E. (1994): Teoretisk vandhygiejne (Water Chemistry). 4th ed., Polyteknisk Forlag, Lyngby, Denmark.

/2/ Henze, M., Grady, C.P.L. Jr., Gujer, W., Marais, G. v. R. and Matsuo, T. (1987): Activated Sludge Model No. 1. IAWPRC, London. (IAWPRC Scientific and Technical Reports No. 1).

/3/ Dott, W. and Wetzel, A. (1984): Microbiologische Untersuchungen einer zweistufigen Adsorptions-Belebungsanlage (Microbiological Investigations of A 2-Step Adsorption Plant). *Z. f. Wasser- u. Abwasser-Forsch.*, **17**, 182-185.

/4/ Grady, C.P.L., Daigger, G. and Lim, H.C. (1998): Biological Wastewater Treatment. 2. ed. Marcel Dekker, Inc. New York, NY.

/5/ Abwassertechnische Vereinigung (1997): Biologische und weitergehende Abwasserreinigung (Biological and advanced wastewater treatment). 4th ed., Ernst & Sohn, Berlin.

/6/ EPA (1977): Process Design Manual. Wastewater Treatment Facilities for Sewered Small Communities. U.S. Environmental Protection Agency, Cincinnati, Ohio. (EPA 625/1-77-009).

Biofilters can use many types of carrier material both natural like stones or artificial like plastic

5 Biofilters

By Poul Harremoës and Mogens Henze

Biofilters are characterized by bacteria being attached to a solid surface in the form of a biofilm. Biofilms are a dense layer of bacteria characterized by their ability to adhere to a solid medium and form a fixed film of polymers in which the bacteria are protected against sloughing off. Biofilters have a short hydraulic retention time and hence free bacteria in the water will be washed out.

The disadvantage of biofilters is the low efficiency of the biomass. The reason is that the substances must be carried through the biofilm to be removed by the bacteria. This transport takes place by molecular diffusion which is often limiting the removal in practice. This phenomenon must be understood in order to understand the functioning of biofilters.

5.1. Biofilm kinetics

Figure 5.1 shows an idealized biofilm which is assumed to be homogenous. In the water, outside the biofilm, the concentration of a substance is assumed to be S. The transport into the biofilm takes place by molecular diffusion using the diffusion coefficient D.

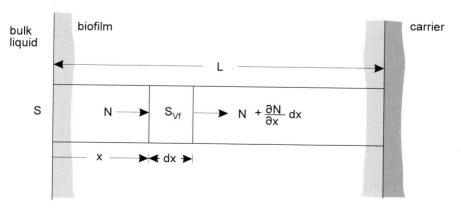

Figure 5.1. Idealized biological film in which the substance is moved by diffusion. The overall process may be limited by diffusion.

Biofilm kinetics

For an infinitesimal section of the biofilm, the following balance can be set up under stationary conditions.

in = out + removed

$$N = N + \frac{\partial N}{\partial x} \cdot dx + r_{Vf} \cdot dx \tag{5.1}$$

where N is the transport through the cross section
r_{Vf} is the volumetric reaction rate of the biofilm.

$$\frac{\partial N}{\partial x} = -r_{Vf}$$

As the transport through the cross section exclusively takes place by diffusion, we have:

$$N = -D \cdot \frac{\partial S_{Vf}}{\partial x}$$

$$\frac{\partial N}{\partial x} = -D \cdot \frac{\partial^2 S_{Vf}}{\partial x^2}$$

$$\frac{\partial^2 S_{Vf}}{\partial x^2} = \frac{r_{Vf}}{D}$$

Quantitatively, this equation can be interpreted in such a way that the second derivative of a concentration distribution expresses a curvature on the distribution. If there is no reaction, the concentration distribution is rectilinear, for example by diffusion in the water phase, see Section 5.3. If there is production, the distribution curves upwards. If there is removal the distribution curves downwards as it appears from the following pages.

This equation is made dimensionless by measuring relative to the characteristic parameters:

$$s_{Vf} = \frac{S_{Vf}}{S} \qquad \xi = \frac{x}{L} \tag{5.2}$$

$$\frac{\delta^2 s_{Vf}}{\delta \xi^2} = \frac{r_{Vf} L^2}{DS}$$

As a solution to this second order differential equation, two cases are considered: A zero and a first order reaction:

First order reaction:

$$r_{Vf} = k_{1Vf} \cdot S_{Vf}$$

where S_{Vf} is the concentration in the biofilm
k_{1Vf} is a first order rate constant with the dimension d^{-1}

$$\frac{\partial^2 s_{Vf}}{\partial \xi^2} = \frac{k_{1Vf} L^2}{D} s_{Vf} = \alpha^2 s_{Vf} \quad , \quad \alpha = \sqrt{\frac{k_{1Vf} L^2}{D}}$$

$$\frac{\partial^2 s_{Vf}}{\partial \xi^2} + 0 \frac{\partial s_{Vf}}{\partial \xi} - \alpha^2 s_{Vf} = 0 \tag{5.3}$$

The characteristic equation of this second order homogenous differential equation with constant coefficients is:

$$R^2 - \alpha^2 = 0$$
$$R = \pm \alpha$$

The complete solution for real values of α is

$$s_{Vf} = A e^{\alpha \xi} + B e^{-\alpha \xi}$$

The constants A and B are determined by these limiting conditions:

$$\xi = 0 \quad s_{Vf} = 1$$
$$\xi = 1 \quad \frac{\partial s_{Vf}}{\partial \xi} = 0$$

from which the complete solution is obtained:

$$s_{Vf} = \cosh \alpha \xi - \tanh \alpha \sinh \alpha \xi \tag{5.4}$$

By using the addition formulas, we get:

$$s_{Vf} = \frac{\cosh(\alpha(1-\xi))}{\cosh \alpha} \quad , \quad \alpha = \sqrt{\frac{k_{1Vf} L^2}{D}} \tag{5.5}$$

The concentration distribution for different values of the dimensionless expression of the geometry and diffusion of the biofilm is shown in figure 5.2.
The transport through the surface of the biofilm is

$$N = -D \left(\frac{\partial S_{Vf}}{\partial x} \right)_{x=0} = -\frac{D}{L} \left(\frac{\partial s_{Vf}}{\partial \xi} \right)_{\xi=0} S$$

By differentiation of 5.5 and substitution, we get:

$$N = \frac{D}{L} (\alpha \tanh \alpha) \quad r_A = k_{1Vf} L \frac{\tanh \alpha}{\alpha} S$$

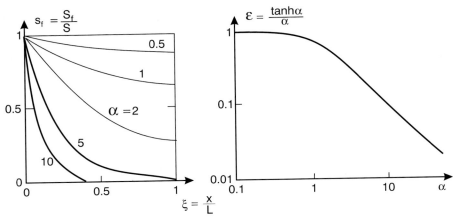

Figure 5.2. The concentration distribution and efficiency factor ε for a biological film in which a first order reaction occurs. The efficiency is reduced due to the limited transport capacity of the diffusion.

from which it can be seen that a first order reaction in the biofilm can be interpreted as a corresponding first order reaction in the water outside the biofilm. The reaction as a transport through the surface of the film is proportional to the concentration just outside the film.

For the reaction per surface area we get:

$$r_A = N = k_{1Vf} \, L \, S \cdot \varepsilon \quad , \quad \varepsilon = \frac{\tanh \alpha}{\alpha} \qquad (5.6)$$

and for the reaction rate per surface area we get:

$$k_{1A} = k_{1Vf} \, L \varepsilon \qquad (5.7)$$

where k_{1A} is a reduced rate constant, ε is an efficiency factor.

In figure 5.2 the efficiency factor ε is shown as a function of the dimensionless expression α for film geometry and diffusion. It appears from the figure that for $\alpha < 1$, $\varepsilon \sim 1.0$, corresponding to a 100 per cent efficiency. The film is fully penetrated. The film is considered to be thick. For $\alpha > 1$ applies

$$\varepsilon \sim \frac{1}{\alpha} = \frac{\sqrt{\frac{D}{k_{1Vf}}}}{L} \qquad (5.8)$$

where $\sqrt{\frac{D}{k_{1Vf}}}$ has the dimension "length" and can be understood as the efficient diffusion path.

$$k_{1A} = k_{1Vf} \cdot L \cdot \frac{\sqrt{\frac{D}{k_{1Vf}}}}{L} = \sqrt{D \cdot k_{1Vf}} \qquad (5.9)$$

for thick biofilms.

Zero order reaction

For the zero order reaction we have a much simpler expression

$$r_{Vf} = k_{0Vf} \qquad (5.10)$$

$$\frac{\partial^2 s_{Vf}}{\partial \xi^2} = \frac{k_{0Vf} L^2}{DS} \qquad (5.11)$$

The complete integral is:

$$s_{Vf} = \frac{k_{0Vf} L^2}{2 DS} \xi^2 + K_1 \xi + K_2$$

For the limiting condition at the surface we get:

$$\xi = 0 \; , \; s_{Vf} = 1 \; \rightarrow \; K_2 = 1$$

If the observed substance can penetrate the film fully, we get the following limiting conditions:

$$\xi = 1 \; , \; \frac{\partial s}{\partial \xi} = 0 \; \rightarrow \; K_1 = \frac{-k_{0Vf} L^2}{DS}$$

$$s_{Vf} = \frac{k_{0Vf} L^2}{2 DS} \xi^2 - \frac{k_{0Vf} L^2}{DS} \xi + 1$$

$$s_{Vf} = \frac{\xi^2}{\beta^2} - 2 \frac{\xi}{\beta^2} + 1 \qquad (5.12)$$

$$\beta = \sqrt{\frac{2 DS}{k_{0Vf} L^2}} \qquad (5.13)$$

The concentration distribution in the biofilm is shown in figure 5.3. For the total transport through the surface of the film we get

$$r_A = N = k_{0Vf} L$$

corresponding to the biofilm being fully efficient, which means that the removal is taking place according to the zero order reaction throughout the full length of the film. This requires, however, that there is something to remove in the very heart of the film, which means that the concentration has to be greater than zero for $\zeta = 1$. The result is the following condition:

$$\beta > 1$$

If this condition is not complied with, the limiting conditions must be changed:

$$\xi = \xi' \quad , \quad s_{vf} = 0$$

$$\xi = \xi' \quad , \quad \frac{\partial s_{vf}}{\partial \xi} = 0$$

From which we get the following solution:

$$s_{vf} = \frac{k_{ovf} L^2}{2 DS} \xi^2 - \frac{k_{ovf} L^2}{DS} \xi' \xi + 1$$

$$\xi' = \sqrt{\frac{2 DS}{k_{ovf} L^2}} = \beta \tag{5.14}$$

$$s_{vf} = \frac{\xi^2}{\beta^2} - 2 \frac{\xi \xi'}{\beta^2} + 1 \tag{5.15}$$

where ξ' is the efficient part of the film. The concentration distribution in the film for this case is also shown in figure 5.3.

From Expression 5.14 it appears that the dimensionless parameter β is an expression of the relative length that a substrate can penetrate the biofilm. Hence β is called the degree of penetration.

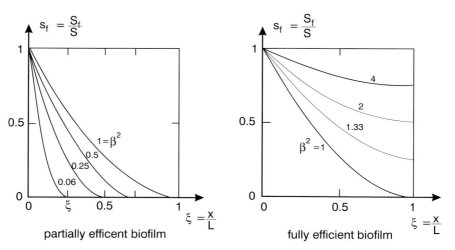

Figure 5.3. The concentration distributions in a biofilm in which a first order reaction occurs. Right: a biofilm in which the substance diffuses to the bottom. Left: a biofilm in which the substance cannot penetrate to the bottom of the film.

Hence, for the overall transport through the surface of the film we get:

$$r_A = N = L \xi' k_{0Vf} = \sqrt{2 k_{0Vf} D} \, S^{1/2} \tag{5.16}$$

It is remarkable that while the zero order reaction in the fully efficient biofilm still resulted in a reaction of zero order - independent of S - the zero order reaction in the partially efficient film results in a half order reaction in the water outside the film. For the surface area specific reaction rates of the material, of which the film forms a part, we get:

A fully efficient biofilm:

$$r_A = k_{0A} = k_{0Vf} L$$

A partially efficient biofilm:

$$r_A = k_{1/2 A} S^{1/2} = \sqrt{2 k_{0Vf} D} \, S^{1/2} \tag{5.17}$$

$$k_{1/2 A} = \sqrt{2 k_{0Vf} D} \tag{5.18}$$

A more detailed study of zero order kinetics in the biofilm can be found in /1/, /2/, /3/ and /4/.

In practise, the concentrations are in the ½-order to 0-order range (bigger than $2 \cdot K_s$, see below). That makes it very convenient to design experiments such that this feature is exploited. If the reaction rate is shown to the ½-power, the reaction rate is a linear function of $S^{1/2}$ and a straight horizontal line in the 0-order range. This is shown in figure 5.4a. The transition from ½- to 0-order, corresponding to the intersection for $\exists = 1$, can be easily determined because it is an intersection between two straight lines. If D is known (e.g. from scientific tables, see below), then L, the effective thickness, of the biofilm can be determined. This is a very funda-

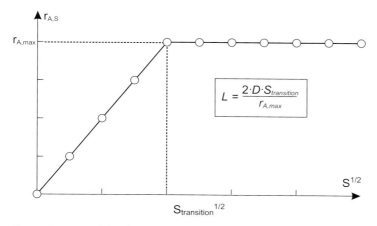

Figure 5.4a Transition from ½- to 0-order of reaction in the bulk water, in a convenient plot for determination of biofilm parameters.

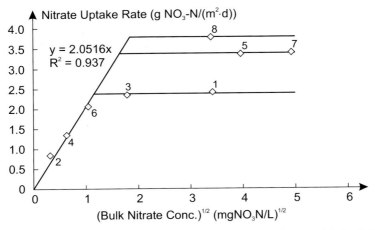

Figure 5.4b. The oxidising agent is nitrate in an experiment with denitrification (see chapter 7) with plenty of organic matter present. Due to growth, the 0-order reaction rate increase with time from the first experiment, no. 1, to the last, no. 8, corresponding to an increasing thickness. The ½-order rate is not affected by thickness, as predicted by theory. The transition from ½- to 0-order reaction rate varies with increasing thickness.

mental experiment to make for verification of biofilm kinetics and for determination of process parameters. Figure 5.4b shows an example of such an experiment.

Example 5.1

At the top of a trickling filter there is a high concentration of a removable, dissolved organic matter and an oxygen concentration near saturation: 8 g/m^3. Both substances diffuse into the biofilms where the organic matter is mineralized while consuming oxygen at the rate of $k_{0f,O2}$ = 200,000 g/(m^3 · d). The diffusion coefficient of the oxygen is set at D_{O2} = 1.7 · 10^{-4} m^2/d, see Tables 5.1 and 5.2.

How far does the oxygen penetrate the biofilm:

$$\beta L = \sqrt{\frac{2DS}{k_{0vf}}} = \sqrt{\frac{2 \cdot 1{,}7 \cdot 10^{-4} \cdot 8}{200{,}000}} = 117 \, \mu m$$

It appears that if the biofilm is thinner than 117μm, the film will be fully penetrated. Conversely, if the film is thicker than 117μm, the film will be partially penetrated and irrespective of the thickness it is a half order process, the reaction rate for oxygen being:

$$k_{½A,O2} = \sqrt{2 k_{0vf} \cdot D} = \sqrt{2 \cdot 200{,}000 \cdot 1{,}7 \cdot 10^{-4}} = 8.2 \, g^{½} \, m^{-½} d^{-1}$$

$$r_{A,O2} = 8.2 \cdot 8^{½} = 23 \, g \, O_2 / (m^2 \cdot d)$$

In practice, the biofilm in a trickling filter is considerably thicker than 117 μm (of the order of 1 mm). Hence partial penetration and half order processes are dominant.

The overall reaction pattern for a biofilm can be summarized as follows:

$$\text{first order: } k_{1A} = k_{1Vf} L \cdot \varepsilon \begin{cases} \varepsilon = \dfrac{\tanh \alpha}{\alpha} \\ \alpha = \sqrt{\dfrac{k_{1Vf} L^2}{D}} \end{cases} \quad (5.19)$$

$$\text{zero order: } \begin{cases} \beta > 1 & k_{0A} = k_{0Vf} L \\ \beta < 1 & k_{\frac{1}{2}A} = \sqrt{2D k_{0Vf}} \end{cases} \quad \beta = \sqrt{\dfrac{2DS}{k_{0Vf} L^2}}$$

An overview of the relationship between the three orders of reaction can be obtained by substituting the following relation between the first and zero orders of reaction:

$$k_{0Vf} = k_{1Vf} K \quad (5.20)$$

where K is the concentration at which the two reaction rates would be identical if there were only first and zero orders. If Monod kinetics are applied $K = K_S$. Hence for first and zero orders only:

$$\text{first order:} \quad \kappa = \dfrac{r_A}{k_{0A}} = \dfrac{k_{1Vf} L \varepsilon}{k_{0Vf} L} S = \varepsilon \cdot \dfrac{S}{K} \quad (5.21)$$

$$\text{zero order:} \begin{cases} \beta > 1 & \kappa = \dfrac{r_A}{k_{0A}} = 1 \\ \beta < 1 & \kappa = \dfrac{r_A}{k_{0A}} = \dfrac{k_{\frac{1}{2}A} S^{\frac{1}{2}}}{k_0 A} = \dfrac{\sqrt{2D k_{0Vf}} S^{\frac{1}{2}}}{k_{0Vf} L} = \beta = \dfrac{\sqrt{2}}{\alpha} \left(\dfrac{S}{K}\right)^{\frac{1}{2}} \end{cases}$$

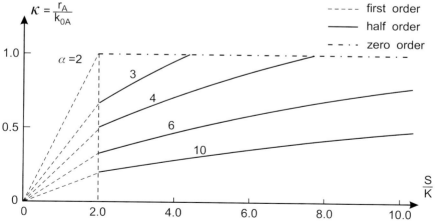

Figure 5.5. A dimensionless representation of the reaction rate as a function of the concentration outside a biofilm. There are three orders of reaction, a zero order, a first order and a half order for a > 2.

This is shown in figure 5.5 in an arithmetic representation. It appears that there is no half order process for $\alpha < 2$. For $\alpha > 2$, the half order process forms a soft transition between the first and zero order reactions. This soft transition has often been confused with Monod kinetics which it resembles. The half order reaction is, however, an expression of a diffusional limitation in the biomass and depends on the diffusion path, whereas Monod kinetics in theory apply to the individual bacterium.

Example 5.2

In the last part of a filter, the concentration of the easily degradable organic matter is reduced to 10 g /m³. For the removal of the organic matter in the biofilm K_S = 10 g/m³ and k_{0Vf} = 300 kg/(m³· d). The biofilm is 200 µm thick. The diffusion coefficient is $0.4 \cdot 10^{-4}$ m²/d.

Find the order of reaction and the rate of reaction.

S/K = 10/10 = 1.0

The change in the order of reaction takes place at S/K = 2; hence it is a first order process. The change from a half order to a first order takes place at S = 20 g/m³.

The internal first order rate constant is

$k_{1Vf} = k_{0Vf}/K = 300 \cdot 10^3/10 = 30{,}000$ d^{-1}

It is estimated that the biofilm is only slightly penetrated because

$$\alpha = L\sqrt{\frac{k_{1Vf}}{D}} = 200 \cdot 10^{-6} \sqrt{\frac{30{,}000}{0.4 \cdot 10^{-4}}} = 5.5$$

$$\varepsilon = \frac{1}{\alpha} = 0.18$$

The removal per surface area without diffusional limitation ($\varepsilon \sim 1$) would be

$k_{1A} = k_{1Vf} \cdot L = 30{,}000$ d$^{-1} \cdot 200 \cdot 10^{-6}$ m = 6 m/d

$r_A = k_{1A} S_{org} = k_{1Vf} L S_{org} = 60$ g/(m² · d) (without limitation)

Due to the diffusional limitation, the removal has been reduced to

$k_{1A} = 6 \cdot 0.18 = 1.1$ m/d

$r_A = 60 \cdot 0.18 = 11.0$ g/(m² · d) (with diffusional limitation)

These theories for diffusional limitation of reactions in biomass, specifically biofilms and bioflocs, can be generalized to a random process of order n. Hence it proves that the generally governing parameter can be written:

$$\Phi = \sqrt{\frac{(n+1)k_{nVf} L^2 S^{n-1}}{2D}} \tag{5.22}$$

For zero order we get:

$$\Phi = \sqrt{\frac{k_{0Vf} L^2}{2DS}} = \frac{1}{\beta} \qquad (5.23)$$

For first order we get:

$$\Phi = \sqrt{\frac{k_{1Vf} L^2}{D}} = \alpha \qquad (5.24)$$

The quantity Φ is called the **Thiele module**.

5.2. Biofilm kinetic parameters

The removal of a substrate in a biofilm takes place through bacterial reactions inside the biofilm where the individual bacterium behaves in the same manner as bacteria in suspension. For the bacterial growth applies:

$$r_{V,XB} = \mu_{max} \frac{S}{S + K_S} \cdot X_B$$

where V is understood as the volume inside the biofilm. This is analogized by the substrate removal. For the substrate removal, the following zero order and first order approximations apply:

$$\begin{aligned} k_{0Vf} &= r_{V,S} = \frac{\mu_{max}}{Y_{max}} \cdot X_B \\ k_{1Vf} \cdot S &= r_{V,S} = \frac{\mu_{max}}{Y_{max}} \cdot \frac{S}{K_S} X_B \end{aligned} \qquad (5.25)$$

In these expressions, kinetic constants as those known from suspended cultures can be used.

Compared with suspended cultures the difference is the concentration of bacteria which is much higher in a biofilm than in a suspension. In activated sludge plants, the concentration varies from 2 to 6 kg VSS/m³. In biofilms the concentration varies from 10 to 60 kg VSS/m³. The only new parameter which comes into use is therefore the diffusion coefficient D. The measurement of this coefficient in the biofilm is uncertain and should not be relied upon. In practice, the diffusion coefficient can be put equal to or a little lower than the molecular diffusion coefficient which can be looked up in standard reference books /5/. Table 5.1 shows relevant diffusion coefficients in pure water. Often a reduction factor of 0.8 is used for biofilms, see Table 5.2; it should be stressed, however, that the variation is wide, depending for example on the surface structure of the biofilm /6/.

BIOFILM KINETIC PARAMETERS

Substance	D	Substance	D	Substance	D
Oxygen, O_2	2.1	Acetate, CH_3COO^-	1.0	Ammonium, NH_4^+	1.7
Carbon dioxide, CO_2	1.6	Glucose, $C_6H_{12}O_6$	0.6	Nitrite, NO_2^-	0.9
Hydrogen carbonate, HCO_3^-	1.0			Nitrate, NO_3^-	1.6
Carbonate, CO_3^{2-}	0.4				

Table 5.1. Diffusion coefficients in pure water at 25°C /5/. Unit 10^{-4} m²/d.

Substance	D 10^{-4} m²/d	$v_{O2,S}$* g COD/g O_2	k_{0vf} kg COD/(m³ · d)
Oxygen	1.0 - 2.1	-	25 - 200
Acetic acid	0.3 - 0.7	2.1	230 - 300
Methanol	0.8 - 4	1.2	40 - 110
Glucose	0.1 - 0.7	2.4	350 - 550
unspec. COD	0.3 - 0.6	1.4 - 2	50 - 500
unspec. BOD	0.3 - 0.6	0.8 - 1.2	25 - 250

* in connection with biological growth.

Table 5.2. Diffusion coefficients for oxygen and organic substrates, stoichiometric conditions, and estimated removal rate internally in biofilms. The estimates are for 25°C and should be handled with care.

The bacterial density is a difficult quantity to predict since it is not a constant, but varies in a rather unknown manner with the growth history of the biofilm: Growth on small substrate concentrations yields a large bacterial density and conversely for high concentrations easily degradable, dissolved substrates.

Example 5.3

In an ideally mixed trickling filter the specific surface is 100 m²/m³. COD is removed in a 1 mm thick biofilm with kinetic constants corresponding to activated sludge plants (Table 3.7):

μ_{max} = 6 d^{-1}

$K_{s,COD}$ = 20 g COD(B)/m³

Y_H = 0.67 g COD(B)/g COD(S)

The concentration of bacteria in the biofilm is 40 kg VSS /m³, corresponding to 40 · 1.4 = 56 kg COD(B)/m³

What is the volumetric concentration of bacteria in the biofilter:

X = 100 m² / m³ · 1 · 10^{-3} m · 40 kg VSS / m³ = 4 kg VSS / m³
= 5.6 kg COD(B) / m³

It appears that the effective biomass in a trickling filter is not bigger than in an activated sludge plant. The bigger specific biomass is obtained in filters with a very large specific surface, for example in fluidized filters.

How big is the removal per surface area when the COD-concentration in the filter is 50 g/m³?

This is a zero order process in the biofilm because the concentration in the water is higher than $2 \cdot K_S$.

$$k_{0vf} = \frac{6}{0.67} \cdot 56 = 500 \text{ kg COD / (m}^3 \text{ biofilm} \cdot \text{d)}$$

The diffusion coefficient for the organic matter is set at $0.4 \cdot 10^{-4}$ m²/d.

$$k_{\frac{1}{2}A} = \sqrt{2 k_{0vf} \cdot D} = \sqrt{2 \cdot 500 \cdot 10^3 \cdot 0.4 \cdot 10^{-4}} = 6.3 \text{ g}^{\frac{1}{2}}\text{m}^{-\frac{1}{2}}\text{d}^{-1}$$

$$r_A = 6.3 \sqrt{50} = 45 \text{ g COD / (m}^2 \cdot \text{d)}$$

This yields $45 \cdot 100$ m²/m³ = 4,500 g COD/(m³ · d) volumetric removal in the filter, corresponding to a **high-rate** trickling filter. (It is assumed that there is sufficient oxygen; see Example 5.5).

5.3. Hydraulic film diffusion

Apart from the diffusional limitation in the biofilm itself, a limitation occurs in practice by transport of substrate from the bulk water to the surface of the biofilm. This transport can be described in a simplified manner by a proportionality between the difference in transport and concentration

$$N = h(S - S_g) \tag{5.26}$$

where h is the transfer coefficient (m/d). The phenomenon is an analogue for the transport of impulse, heat and substance. Much literature is available on tubes, plane surfaces, etc.; but it may be difficult to transfer the information to a biofilm whose surface structure is often badly defined in respect of hydrodynamics. Very few figures from experience are available for biofilms.

In the case where there is a diffusional limitation in the hydraulic film and in the biofilm, the result for a zero order reaction in the biofilm is:

$$N = h(S - S_g) = -D\left(\frac{dS_{Vf}}{dx}\right)_{x=0} \tag{5.27}$$

because the flux across the interface must be the same, both in relation to the hydraulic film as well as the biofilm. The concentration profile from the concentration S in the bulk water to $S_f = 0$ in the biofilm is shown in figure 5.6.

By integration of the differential equation for biofilm diffusion:

$$s_{Vf} = \frac{k_{0Vf} L^2}{2 DS} \xi^2 + K_1 \xi + K_2$$

where $s_{Vf} = \frac{S_{Vf}}{S_g}$, $\xi = \frac{x}{L}$

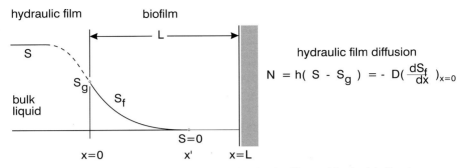

Figure 5.6. Concentration distribution in the hydraulic film and in the biofilm for a partially penetrated biofilm.

The limiting conditions are:

$$\xi = 0 \begin{cases} s_{vf} = s_g \\ \dfrac{ds_{vf}}{d\xi} = \dfrac{hL}{D}(s_g - 1) \end{cases} \qquad \xi = \xi' \begin{cases} s_{vf} = 0 \\ \dfrac{ds_{vf}}{d\xi} = 0 \end{cases}$$

h is the transfer coefficient for the hydraulic film diffusion. The solution of the equations is:

$$\text{or} \quad \begin{rcases} \dfrac{r_A}{k_{\frac{1}{2}A} S^{\frac{1}{2}}} = \sqrt{1 + \dfrac{1}{4\lambda^2}} - \dfrac{1}{2\lambda} \\ \dfrac{r_A}{h \cdot S} = \dfrac{1}{2\lambda^2}(\sqrt{1 + 4\lambda^2} - 1) \end{rcases} \quad \lambda = \dfrac{hS}{k_{\frac{1}{2}A} S^{\frac{1}{2}}} \qquad (5.28)$$

where λ is a dimensionless quantity which expresses the ratio between hydraulic film diffusion and biofilm diffusion corresponding to:

$$\lambda \to 0 \quad \to \quad r_A \to hS$$
$$\lambda \to \infty \quad \to \quad r_A \to k_{\frac{1}{2}A} S^{\frac{1}{2}}$$

The result is shown in figure 5.7. Hence there is a soft transition between the pure hydraulic film diffusion, which corresponds to a first order reaction, and the half order reaction for the biofilm.

For the first order reaction in the biofilm, a corresponding condition for the flux through the area can be set up:

$$N = h(S - S_g) = r_A = k_{1A} S_g$$

$$\left.\begin{array}{l}\dfrac{r_A}{k_{1A}S} = \dfrac{\lambda}{\lambda+1} \\ \dfrac{r_A}{hS} = \dfrac{1}{\lambda+1}\end{array}\right\} \quad \lambda = \dfrac{h}{k_{1A}} \qquad (5.29)$$

but the order of reaction will in any case be a reaction of first order, only reduced by the diffusional limitation. In this case it may thus be difficult to distinguish between the two kinds of diffusional limitation.

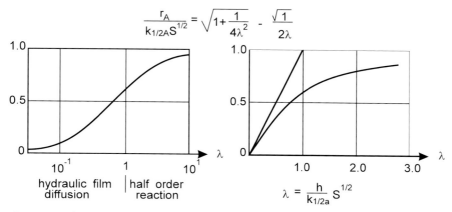

Figure 5.7. The transition between dominant hydraulic film diffusion and dominant half order heterogeneous biofilm reaction.

Example 5.4

By interpretation of tests it is necessary to take into consideration the possibility that both hydraulic film diffusion and diffusional limitation in the biofilm have an effect. Figure 5.8 shows a number of measuring points for the removal rate per surface area of nitrate which is removed by denitrifying bacteria in a thick biofilm. Apparently the following relationship is obtained:

$r_A = 0.09 \cdot S^{0.64}$

hence a deviation from the expected ½ order.

By substitution

h = 0.15 m/h
$k_{½A}$ = 0.13 g^{½}m$^{-½}$h^{-1}

the full line shown in figure 5.8 is obtained. For S equal to approx. 50 g $NO_3^- - N/m^3$, the reaction is half order. Hence the exponent found (n = 0.64) can be attributed to the fact that the data are located in the transfer zone between a half order (n = 0.5) and a first order (n = 1.00); but the data do not have a sufficient span to distinctly show the first order reaction. Therefore it is sometimes tempting to make a linear regression on something which is, in fact, a curved graph.

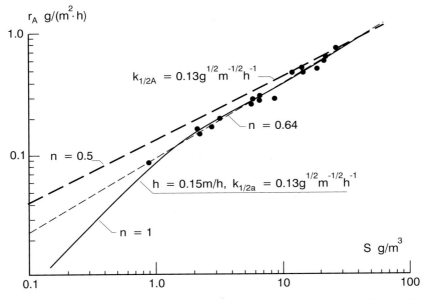

Figure 5.8. Surface area specific reaction rates of nitrate in a rotating disc as a function of the nitrate concentration in the water phase.

5.4. Two-component diffusion

Biological removals are almost always redox-processes which require two substrates: an oxidant and a reductant; for example oxygen and organic matter. One of the most significant results of the biofilm kinetics is the determination as to which substrate is limiting for the removal. This is regulated by the removal rate and the diffusional rate for each of the two substrates:

Oxidant:

$$\frac{d^2 S_{Vf,ox}}{dx^2} = \frac{k_{0,ox}}{D_{ox}}$$

$$S_{Vf,ox} = \frac{1}{2}\frac{k_{0,ox}}{D_{ox}}x^2 + K_{1,ox}\, x + K_{2,ox}$$

$$x = 0 \quad S_{Vf,ox} = S_{ox} = K_{2,ox}$$

$$x = x_1 \quad \frac{dS_{Vf,ox}}{dx} = 0 \,,\quad K_{1,ox} = -\frac{k_{0,ox}}{D_{ox}}x_1$$

$$S_{Vf,ox} = \frac{1}{2} \frac{k_{0,ox}}{D_{ox}} x^2 - \frac{k_{0,ox}}{D_{ox}} x_1 \cdot x + S_{ox} \tag{5.30}$$

where the index "ox" stands for oxidant. x_1 is the point in the biofilm beyond which no removal takes place because one of the substrates has been used up. This is illustrated in figure 5.9.

By analogy we correspondingly get for:

Reductant:

$$S_{Vf,red} = \frac{1}{2} \frac{k_{0,red}}{D_{red}} X^2 - \frac{k_{0,red}}{D_{red}} X_1 \cdot X + S_{red} \tag{5.31}$$

If it is assumed that there is sufficient reductant (figure 5.9, left), the oxidant will be used up for $x = x_1$.

$$x = x_1 \quad S_{Vf,ox} = 0 \quad x_1 = \beta_{ox} \cdot L = \sqrt{\frac{2 D_{ox} S_{ox}}{k_{0,ox}}}$$

$$r_{A,ox} = \sqrt{2 D_{ox} k_{0,ox}} \cdot (S_{ox})^{1/2}$$

as previously derived for a limiting substrate.

For the reductant we then get the removal per surface area of the biofilm

$$r_{A,red} = k_{0,red} \cdot x_1 = k_{0,red} \sqrt{\frac{2 D_{ox} S_{ox}}{k_{0,ox}}}$$

The removal of the reductant is controlled by the penetration of the oxidant. The reductant is present throughout the film, but cannot be removed for $x > x_1$ because no oxidant is present (figure 5.9, left).

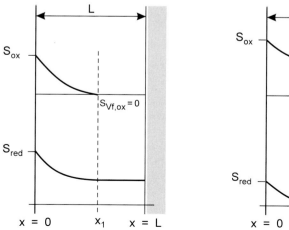
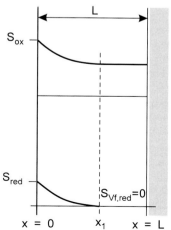

Figure 5.9. Illustration of partial penetration and rate limitation for two components.

Two-component diffusion

The substrate which is limiting for the removal is determined by the substrate which penetrates the shortest distance into the biofilm:

$\beta_{ox} < \beta_{red}$ the oxidant is limiting
$\beta_{red} < \beta_{ox}$ the reductant is limiting

$$\frac{\beta_{ox}}{\beta_{red}} = \frac{\sqrt{\frac{2 D_{ox} S_{ox}}{k_{0,ox}}}}{\sqrt{\frac{2 D_{red} S_{red}}{k_{0,red}}}} \gtreqless 1$$

On the basis of the above, the following criterium is developed:

$$\frac{S_{ox}}{S_{red}} \gtreqless \frac{D_{red}}{D_{ox}} \cdot \frac{k_{0,ox}}{k_{0,red}} = \frac{D_{red}}{D_{ox}} \cdot \frac{1}{v_{ox,red}} \qquad (5.32)$$

where $v_{ox,red}$ is the stoichiometric coefficient.

If the sign of inequality > applies, the reductant is potentially limiting for the removal. Conversely for <, the oxidant will be potentially limiting because it has the shortest penetration depth.

Example 5.5

A biofilter treats wastewater where the organic matter can be assumed to be soluble, that is, diffusible in the biofilm. The influent concentration is 150 g COD/ m³.

For the kinetic constants it can be estimated (Table 5.2):

$D_{O2} = 1.7 \cdot 10^{-4} m^2/d$

$D_{COD} = 0.4 \cdot 10^{-4} m^2/d$

$v_{O2,COD} = 1.7$ g COD/g O_2

Controlling for the removal rate applies:

$$\frac{S_{O2}}{S_{COD}} \gtreqless \frac{D_{COD}}{D_{O2}} \cdot \frac{1}{v_{O2,COD}} = \frac{0.4 \cdot 10^{-4}}{1 \cdot 7 \cdot 10^{-4} \cdot 1.7} = 0.14 \text{ g } O_2 / \text{ g COD}$$

If the oxygen concentration is less than 0.14 S_{COD}, oxygen is potentially limiting for the removal. At the influent, the oxygen concentration is less than 0.14 g O_2/g COD · 150 g COD/m³ = 21 g O_2/m³ because this is greater than the saturation with the air: 10 g O_2/m³ at 15°C.

If the oxygen concentration in the filter just ahead of the outlet is assumed to be 2 g O_2/m³, the COD-concentration should be less than S_{O2}/0.14 = 2/0.14 = 14 g COD/m³ in order to be limiting.

The conclusion is that in a filter treating dissolved organic matter with oxygen as the only oxidant, the oxygen will be limiting for the removal in the major part or all through the filter. Only if at some point the COD gets below 14 g COD/m³, COD will become limiting.

5.5. Filter kinetics

For a given medium in a filter, the analysis of the filter processes can be made in two stages as shown in figure 5.10:

Removal in the biofilm itself is here assumed to be of zero order

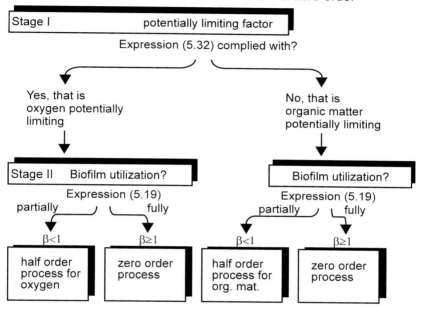

Figure 5.10. Biofilm kinetic user guide. The calculation procedure is shown which is used to determine the kinetics in relation to the wastewater (bulk phase).

I. It is determined whether oxygen or organic matter is the potentially limiting factor and hence controls the removal rate and the kinetics.

II. It is determined whether the whole biofilm is active (that is, whether the potentially limiting matter is found throughout the biofilm, or whether only the outer part of the biofilm removes substances). This is important for the order of reaction.

The calculation path is shown in figure 5.10.

Figure 5.11 shows an ideally mixed tank in which a heterogeneous process takes place without hydraulic film diffusion. For each of the three orders of reaction we get zero order:

$$r_A = k_{0A}$$
$$S_3 = S_1 - \frac{k_{0A} A_{2^*}}{Q} \tag{5.33}$$

FILTER KINETICS

Figure 5.11. Ideally mixed reactor in which a heterogeneous process occurs.

half order:

$$r_A = k_{½A} \cdot S_3^{½}$$

$$S_3^{½} = \sqrt{\left(\frac{k_{½A} A_{2^*}}{2Q}\right)^2 + S_1} - \frac{k_{½A} A_{2^*}}{2Q} \qquad (5.34)$$

first order:

$$r_A = k_{1A} S_3$$

$$S_3 = \frac{QS_1}{Q + k_{1A} A_{2^*}} \qquad (5.35)$$

Notice that the volume of the reactor, and hence the retention time, is not included in the expressions at all. These expressions can be used for simulation of processes in filters where the whole reactor can be considered to be ideally mixed. The data given in Example 5.4 have been obtained through measurement of nitrate removal on a rotating disc with an ideal mix.

The flow through a submerged filter (that is, a filter which is fully submerged in the water) can, with reasonable approximation, be considered to be plug flow. Figure 5.12 shows a submerged filter. For a infinitesimal section, the following balance can be set up under stationary conditions:

$$\text{in} - \text{removed} = \text{out}$$
$$QS - r_{V,s} A dy = Q(S + \frac{\partial S}{\partial Y}) dy \qquad (5.36)$$

where Q is the flow through the filter
S is the concentration
A is the cross sectional area of the filter
$r_{V,S}$ is the volumetric removal rate of the filter

$$\frac{\partial S}{\partial y} = -r_{V,s} \frac{A}{Q} \qquad (5.37)$$

or dimensionless:

176

$$s = \frac{S}{S_1}, \quad \eta = \frac{y}{H}, \quad HA = V, \quad \theta = \frac{V}{Q}$$

$$\frac{\partial s}{\partial \eta} = -\frac{r_{V,S}}{S_1} \theta$$
(5.38)

where θ is the hydraulic retention time (corresponding to an empty filter).

Here $r_{V,S} = \omega \cdot r_{A,S}$ can be substituted where ω is the surface area of the filter medium per unit volume of the filter. For $r_{A,S}$ the three cases observed during the biofilm diffusion can be substituted:

Zero order reaction:

$$\frac{\partial s}{\partial \eta} = -\frac{k_{OV}}{S_1} \theta$$
(5.39)

$$s = 1 - \frac{k_{OV}}{S_1} \theta \cdot \eta$$
(5.40)

$$E = \frac{S_1 - S}{S_1} = 1 - s = \frac{k_{OV}}{S_1} \theta \cdot \eta$$
(5.41)

where E is the treatment efficiency. This is shown in figure 5.12. The expression applies, of course, only to

$$0 < \frac{k_{OV}}{S_1} \cdot \theta < 1$$

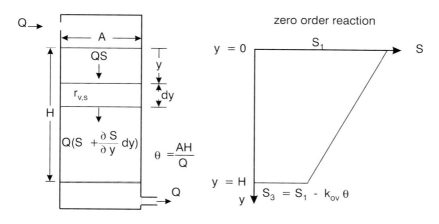

Figure 5.12. Mass balance for a submerged filter in which a heterogeneous process takes place, and the concentration distribution in the filter if a zero order reaction occurs.

Half order reaction:

$$r_V = \omega\, r_A = \omega\, k_{\frac{1}{2}A}\, S^{\frac{1}{2}} = k_{\frac{1}{2}V}\, S^{\frac{1}{2}} = k_{\frac{1}{2}V}\, S_1^{\frac{1}{2}} s^{\frac{1}{2}} \tag{5.42}$$

$$\frac{\partial s}{\partial \eta} = -\frac{k_{\frac{1}{2}V}}{S_1^{\frac{1}{2}}}\, \theta^{\frac{1}{2}} \cdot s^{\frac{1}{2}} \tag{5.43}$$

$$s = \left(1 - \frac{1}{2}\frac{k_{\frac{1}{2}V}}{S_1^{\frac{1}{2}}}(\theta\,\eta)\right)^2 \tag{5.44}$$

$$E = 1 - \left(1 - \frac{1}{2}\frac{k_{\frac{1}{2}V}}{S_1^{\frac{1}{2}}}(\theta\,\eta)\right)^2 \tag{5.45}$$

First order reaction:

$$\frac{\partial s}{\partial \eta} = -k_{1V}\,\theta\, s \tag{5.46}$$

$$s = e^{-k_{1V}\,\theta\cdot\eta} \tag{5.47}$$

Each of these expressions for the relationship between treatment efficiency and retention time for different values of the respective reaction rates and influent concentrations can be used independently. A special case is, however, a filter with biofilm diffusion because all three cases can occur at different levels in the same filter. At the influent, the concentration may be sufficiently big for the medium to diffuse to the bottom of the biofilm - a zero order reaction. After some treatment, the reduced concentration means that the biofilm becomes partially efficient - a half order reaction. Further down in the filter the concentration outside the biofilm will drop from values corresponding to a reaction of half order to values corresponding to a reaction of first order. The whole biofilm will then behave as having a first order reaction with reduced efficiency; see Section 5.1.

The transition between the zero and half order reaction can be directly determined. The transition criterium is, according to the previous section:

$$S_0 = \frac{L^2 k_{0Vf}}{2D} \quad \beta = 1 \quad \begin{cases} S > S_0 & k_{0V} = \omega\, k_{0A} \\ S < S_0 & k_{\frac{1}{2}V} = \omega\, \sqrt{2 k_{0Vf}\, D} \end{cases}$$

The transition from half order to first order can be determined correspondingly

$$S_1 = 2K_s \quad \begin{cases} S > S_1 & k_{\frac{1}{2}V} = \omega\,\sqrt{2k_{0Vf}\,D} \\ S < S_1 & k_{1V} = \omega \cdot k_{1Vf}\cdot L\cdot\varepsilon \end{cases}$$

Example 5.6

A 4 m high submerged filter filled with plastics packing with a specific surface of 100 m²/m³ has a nominal filter velocity of 1 m/h. The wastewater has a concentration of 500 g COD/m³. The same process constants as in Examples 5.3 and 5.5 are used.

Air is constantly bubbled into the filter resulting in an efficient aeration so that the whole water column on the whole depth is assumed to have an oxygen concentration of 2 g O₂/m³. According to Example 5.5, the filter will be oxygen limited from the top to the position in the filter where S_{O2}/S_{COD} = 0.14 g O₂/g COD, that is, when S_{COD} has been reduced to 14 g COD/m³.

The biofilm is sufficiently thick to make it partly penetrated in all practical ranges of concentrations. In the upper part of the filter, the removal is a half order reaction, limited by the oxygen concentration. This was shown in Example 5.5. However, because the oxygen concentration is assumed to be constant down through the filter, the removal of organic matter will be seen as a zero order reaction, until at some point, where the COD will be limiting, resulting in a potential half order reaction with respect to COD.

$$k_{ov} = \omega \cdot k_{½A,O2} \cdot S_{O2}^{½}$$

$$k_{½A,O2} = \sqrt{2 D_{O2} \cdot k_{0f,O2}} \qquad (5.48)$$

$$k_{0f,O2} = \frac{\mu_{max}}{Y_H} \cdot X \cdot \frac{1}{\nu_{O2,COD}}$$

$$k_{0f,O2} = \frac{6}{0.67} \cdot 56 \cdot \frac{1}{1.7} = 295 \; gO_2 / (m^3 \cdot d)$$

$$k_{½A,O2} = \sqrt{2 \cdot 1.7 \cdot 10^{-4} \cdot 295 \cdot 10^3} = 10 \; (g\,O_2)^{½} m^{-½} d^{-1}$$

$$k_{ov} = 100 \cdot 10 \cdot \sqrt{2} = 1400 \; g\,O_2 / (m^3 \cdot d) = 2400 \; g\,COD / (m^3 \cdot d)$$

which corresponds to a high-rate filter.

(Hence the result in Example 5.3 was not realistic because oxygen is limiting for the removal.)

The filter is loaded with 1 m/h corresponding to 24 m/d and a concentration of 500 g COD/m³, or 500 · 24 = 12,000 g COD/(m² · d). The filter removes 2400 g COD/(m³ · d) over 4 m, that is, 9600 g COD/(m² · d). Residue: 2400 g COD/(m² · d), corresponding to an effluent concentration of 2400/24 = 100 g COD/m³. Hence a COD-limited removal (100 g COD/m³ > 14 g COD/m³) does not occur.

5.6. Mass balances for biofilters

Normally it is not necessary to recycle sludge over a biofilter plant, as the sludge concentration becomes sufficiently large by means of the biofilm and a sufficiently large specific solid surface (on the carrier material). As recycle of sludge is not necessary, we can, in certain cases, leave out a secondary settling tank - see Section 6 about nitrification. Normally, however, the sludge present in the water which has passed the biofilter needs a subsequent settling. The sludge comes from sloughed off biofilm and from the content of suspended solids in the influent water.

5.6.1. Biofilters without recycle

A mass balance for material over the biofilter itself looks as follows:

$$Q_1 \cdot C_1 - r_{V,S} \cdot V_2 = Q_3 \cdot C_3 \tag{5.49}$$

The removal can also be expressed by means of the sludge concentration, X_2, as

$$r_{V,S} \cdot V_2 = r_{X,S} \cdot V_2 \cdot X_2 \tag{5.50}$$

As X_2 should only include the active biomass which is usually unknown, it is frequently common practice to relate the removal to the volume or the surface of the carrier.

If the removal is expressed per unit area of carrier, Expression (5.50) will look as follows:

$$Q_1 \cdot C_1 - r_{A,S} \cdot A_{2^*} = Q_3 \cdot C_3 \tag{5.51}$$

where $r_{A,S}$ is the removal rate per unit area of carrier
(unit for example kg COD/(m² · d)),
 A_{2^*} is the overall area of the carrier (for example m²).

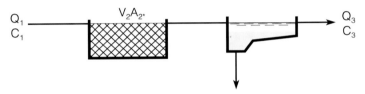

Figure 5.13. Biofilter without recycle. Separation here shown as a settling tank can be with any method of separation (e.g. settling, flotation, centrifuge, membranes etc.)

Biofilters with recycle

Most biofilter plants have recycle which is normally made by recycling directly over the filter (figure 5.14). The recycle ratio, R, is defined as

$$R = Q_6/Q_1 \tag{5.52}$$

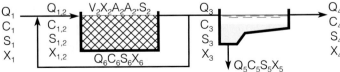

Figure 5.14. Biofilter with recycle. Separation can be by settling, flotation, centrifuge, membranes etc.

The purpose of recycling is to ensure a suitable flow of water through the filter. This is important for most types of biofilter plants. Also, the recycle lowers the influent concentration to the biofilter and can therefore affect the process, both through the order of reaction and as to whether oxygen is limiting or not.

The equation for the removal of substrate will be the same as for biofilters without recycle, but the reaction term will be influenced by the recycle.

5.7. Concepts and definitions

The concepts and definitions for biofilters are in most cases identical to those used in activated sludge plants.

Treatment efficiency is defined as (see figure 5.14)

$$E = (C_1 - C_4)/C_1 \qquad (5.53)$$

Similar to the treatment efficiency in activated sludge plants, the treatment efficiency includes the effect of the secondary settling.

Recycle ratio is defined in Expression (5.52).

Volumetric loading rate is defined as

$$B_V = Q_1 \cdot C_1/V_2 \qquad (5.54)$$

The amount of material added per unit volume per unit time, $Q_1 \cdot C_1$, is calculated at a position prior to a possible recycle. The volumetric loading rate cannot be increased by an internal recycle in the plant.

Biofilm surface area has two meanings, the area of the filter tank in a horizontal section and the surface area of the carrier, the latter area normally being much higher than the horizontal section area. The surface area of the carrier is normally calculated from Expression (5.55)

$$A_{2^*} = \omega \cdot V_2 \qquad (5.55)$$

where ω is the specific surface of the carrier (for example m^2/m^3).

The surface area in a horizontal section, A_2, is used for the calculation of the hydraulic surface loading rate.

Organic surface loading rate is an expression of the load of the surface of the carrier in the filter:

$$B_{A,C} = \frac{Q_1 \cdot C_1}{A_{2^*}} \qquad (5.56)$$

This is the figure which comes closest to the kinetic conditions in the filter.

Hydraulic surface loading rate (or nominal filter rate) is defined by the symbols in figure 5.13 as

$$B_{A,V} = Q_{1,2}/A_2 = (Q_1 + Q_6)/A_2 \qquad (5.57)$$

This parameter is a simple expression of the flow passing the biofilm and hence coupled to the hydraulic erosion which is considered to be crucial for the control of the thickness of the biofilm. (In German literature Q_1 is set at the flow during the daytime, for example calculated as the diurnal flow distributed over 18 hours for large plants, see figure 1.6).

Sludge production, F_{SP}, is the mass of sludge leaving the plant per unit time. According to figure 5.14 it is

$$F_{SP} = Q_5 \cdot X_5 + Q_4 \cdot X_4 \qquad (5.58)$$

For further details concerning sludge production, see Chapter 4, Section 4.2. The comments made in that chapter also apply to biofilter plants. Hence the sludge production can be calculated from Expression (4.13). Note that yield constants for biofilters deviate from yield constants for activated sludge plants.

Surplus sludge production is the part of the sludge production to be further treated in the plant, that is, $Q_5 \cdot X_5$, see figure 5.14.

5.8. Types of plants

Aerobic biofilters must be able to carry out four functions to be able to treat the water satisfactorily. The tank should be designed so that:

- the bacteria necessary for the required process can be attached to a carrier (filter medium),
- the water gets into efficient contact with the sludge attached to the carrier (the biofilm/slime),
- the growth of the biofilm can be controlled so that no clogging will occur,
- oxygen is supplied to the water for the degradation of the organic matter.

Bacteria have a remarkable ability to adhere to solid surfaces, whether it concerns stone, wood or plastics which are the three types of media used in practice. The engineering control of the adhesion of such bacteria which carry out the required process is very practical in water and wastewaster treatment - as opposed to industrial production by means of biological reactors which are often based on carefully selected types of bacteria. In water and wastewater treatment such conditions are established which from experience are known to favour the bacteria which carry out the required process. This also applies to biofilters: the bacteria which cannot adhere to a surface will be washed out of the system. The bacteria which cannot degrade the material concerned cannot compete, see Chapter 3, Section 3.1.2. This may occasionally cause difficulties when starting up biofilter plants.

To achieve an efficient removal of the materials in the water, the water must pass the surface of the biofilm in order that a continuous renewal of water at the biofilm surface takes place. For this purpose the medium is either solid and the water passes the medium as is the case in the conventional biofilters (trickling filters), or the medium is continually carried through the water as the case is in rotating discs. The efficiency in the plant type depends to a large extent on how this contact between biofilm and water is arranged.

Control of the growth of the biofilm and the conditions concerning the oxygen supply are further discussed in Sections 5.10.2 and 5.10.1.

5.8.1. Trickling filters

The trickling filter is the conventional biofilm reactor and many different types of trickling filters have been used during the last century. Figure 5.15 shows the characteristic type which is a typical three-phase system. The filter medium is stationary, typically consisting of 5-20 cm large stones. Any other conceivable materials have been tried out - only plastics are, however, widely used as an alternative to stone.

The wastewater is distributed over the filter and trickles down over the stones to be collected under the stones and carried out. The ventilation through the bottom of the filter is made efficient so as to ensure a steady air current through the filter. In the course of the development it was believed to be necessary to use forced ventilation through such filters, but in practice it has proved that the variation in temperature between filter medium, wastewater and the surrounding air is sufficient to ensure air renewal and reaeration of the water during the trickling from top to bottom. The trickling filter is very efficient with respect to adhesion of bacteria, contact between water and biofilm, and reaeration of the water. The most essential problem with trickling filters is the control of the growth of the biofilm. This is the function making the greatest demand on design and operation. In the oldest, very low-rate plants, the control is biological. The biofilm is allowed to develop uninhibited which leads to local cloggings. The clogging will prevent that oxygen is supplied to the biomass, and it will putrefy locally and be degraded, until there is passage again. Higher animals, such as worms and larvae, also contribute to the degradation of the biomass and sloughing of the biofilm. Thus the filter can become a hatching plant for insects, especially filter flies, and they can be such a nuisance that this is, in fact, the reason why the low-rate trickling filter is not commonly used. To this should be added that due to the low rate the demand for land space is so big that it can only be satisfied for very small plants.

Increasing the loading (often simultaneously hydraulically and with respect to organic matter) on a low-rate plant might result in process failure as the cloggings

Figure 5.15. Classic trickling filter design. In the lower picture the filter suffers from partial clogging (ponding).

spread at a higher rate than the rate at which the cloggings can be degraded. But if the load is increased beyond a certain point, something new happens: From being exclusively biologically controlled, the thickness of the biofilm will now mainly be controlled by hydraulic sloughing off. This is the type of biological trickling filter which is still found in great numbers. In recent years, the trickling filter has experienced a renaissance because plastics have made it possible to create a large volumetric surface in filters of little weight, and these can therefore be built high without incurring considerable costs. Such filters are typically used for the treatment of wastewater with high concentrations as a pretreatment prior to any other treatment.

5.8.2. Submerged filters

Submerged filters are filters where the filter medium is located under the surface of the water. Among many different characteristic designs of filters, the most important characteristic is a stationary or movable filter medium.

Stationary filter medium

Figure 5.16 shows a filter design in principle. What is especially important for the design is: The control of the thickness of the biofilm and the supply of oxygen to the water.

In respect of the thickness of the biofilm and problems with clogging, there are only two solutions to the problem: 1) Spacious filter media with sufficient space and hydraulic conditions, which prevent excessive thicknesses of the film, can be used. With such filter media it is not possible to obtain a large specific surface area, and these plants take up correspondingly more space. 2) The alternative is to backwash the filter. This is done by applying a high flow of water. In filters with a filter medium consisting of small objects, usually sand, the particles are lifted free of each other and whirled around in the high flow whereby the biofilm is scoured by the particles - just as it is known from the traditional filters at waterworks. In filters containing an immobilized filter medium, the washing effect is reached through the flow of water eroding the biofilm off the material.

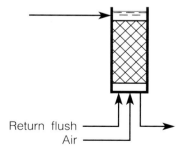

Figure 5.16. Submerged biofilter with stationary filter medium such as small-size filter media (5-10 cm in diameter) or larger surfaces in ingenious patterns to ensure a smooth distribution of water in the filter.

In filters for the removal of organic matter to which oxygen must be supplied, the oxygen can be blown in at the bottom of the filter. When the bubbles rise, the water in the filter is oxidized.

Water can be supplied to the submerged, fixed filter, both at the top and at the bottom, and we will often see that this is used to characterize the filter. However, it is in fact of no importance for the functioning of this type of filter, whether it is an "upflow" or a "downflow" filter.

Movable filter medium

Whereas trickling filters and submerged filters with a fixed filter medium are well known, filters with a movable filter medium were developed during the 70s. The three possibilities, which are in principle different, are illustrated in Figs 5.17, 5.18 and 5.19.

A filter with an upflow through a loose filter medium, typically sand, will have a stationary filter medium as long as the filter medium is not lifted off the pressure gradient from bottom to top. If the pressure in the bottom of the filter equals the weight corresponding to the weight of the filter above, a lifting of the filter medium will occur. The contact between the individual sand particles is broken and the particles become movable. A filter can be operated at the point just where the particles are released from each other. This is called an **expanded filter** - see figure 5.17. The particles constantly grate against each other and keep the biofilm thin. So far, our knowledge is not sufficient to establish how safe this biofilm control is in practice.

If the upflow in the filter is increased beyond the point where a lifting occurs, the filter medium will expand and for a given rate through the filter a balance between the upflow rate and the settling velocity of the particles will occur which depends on the density of the particles. To a given rate corresponds therefore a given degree of expansion. The individual particles of the filter medium will be separated from each other and whirled around in the turbulent upflow. This is called a **fluidized filter** - see figure 5.18. The fluidization creates a very fine hydraulic contact between the water and the biofilm, but it does not facilitate an automatic biofilm control. This must be established separately and takes place by taking out a separate flow of the fluid mix of water and filter medium coated with biomass.

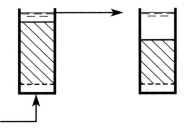

Figure 5.17. Expanded filter in operation and at a standstill. The expansion of the filter layer in operation is typically 30-40 per cent.

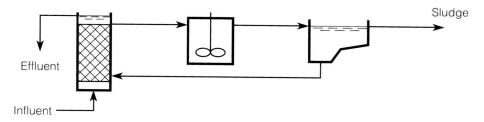

Figure 5.18. Fluidized filter in operation. The filter layer is typically expanded 100 per cent. There are several ways to check biofilm, one of the possibilities being the one shown here where the biofilm on the biofilm particles is sloughed off during vigorous stirring after which the biofilm and the carrier are separated by settling.

Fluidized filters are the most volumetric efficient biological reactors available - with a biomass in the filter of up to 40 kg VSS/m^3. The reactor is, however, not simple to operate in practice, and we have not yet sufficient experience with long-term operation, but this type of plant no doubt has a potential and will find its special niche of application.

In the fluidized filter, the filter medium is kept in suspension in the upflow turbulence. This turbulence can, of course, be established without upflow, and hence we have the **suspended biofilm reactor**, see figure 5.19. Consequently the connection to the activated sludge plant has been established. The difference is just that for the aeration tank, an inert carrier is intentionally added to which the bacteria adhere. In this way the separation properties can be improved either on account of the weight which the filter medium adds to each individual particle, or because the bigger biofilm particles are easier to retain in the plant (due to a higher rate of settling). It may, on the other hand, be necessary to supply more energy in the form of turbulence to the aeration tank to keep the material suspended. At the same time it is a complication that the added filter medium must be handled in the sludge part of the plant. The intentional addition of filter medium to the "activated plants" is as yet at the experimental stage.

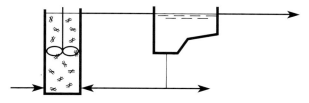

Figure 5.19. Movable filter with stirring. The filter medium is typically different plastics materials, both solid and porous.

5.8.3. Rotating discs

As early as at the turn of the century, experiments were made to solve the three-phase problem by rotating a filter medium alternately through water and air. All kinds of filter media were used, for example branches fixed to a rotating shaft. Today this principle is solely based on plastics in the form of semi-submerged, rotating discs or a rotating tank containing a filter medium. This is illustrated in figure 5.20. These discs or bodies of plastics can be shaped so that they give a very large specific surface and hence a large volumetric efficiency. With a remarkably low energy consumption, the rotation creates an efficient aeration of the water and an efficient contact between water and biofilm. The thickness of the biofilm is controlled by the turbulence which is controlled by the rotational speed. In Germany and Switzerland, rotating discs are widespread in the small treatment plants. In the USA this type of plant has been used in all sizes of plants up through the 70s.

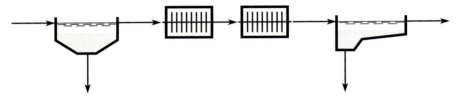

Figure 5.20. Rotating disc plant. The discs are often divided into sections which are operated in series. Similar to all other filter plants, a good mechanical treatment is needed before the water enters the filter.

5.9. Design of biofilters

Based on experiences from many years' operation of conventional biofilters, a set of design criteria has been established which is generally used for the practical design of these plants. These criteria are based on the idea that a given type of plant can be loaded hydraulically or with organic matter up to a limit which can be determined for ordinary use in the design phase. This may sometimes be combined with requirements of the retention time in the system. These design criteria are not based on any special understanding of the phenomena on which the treatment is based, but they are exclusively based on a systematizing of experiences from an endless number of thoroughly measured plants. The criteria are very simple, just like it is remarkable that in spite of this simplicity they have actually been most useful for decades. Likewise there is no doubt that they have given rise to design errors because the circumstances change from one plant to another to a greater extent than it can be expressed in such simple relationships as it will be illustrated for trickling filters and for rotating discs.

5.9.1. Design of trickling filters

According to the classic design criteria for trickling filters the treatment result exclusively depends on the organic volumetric loading rate and the hydraulic surface loading rate.

Based on these parameters and many years' of experience, German design criteria for trickling filters are shown in Table 5.3.

Treatment	No nitrification		Nitrification	
Media	Rock	Plastic (100-200 m^2/m^3)	Rock	Plastic (100-200 m^2/m^3)
Organic volumetric loading rate kg BOD/($m^3 \cdot$d)	0.4	0.4-0.8	0.2	0.2-0.4
Hydraulic loading (inclusive recycle) m/h	0.5-1.0	0.8-1.8	0.4-0.8	0.6-1.5

Table 5.3 German loading guidelines for municipal wastewater treatment/14/.

The experience values in Table 5.3 only apply to ordinary municipal wastewater which is considered to have a relatively homogenous composition. All the curves of treatment results (whether expressed as treatment efficiency or as concentration in the effluent), which can be found in great numbers in the literature, are characterized by a very large spread. This expresses that these simple loading rules cannot take into account the numerous special circumstances which occur in practice. In the above, for example, such basic conditions as: Filter medium, surface area, influent concentration and the special property of the organic matter have not been taken into consideration. We have to be very cautious about using the above loading rules uncritically - although these rules have formed the basis of the successful design of numerous full-scale trickling filters.

Example 5.7

A trickling filter with plastic media has a diameter of 10 m and a height of 2 m and is loaded with 235 m^3/d of mixed domestic and industrial wastewaters. The concentration of BOD is 500g/m^3.

Find the volumetric loading rate and the necessary recycle.

According to Expression (5.54), the volumetric loading rate is:

$$B_V = Q_1 \cdot C_1 / V_2 = Q \cdot C_1/(\pi \cdot r^2 \cdot h) \qquad (5.59)$$

Substitution gives:

$B_V = (235 \, m^3/d) \cdot (500 \, g \, BOD/m^3)/(\pi \cdot (5m)^2 \cdot 2 \, m)$
$B_V = 748 \, g \, BOD/(m^3 \cdot d)$

Hence the filter loading is in the normal range for non-nitrifying filters.

The necessary recycle is controlled by the requirement of the hydraulic surface loading rate which, based on Table 5.3, in this case is estimated at 1.2 m/h.

From Expression (5.57) the necessary recycle, Q_6, can be determined:

$$B_{A,V} = (Q_1 + Q_6)/A_2 \qquad (5.60)$$

Substitution gives:

$1.2 \text{ m/h} \cdot 24 \text{ h/d} = (235 \text{ m}^3/\text{d} + Q_6)/(\pi \cdot (5\text{m})^2)$

$Q_6 = 2{,}027 \text{ m}^3/\text{d}$

that is, a recycle ratio $R = Q_6/Q_1 = (2{,}027 \text{ m}^3/\text{d})(235 \text{ m}^3/\text{d})^{-1} = 8.6$

This is a high recycle.

Figure 5.21. Primary settling, the Lynetten wastewater treatment plant (Copenhagen, Denmark). Primary settling reduces a secondary biological step by approx. 30 per cent. At the same time the energy consumption is reduced, while the sludge production increases. For most types of biofilm processes, primary settling is needed in order tol avoid clogging

5.9.2. Design of discs

Some of the rules for rotating discs are very good examples to show how different the loading rules may be. Table 5.4 lists North American and European examinations which have resulted in recommended loadings. It clearly appears that there is not agreement and that the variation is quite considerable: 5-26 g BOD/(m² · d). It also appears that the more optimistic figures are those given by companies. Notice that the allowable load is stated per m² surface of the discs which is a theoretical, much more relevant size to relate the treatment to, than to m³ of the tank in which the discs rotate.

Recommended loading g BOD/(m² · d)	Year	References
14	1974	Steels /9/
8,5	1978	Murphy and Wilson /10/
26	1976	Autotrol Corp. /11/
13	1977	Envirodisc Corp. /12/
4,9	1974	Ontario Ministry of Environment /13/
8-10	1997	Abwassertechnische Vereinigung /14/

Table 5.4. List of references stating allowable loads for rotating disc plants, applicable for ordinary municipal wastewater at approx. 15°C, to obtain an expected effluent concentration of approx. 15 g BOD/m³ or less. The influent BOD is based on diurnal measurements.

The biofilm control on rotating discs is regulated by the rotational speed which is usually stated as a requirement of a peripheral speed on the discs of not less than 0.3 m/s. To this should be added a minimum requirement of the distance between the discs which is usually stated at 1.5-2.5 cm.

5.9.3. Rules for other types of filter

None of the submerged filter types have gained such common practical use that actual recommendations are available in respect of loading. The biggest volumetric loading has been achieved with fluidized filters which could be loaded with 10 kg BOD/(m³ · d).

5.9.4. Design of biofilters for dissolved organic matter

Today there are no design criteria for biofilters for the removal of organic matter, based on the biofilm theory. In the first section of this chapter it was suggested how a biofilter for the removal of organic matter from wastewater can be designed. The design takes place in 4 stages. (The calculation procedure for stages I and II are shown in figure 5.10).

Stage 0:   Determine the order of reaction in the biofilm itself. It is typically zero order.

Stage I: Calculate, using Expression (5.32), whether oxygen or organic matter is potentially limiting.

Stage II: Calculate whether the biofilm is fully active ($\beta \geq 1$) or only partially active ($\beta < 1$). Often we will find that the biofilm is only partially active, that is, b <1.

Stage III: The removal in the filter can be calculated by means of a mass balance, for example Expression (5.50), for an ideally mixed plant. The removal rate for organic matter, $r_{A,S}$, or oxygen, $r_{A,O2}$, is calculated using one of the 4 expressions below.

For $\beta \geq 1$ the consumption of substrate

$$r_{A,S} = k_{0Vf,S} \cdot L \qquad (5.61a)$$

or if, instead, we look at the oxygen consumption

$$r_{A,O2} = k_{0Vf,O2} \cdot L \qquad (5.61b)$$

$$\text{where } k_{0Vf,O2} = \frac{k_{0Vf,S}}{v_{O2,S}} \qquad (5.61c)$$

For $\beta < 1$ and organic matter as a limiting factor

$$r_{A,S} = (2 \cdot D_{S,2} \cdot k_{0Vf,S} \cdot S_2)^{1/2} = k_{1/2A,S} \cdot S_2^{1/2} \qquad (5.61d)$$

For $\beta < 1$ and oxygen as a limiting factor

$$r_{A,O2} = (2 \cdot D_{O2,2} \cdot k_{0Vf,O2} \cdot S_{O2,2})^{1/2} = k_{1/2A,O2} \cdot S_{O2,2}^{1/2} \qquad (5.61e)$$

A mass balance for organic matter over the plant with for example organic matter as the limiting factor is then (with symbols from figure 5.14):

$$Q_1 \cdot C_1 - k_{1/2A,S} \cdot S_2^{1/2} \cdot A_{2^*} = Q_4 \cdot C_4 + Q_5 \cdot C_5 \qquad (5.62)$$

In Table 5.5, half order rate constants for both removal of organic matter and nitrogen are stated.

Example 5.8

A filter is loaded with 100 m³ of industrial wastewater per day, containing 500 g/m³ acetic acid, HAc. For practical reasons, the filter plant is constructed by units of a surface area of 500 m² each. Experience has shown that each unit can, at a full loading rate, sustain 4 g/m³ oxygen in the water. Each unit is ideally mixed. The biofilm is very thick and is considered to be partially penetrated, $k_{1/2A,O2} = 3.5$ g$^{1/2}$m$^{-1/2}$d^{-1}. How many units are needed in order that the effluent concentration is less than $2 \cdot K_{S,HAc}$, where $K_{S,HAc} = 2$ g HAc/m³?

According to the examples given earlier, the removal will be controlled by oxygen in the first units:

$$r_{A,O2} = 3.5 \text{ g}^{1/2} \text{ m}^{-1/2} \text{ d}^{-1} \cdot (4 \text{ g O}_2 / \text{m}^3)^{1/2} = 7 \text{ g O}_2 / (\text{m}^2 \cdot \text{d})$$

Process	Equipment	Substrate	$k_{1/2A}*$ $g^{1/2}m^{-1/2}d^{-1}$
Oxidation of org. matter	Lab. rot. disc	Acetic acid	3.5-6.2
		Oxygen	3.2-4.1
	-	Methanol	1.4-1.8
	-	Oxygen	4.2
	-	Glucose	3.2-3.8
	-	Oxygen	3.3
Nitrification	Rot. discs	Ammonium	1.5
	-	Oxygen	3.8
	Lab. filter	Oxygen	1.4
	Lab. rot. disc	Ammonium	5.6
	-	Ammonium	4.5
	-	Nitrite	5.1
Denitrification	-	Methanol	2.8-5.4
	-	Nitrate	0.6-3.7
	-	Nitrate	3.1

$k_{1/2A,S} = (2 \cdot D_{S,2} \cdot k_{0Vf})^{1/2}$

Table 5.5. List of half order rate constants for biofilters. Data from /14/,/16/,/24/,/8/,/19/.

$r_{A,O2} \cdot A_2* = 7 \cdot 500 = 3500$ g O_2 / d.

From Example 3.2 we get the stoichiometric conditions for the biological removal of HAc: 1 mole of HAc ~ 60 g HAc consumes 0.9 mole of O_2 ~ $0.9 \cdot 32 = 29$ g O_2

Hence

$$v_{O2,HAc} = \frac{60 \text{ g HAc}}{29 \text{ g } O_2} = 2.1 \text{ g HAc/ g } O_2$$

and thus

$r_{A,HAc} \cdot A_{2*} = 3500 \cdot 2.1 = 7.3$ kg HAc/d

The flow of organic matter $C_1 = S_1 = 500$ g/m³

$Q_1 C_1 = 100 \cdot 500 = 50$ kg HAc/d

Hence at least 7 units must be arranged in a series. The first six will work identically because the HAc-concentration is greater than the limit for being rate limiting in relation to oxygen. In the first six units $6 \cdot 7.3 = 44$ kg HAc/d are removed, and hence there are 6 kg HAc/d left corresponding to an influent concentration of 60 g HAc/m³.

In the seventh unit, HAc is expected to be rate limiting.

$k_{1/2,HAc} = (2 \cdot D_{HAc} \cdot k_{0Vf,HAc})^{1/2}$

$$k_{\frac{1}{2},O2} = (2D_{O2} \cdot k_{0Vf,O2})^{\frac{1}{2}}$$

$$\frac{k_{0Vf,HAc}}{k_{0Vf,O2}} = v_{O2,HAc} = 2.1 \text{ g HAc/gO}_2$$

$$\frac{k_{\frac{1}{2},HAc}}{k_{\frac{1}{2},O2}} = \left(\frac{2 \cdot D_{HAc} \cdot k_{0Vf,HAc}}{2 \cdot D_{O2} \cdot k_{0Vf,O2}}\right)^{\frac{1}{2}} = \left(\frac{D_{HAc}}{D_{O2}} \cdot v_{O2,HAc}\right)^{\frac{1}{2}}$$

$$D_{HAc} = 0.7 \cdot 10^{-4} \text{ m}^2/\text{d}$$
$$D_{O2} = 1.7 \cdot 10^{-4} \text{ m}^2/\text{d, see Table 5.2}$$

$$k_{\frac{1}{2},HAc} = k_{\frac{1}{2},A,O2}\left(\frac{D_{HAc}}{D_{O2}} \cdot v_{O2,HAc}\right)^{\frac{1}{2}} = 3.5 \cdot \left(\frac{0.7}{1.7} \cdot 2.1\right)^{\frac{1}{2}} = 3.3 \text{ (g HAc)}^{\frac{1}{2}}\text{m}^{-\frac{1}{2}}\text{d}^{-1}$$

The balance for HAc:

$$QS_1 - r_{A,HAc} \cdot A_2^* = QS_3$$

$$6000 \text{ g HAc/d} - 3.3 \cdot 500 \cdot S_3^{\frac{1}{2}} = 100 \cdot S_3$$

This gives the solution:

$$S_3^{\frac{1}{2}} = 3.1 \quad \left(\text{or } S_3^{\frac{1}{2}} = -19\right)$$
$$S_3 = 9.6 \text{ g HAc/m}^3$$

At this concentration HAc will be limiting for the removal:

$$\frac{D_{red}}{D_{ox} \cdot v_{ox,red}} = \frac{0.7}{1.7 \cdot 2.1} = 0.20$$

$$\frac{S_{ox}}{S_{red}} = \frac{4 \text{ g O}_2/\text{m}^3}{9.6 \text{ g HAc/m}^3} > 0.20 \text{ g O}_2/\text{g HAc}$$

Notice: The reductant is thus limiting for the process, see Expression (5.32). HAc is limiting when S_{HAc} is less than $4/0.20 = 20$ g HAc/m^3.

An eighth unit is necessary as the concentration is still higher than $2 \cdot K_S$. In the eighth unit, the removal follows a first order process as the concentration is less than a $2 \cdot K_S = 4$ g HAc/m^3.

Expression (5.20) gives:

$$k_{1Vf,HAc} = k_{0Vf,HAc}/K_{S,HAc} = \frac{(k_{\frac{1}{2},HAc})^2}{2 D_{HAc} \cdot K_{S,HAc}}$$

$$k_{1Vf,HAc} = \frac{3.3^2}{2 \cdot 0.7 \cdot 10^{-4} \cdot 2} = 3.9 \cdot 10^4 \text{ d}^{-1}$$

As the biofilm is considered to be very thick, we have from Expresssion (5.9):

$$k_{1A,HAc} = \sqrt{D_{HAc} \cdot k_{1Vf,HAc}} = (0.7 \cdot 10^{-4} \cdot 3.9 \cdot 10^4)^{\frac{1}{2}} = 1.7 \text{ m/d}$$

Balance:

$$QS_1 - k_{1A} \cdot A_2^* \cdot S_3 = QS_3$$
$$960 \text{ g HAc/d} = (1.7 \cdot 500 + 100) S_3$$
$$S_3 = 1 \text{g HAc/m}^3$$

5.10. Technical conditions concerning biofilters

A number of important conditions will be briefly discussed below. The conditions are important for the functioning of biofilter plants and include

- aeration
- growth and sloughing off of the biofilm

5.10.1 Aeration in biofilters

As it appears from the above, the supply of oxygen to biofilters is of greater importance than for other biological plants because the oxygen supply is not just stoichiometrically decisive, but frequently it is also governing for the reaction rate. The following types of aeration can be used:

Separate aeration of the flow of recycle
This type of aeration is not different from the aeration which is required in activated sludge plants. The theories gained from the activated sludge plants can be used directly /15/.

Aeration in the filter itself
The theory for this type of aeration is in principle not different either from the aeration in an activated sludge plant, but of course the presence of the filter medium has a certain effect.

Aeration of the biofilm
In discs the filter medium is covered by a thin water film on the surface of the biofilm. Hence the conditions for the transfer of oxygen from air to biofilm are very special. More details on the subject can be found in /16/.

5.10.2 Growth and sloughing off of the biofilm

Biofilters are based on the ability of the bacteria to attach and develop on a solid medium. Technologically it is desirable that a stable state with equilibrium between growth and sloughing off is developed. No matter how fundamental these conditions might be it is, however, a fact that too little is known about these conditions, apart from the very technological empiricism.

Adhesion

In practice, bacteria will adhere to any surface - they have in fact a partiality for adhesion - irrespective of the type of surface (except where countermeasures have been taken, for example marine coating). The start of a biofilm is simply that the medium concerned is in contact with water, in this case with the wastewater. In practice, the development of a functional film takes approx. 14 days under aerobic conditions. The building up is commenced faster on surfaces with earlier growth than on completely clean surfaces. The building up is selective and the bacteria which are not attached will simply be washed out of the plant. For special substances and processes the starting up may be a big problem.

Types of biofilm

The characteristics of the biofilm are of course a function of the characteristics of the bacteria which form the biofilm. No applicable examinations are available whatsoever of the relationship between microbiological determination of species and measured biofilm characteristics. It can only be concluded that two types of biofilm generally occur: The dense film and the filamentous film. The dense film is a coherent, immobilized biomass. In its ideal form, it has a plane, smooth surface. For this type of biofilm there is clear experimental proof of the good, theoretical description of the removal as described earlier in the biofilm kinetics. The filamentous biofilm is dominated by filamentous bacteria which grow in continuation of each other and form filaments attached to the filter medium, typically the Chlamydobacteriales Sphaerotilus Natans, (also known from highly polluted streams under the name of: Sewage fungus). In its extreme form, this type of film forms a mat whose filaments are moved by the liquid flow. This causes turbulent movements around the filaments for which reason the transport of substrate into the biofilm is increased compared with molecular diffusion. These films will therefore remove material according to a zero order reaction at lower concentrations than calculated according to the biofilm kinetics as discussed above.

The bacterial density in the film is decisive for the removal rate. What is interesting is the mass of the bacteria which carry out the process observed. It is difficult to determine this quantity by measurement, but it can be calculated as a fraction of the total mass in the same way as for the bacteria in the sludge in an activated sludge plant. The total mass may vary within wide ranges: 10-100 kg VSS/m^3; but data primarily fall in the range 40-60 kg VSS/m^3.

The sloughing off of the biofilm

The removal of organic matter corresponds to a continuous growth of bacteria and hence a growth of the thickness of the biofilm. If this is not balanced by a corresponding sloughing off, the result must necessarily be clogging. The following conditions are important for the sloughing off and are used technologically, intentionally or unintentionally, as part of the control of the biofilm:

- Hydraulic erosion acts continually on the surface of the biofilm and leads to a steady sloughing off on the outer side. In its extreme version it is the total sloughing off of the biofilm on the sand in the recycle flow of a fluidized filter in a tank where a stirrer creates a very high turbulence level. The requirements of the hydraulic loading of trickling filters and of the rotational speed of rotating discs are empirically dictated from a desire to cause sloughing off, but the flow of water is not so strong that it can by itself cause the sloughing off. Other factors release the film so that it can be sloughed off hydraulically.
- Degradation of starved out bacteria in the bottom of biofilms may cause a weakening of the adhesion. Biofilms will in practice have a tendency to grow to a thickness where they are only partially penetrated with substrate. In the aerobic filter, whose reaction rate is controlled by the oxygen, anaerobic conditions will occur in the bottom of the film. This will degrade the bacteria in the bottom and destroy the adhesion. Sufficiently weakened, the film will be sloughed off completely over a smaller area by hydraulic erosion.
- Super saturation and bubble formation in the bottom of the biofilm may destroy the adhesion - for example methane production under anaerobic conditions and the production of pure nitrogen by denitrification.

In the last two cases the film is sloughed off completely, and a naked area is left on the filter medium where new growth starts. The filter is thus continuously in a state of sloughing off and regrowth. Therefore the biofilm never has a well-defined thickness in practice which applies to the whole film.

A special phenomenon is the grazing of higher animals on the bacterial mass in the filter. Very little is known about this, but it is undoubtedly of importance and is for example reflected in seasonal variations in the biomass in a trickling filter and in sudden changes in the biofilm in for example nitrifying plants /17/. A conventional trickling filter will contain most biomass in the early spring which makes the risk of clogging correspondingly higher.

5.11. Removal of particulate organic matter

As discussed in the previous sections, we do not today have an applicable hypothesis for the removal of organic matter which cannot diffuse into a biofilm. Until such a hypothesis has been found, we have to rely on emperical load norms, but new knowledge is on its way.

A distinction should be made between two types of substrates:

- diffusible
- non-diffusible

Figure 5.22. Real-life biofilters must also be able to remove particulate organic matter. The photo shows a biofilter plant (Biocarbone) in Meyzieu (France).

Diffusible means that the substrate can be carried into the biofilm by molecular diffusion. There is no well-defined analyzeable limit, but genuine soluble substances such as oxygen, acetic acid, methanol and glucose clearly belong to the diffusible category whereas suspended solids as defined for the filtering through a filter (GF/A filter, pore size 1-1.6 µm) definitely belong to the non-diffusible category. In this connection it is misleading that the substance running through the filter mentioned above is traditionally called "dissolved" (Chapter 1 and 2).

A major part of organic matter in municipal wastewater consists of fractions which pass a 1-1.6 µm filter, but which are not genuinely soluble.

Molecules with a molecular weight of less than 10^3 can be absorbed through the cell membrane on the bacteria. It is not known where the diffusion limit into the biofilm lies. 40-80 per cent of COD in municipal wastewater are particles bigger than 0.1 µm /18/. Very little material in the size fraction 0.01 - 0.1µm has been found. However, there is a significant fraction in the range 0.1-1 µm which is not included in suspended solids, but in the "dissolved" fraction which certainly cannot diffuse into the biofilm.

It is a condition for the removal of a non-diffusible material that it is degraded extracellularly in a diffusible form in order to subsequently diffuse into the biofilm where it is mineralized while consuming oxygen. The collective name for the extracellular degradation is hydrolysis (see Chapter 3, Section 3.2.2.).

There are two different theories of the mechanism for hydrolysis in biofilm reactors:

Particle adsorption. According to this thesis the particles will be adsorbed to the surface of the biofilm. This can be done by diffusion from the bulk water to the surface, by settling and by entrapment on a rough (or filamentous) biofilm surface /20/. Only qualitative determinations are available. We have no verification nor any operational formulation of kinetic expressions.

External hydrolysis. From the biofilm extracellular enzymes are released to the bulk water in which the hydrolysis takes place. The hydrolyzates should subsequently diffuse into the biofilm and be degraded there. This mechanism is experimentally demonstrated for starch as a non-diffusible substance, and kinetic expressions are available, see the simplified representation below:

Balance for non-diffusible matter (for example starch) in an ideally mixed tank:

$$Q \, X_{R,1} = r_{V,XR} \, V_2 + Q \, X_{R,3} \tag{5.63}$$

where X_R is the non-diffusible matter. $r_{V,XR}$ is the volumetric hydrolysis rate in the reactor. The process is assumed to be a first order process in both X_S and S_E (the concentration of enzyme in bulk water)

$$r_{V,XR} = k_E \, S_{E,2} \, X_{R,2} \tag{5.64}$$

where k_E is the hydrolysis constant.

The enzymes are produced in the biofilm at a surface area specific production rate $r_{A,E}$. It is assumed that there is no significant enzyme in the influent and that no removal of the enzyme takes place in the reactor. This leads to the following **enzyme balance**:

$$Q_1 \, S_{E,2} = r_{A,E} \cdot A_{2*} \tag{5.65}$$

In an ideally mixed reactor $X_{R,2} = X_{R,3}$, $S_{E,2} = S_{E,3}$.

$$Q_1 \, X_{R,1} = k_E \cdot r_{A,E} \, X_{R,3} \, A_{2*} \frac{V_2}{Q_1} + Q_1 \, X_{R,3} \tag{5.66}$$

$$D_H = \frac{X_{R,1} - X_{R,3}}{X_{R,1}} = \frac{1}{1 + \frac{Q_1^2}{k_E \, r_{A,E} \, A_{2*} \, V_2}} \tag{5.67}$$

where D_H is called the degree of hydrolysis, which means the part of the non-diffusible substrate which is hydrolyzed to diffusible substrate.

Notice that the retention time $= V_2/Q_1$ and that the hydraulic surface loading rate of the biofilm Q_1/A_{2*} cannot be used individually to describe a loading. It is the combination $Q_1^2/(A_{2^*} \cdot V_2)$ which we can call the combined load. Figure 5.23 shows the degree of hydrolysis as a function of these quantities.

It is now simple to set up the **balance for diffusible matter**, S_D:

$$Q_1 \cdot S_{D,1} + Q_1 X_{R,1} \cdot D_H = k_{\frac{1}{2}A,D} A_{2^*} \cdot S_{D,2}^{\frac{1}{2}} + Q_1 S_{D,3} \tag{5.68}$$

Apart from the loading with diffusible substrate in the influent, the reactor is loaded with the diffusible matter which is produced by hydrolysis. It is assumed that the biofilm is thick and that the diffusible matter is removed by a half order process.

The filter kinetics from Section 5.5 can be directly used if the influent concentration is understood as

$$S_1 = S_{D,1} + X_{R,1} \cdot D_H \tag{5.69}$$

$$S_{D,3}^{\frac{1}{2}} = -\frac{k_{\frac{1}{2}A,D} \cdot A_{2^*}}{2Q_1} + \sqrt{\left(\frac{k_{\frac{1}{2}A,D} A_{2^*}}{2Q_1}\right)^2 + \left(S_{D,1} + X_{R,1} \cdot D_H\right)} \tag{5.70}$$

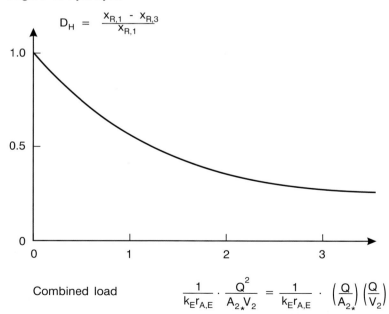

Figure 5.23. The degree of hydrolysis as a function of the water flow, the biofilm surface and the water volume. /21/

$$E = \frac{C_{T,1} - C_{T,3}}{C_{T,1}} = \frac{(S_{D,1} + X_{R,1}) - (S_{D,3} + X_{R,3})}{(S_{D,1} + X_{R,1})} \qquad (5.71)$$

as $C_T = S_D + X_R$

The result is shown in figure 5.24 for the degree of hydrolysis 1.0.

In a biofilm reactor the removal only depends on the accessible biofilm area, and not on the volume. By hydrolysis, the volume is included as a modifying factor because the enzymes are washed off the reactor. If only the volume is reduced, the degree of hydrolysis will decrease, $S_{D,3}$ will decrease and the surface area specific removal will decrease.

This effect is reinforced by the fact that the enzyme production depends on the removal in the biofilm:

$$r_{A,E} \propto \beta L$$

By reduced volume the degree of hydrolysis decreases further because $r_{A,E}$ decreases by reduced removal.

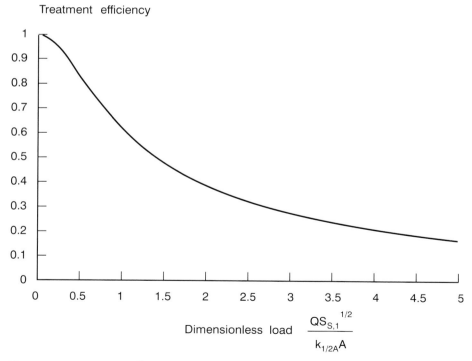

Figure 5.24. Treatment efficiency as a function with the load for a half order process witha degree of hydrolysis of 1.0. /20/

Example 5.9

A filter plant is used for the treatment of wastewater from a potato flour factory whose wastewater is assumed only to contain starch:

Flow of wastewater: $Q = 500$ m³/d
Starch concentration: $C_1 = 100$ g COD/m³
Biofilm area: $A_{2*} = 500$ m²
Tank volume: $V_2 = 5$ m³
Oxygen concentration: $C_{O2,2} = 1$ g/m³

and as a preliminary assumption, the oxygen is assumed to be limiting for the removal:

Internal oxygen removal $k_{0Vf,O2}$ = 200 kg O_2 / (m³ · d)
Diffusional coeff. for oxygen D_{O2} = 1.7 · 10⁻⁴ m² / d
Stoichiometric coefficient $v_{O2,COD}$ = 2.4 g COD / gO_2

$$\beta L = \left(\frac{2 D_{O2} \cdot C_{O2,2}}{k_{0Vf,O2}}\right)^{1/2} = \left(\frac{2 \cdot 1.7 \cdot 10^{-4} \cdot 1}{200,000}\right)^{1/2} = 41 \, \mu m$$

The starch is hydrolyzed to glucose. The removal rate for glucose is:

$$\begin{aligned} r_{A,glu} &= \beta L \cdot k_{0Vf,O2} \cdot v_{O2,COD} \\ &= 41 \cdot 10^{-6} \cdot 200,000 \cdot 2.4 \\ &= 20 \text{ g COD} / (m^2 \cdot d) \end{aligned}$$

If all of the starch is hydrolyzed: $D_H = 1.0$, the glucose balance would look as follows:

$$Q_1 \cdot C_1 \cdot D_H = r_A \cdot A_{2*} + Q_1 \cdot S_3$$

$$S_3 = C_1 \cdot D_H - r_A \frac{A_{2*}}{Q_1} = 100 \cdot 1.0 - 20 \cdot \frac{500}{500} = 100 - 20 = 80 g/m^3$$

which is such a large concentration of glucose that oxygen is limiting for the removal. Empirical constants for the volumetric enzyme production of biofilm and hydrolysis:

Enzyme prod. $r_{V,E}$ = 300 · 10⁶ IU / (m³ biofilm · d)
Hydrolysis k_E = 0.05 m³ / (IU · d)

where IU is the "International Unit" for enzymes.

$$\begin{aligned} r_{A,E} &= \beta L \cdot r_{V,E} = 41 \cdot 10^{-6} \cdot 300 \cdot 10^6 = 12,300 \text{ IU} / (m^2 \cdot d) \\ k_E \cdot r_{A,E} &= 0.05 \cdot 12,300 = 615 \, m/d^2 \end{aligned}$$

(These quantities are still very uncertain.)

Substitution in the formula for the degree of hydrolysis (5.67) gives:

$$D_H = \frac{1}{1 + \frac{1}{k_E r_{A,E}} \cdot \frac{Q_1^2}{A_{2*} \cdot V_2}} = \frac{1}{1 + \frac{1}{615} \cdot 100} = 0.86$$

where the "combined load" is $Q_1^2 / A_{2*} \cdot V_2 = 500^2 / 500 \cdot 5 = 100 \, m/d^2$.

Starch concentration in the effluent is

$$X_{R,3} = C_1(1 - D_H) = 100 \cdot 0.14 = 14 \, g/m^3$$

Glucose concentration in the effluent is

$$S_3 = 100 \cdot 0.86 - 20 \cdot \frac{500}{500} = 86 - 20 = 66 \, g/m^3$$

If the filter is made more efficient, for example by using a fluidized filter with a carrier of fine particles and by increasing the specific volume to 1000 m²/m³, the volume is reduced to 0.5 m³.

$$\frac{Q_1^2}{A_2 \cdot V_2} = \frac{500^2}{500 \cdot 0.5} = 1000 \text{ m/d}$$

$$\frac{1}{1 + \frac{1000}{615}} = 0.38$$

$$X_{R,3} = 100 \cdot 0.62 = 62 \text{ g/m}^3$$

$$S_3 = 100 \cdot 0.38 - 20 \frac{500}{500} = 38 - 20 = 18 \text{ g/m}^3$$

It is the hydrolysis which is the rate limiting step for the process.

5.12. Detailed model

So far the description of the processes in the biofilm has been influenced by such simplifying assumptions that the equations can be solved analytically and that the interaction of parameters is relatively easily ignored. Such a description has its advantage in that it facilitates the understanding and gives a general view; but the more complicated the interaction of processes is, the more difficult it is to find suitable simplifications and analytical solutions. This interaction can, on the other hand, be described by detailed, deterministic modelling by which the door is opened to a alienating world of complicated differential equations, complicated functional expressions and solutions which are difficult to understand and which can only be obtained by computer calculations. However, there is no doubt about the need for the increased realism in the description and the need for such models for scientific examinations and soon also for the practical design.

The following description mainly follows the description in /22/. Figure 5.24 shows an element in the biofilm and the corresponding definitions. The biofilm consists of several phases such as water, solids (and possibly air). The concentration, C_{ki}, of a substance is stated as the amount of a given substance in a unit volume of the phase concerned. The volume fraction, ε_k, of a solid k is stated by the volume of the solid in a unit volume of the biofilm containing all the phases k.

For the phases it applies:

$$\sum_k \varepsilon_k = 1 \tag{5.72}$$

For the concentration it correspondingly applies that the concentration distributed on a total unit volume is $C_{fi} = \varepsilon_k \cdot C_{ki}$, where C_{ki} is the definition of the model concentration which has been used so far. It is necessary to make this distinction when there are many phases; for example several phases of solids and different types of bacteria.

The differential equation for an element in the biofilm can be written as:

$$\frac{\partial \varepsilon_k C_{ki}}{\partial t} = \varepsilon_k \frac{\partial C_{ki}}{\partial t} + C_{ki} \cdot \frac{\partial \varepsilon_k}{\partial t} = -\frac{\partial j_{ki}}{\partial x} + r_{V,ki} \qquad (5.73)$$

where j_{ki} is the flux of the substance i in phase k
 $r_{V,ki}$ is the removal rate for the substance i in phase k per unit volume
 ε_k is the volume fraction of phase k
 C_{ki} is the concentration of substance i in phase k.

The first term ahead of the equals sign is the local time derivative from the mean concentration which, just after the first equals sign, is split into a time derivative for the concentration and one for the volume fraction of a phase. The last term but one is the local, geometrical derivative of the flux through a unit cross section. The last term expresses the removal in the biofilm.

$$\frac{\partial \varepsilon_k C_{ki}}{\partial t} = -\frac{\partial j_{ki}}{\partial x} + r_{V,ki} = 0 \qquad (5.74)$$

$$j_{ki} = D_i \frac{\partial \varepsilon_k C_{ki}}{\partial x}$$

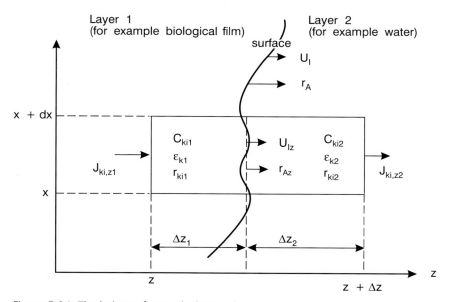

Figure 5.24. The balance for a calculation element in the interface between water phase and biofilm. /22/

which yields the well-known expression

$$D_i \frac{\partial^2 \varepsilon_k C_{ki}}{\partial x^2} = D_i \frac{\partial^2 C_{fi}}{\partial x^2} = r_{V,ki} \tag{5.75}$$

corresponding to the derived equations in Section 5.1. The well-known equations have thus been generalized in respect of phase and component, as well as time and place.

The processes are described by the well-known process matrix, for example Table 11.1.

In principle, all that is left now is to define the limiting conditions:

$$u_1 (\varepsilon_{k1} \cdot C_{ki1} - \varepsilon_{k2} \cdot C_{ki2}) = j_{ki1} - j_{ki2} + r_{Aki} \tag{5.76}$$

where u_1 is the rate at which the interface moves (for example due to growth)
       1   is on the one side the interface
       2   is on the other side of the interface

The first term describes the transport of substance by a movement of the interface. The second term describes the flux to and from the interface and the last term is a surface reaction rate (for example adsorption of particles to the surface (positive) or erosion from the surface). Simple cases will illustrate the limiting condition:

Back wall: $u_1 = 0$, $j_{ki1} = 0$, $r_{A,ki} = 0$, which results in

$$j_{ki2} = 0 \; ; \quad D_i \frac{\partial \varepsilon_k C_{ki}}{\partial x} = 0 \tag{5.77}$$

The concentration profile has a horizontal tangent as already described.
Stationary surface without reaction:

$$u_1 = 0, \quad r_{A,ki} = 0$$

$$j_{ki1} = j_{ki2}$$

The diffusion of a substance must be the same on both sides of the area because no accumulation can take place in the interface with the given assumptions.

With these equations all processes can be described if we know the kinetic expressions for the processes; for variations in time and place, for random composition of the phase (types of bacteria) and for selected limiting conditions. On the other hand it is not easy to solve these differential equations under the given practical conditions. However, today we have computer programs which are tailored for the solution of these equations for conditions corresponding to the biofilm /23/. Such programs will be increasingly used for analysis, operation and design.

In the next chapter an example is given which is based on a solution with this program.

References

/1/ Harremoës, P. (1976): The Significance of Pore Diffusion to Filter Denitrification. *J. WPCF*, 48, 377-388.

/2/ Harremoës, P. and Riemer, M. (1977): Pilot-Scale Experiments on Down-Flow Filter Denitrification. *Prog. Water Technol.*, **8**, (4/5), 557-576

/3/ Harremoës, P. and Riemer, M. (1975): Report on Pilot Studies of Denitrification in Down-flow Filters. Department of Environmental Engineering, Technical University of Denmark, Lyngby, Denmark

/4/ Harremoës, P. (1978): Biofilm kinetics. Chapter 4 in: Mitchell, R. (ed.): Water Pollution Microbiology, 2, pp. 71-109. John Wiley & Sons, New York, N.Y.

/5/ Perry, J. (1963): Chemical Engineers Handbook, McGraw-Hill, New York, N.Y.

/6/ Siegrist, H. and Gujer, W. (1985): Mass transfer Mechanisms in a Heterotrophic Biofilm. *Water Res.*, **19**, 1369-1378.

/7/ Jansen, J. la Cour and Kristensen, G.H. (1980): Fixed Film Kinetics - dentrification in fixed films. Department of Environmental Engineering, Technical University of Denmark, Lyngby. (Report 80-59).

/8/ Onuma, M. and Omura, T. (1982): Mass-transfer Characteristics with Microbial Systems. *Water Sci. Technol.*, **14**, (6/7), 553-568.

/9/ Steels, I.H. (1974): Design Basis for the Rotating Disc Proces. *Effluent Water Treatment J.*, **14**, 431-445.

/10/ Murphy, K.L. and Wilson, R.W. (1978): Pilot Plant Studies of Rotating Biological Contactors Treating Municipal Wastewater, Prepared for the Central Mortgage and Housing Corporation. International Environmental Consultans Ltd., Toronto, Ontario.

/11/ Autotrol Corporation (1976): Wastewater Treatment Systems. Manufacturers Design Manual, Autotrol Corp., Section C. South Lake Tahoe, NV, USA.

/12/ Envirodisc Corporation (1977): Rotating Disc Wastewater Treatment. Manufacturers Design Manual. USA.

/13/ Ahlberg, R.N. and Kwong, T.S. (1974): Process Evaluation of a Rotating Biological Contactor for Municipal Wastewater Treatment. Ontario Ministry of Environment, Wastewater Treatment Section, Pollution Control Plannning Branch, Ottawa, Ontario. (Research paper no. W2041).

/14/ Abwassertechnische Vereinigung (1997): Biologische und weitergehende Abwasserreinigung (Biological and advanced wastewater treatment). 4th ed. Ernst & Sohn, Berlin.

/15/ Harremoës, P., Henze, M., Arvin, E. and Dahi, E. (1994): Teoretisk vandhygiejne. (Water Chemistry) 4th ed. Polyteknisk Forlag, Lyngby, Denmark.

/16/ Gönenç, E. (1982): Nitrification on Rotating Biological Contactors. Department of Environmental Engineering, Technical University of Denmark, Lyngby, Denmark.

/17/ Iwai, S., Oshino, Y. and Tsukada, T. (1990): Design and Operation of Small Wastewater Treatment Plants by the Microbial Film Process. *Water Sci. Technol.*, **22**, (3/4), 139-144.

/18/ Levine, A.D., Tchobanoglous, G. and Asano, T. (1985): Characterization of the Size Distribution of Contaminants in Wastewater Treatment and Reuse Implications. *J. WPCF*, **57**, 805-816.

/19/ Jansen, J. la Cour (1982): Fixed Film Kinetics - kinetics of soluble substrates. Ph.D. Thesis. Department of Environmental Engineering, Technical University of Denmark, Lyngby. (Report 81-35).

/20/ Bouwer, E.J. (1987): Theoretical Investigation of Particle Deposition in Biofilm Systems. *Water Res.*, **21**, 1489-1498.

/21/ Larsen, T.A. (1991): Degradation of Colloidal Organic Matter in Biofilm Reactors. Ph. D. Thesis. Department of Environmental Engineering, Technical University of Denmark, Lyngby, Denmark.

/22/ Characklis, W.G., et al (1989): Modelling of Biofilm Systems. IAWPRC Task Group on Modelling of Biofilm Systems, Institute of Process Analysis, Montana State University, Bozeman, MT.

/23/ Reichert, P., Ruchti, J. and Wanner, O. (1989): Biosim Interactive Program for the Simulation of Dynamics of Mixed Culture Biofilm systems. Swiss Fed. Inst. Wat. Resources and Wat. Poll. Control., Duebendorf, Switzerland.

/24/ Watanabe, Y. and M. Ishiguro (1978): Dentrification Kinetics in a Submerged Rotating Biological Disk Unit. *Prog. Water Technol.*, **10**, (5), 187-195.

/25/ Ekama, G.A. et al. (1997): Secondary settling tanks:Theory, Modelling, Design and Operation. IAWQ London. (IAWQ Scientific and Technical Report No.6).

References for general information about biofilms:
Bernard, J. (ed) (1990): Technical Advances in Biofilm Reactors, *Water Sci.Technol.*, **22**, No. 1/2.
Characklis, W.G. and Marshall, K.A. (1990): Biofilms. John Wiley & Sons, New York, N.Y. 800 pp.
Characklis, W.G. and Wilderer, P.A. (1989): Structure and Function of Biofilms. Report of the Dahlem Workshop Berlin 1988. John Wiley & Sons, New York, N.Y. (Life Sciences Research Report 46). 387 pp.
Harremoës, P. and Arvin, E. (eds.) (1990): Biofilm Kinetics and Fixed Film Reactors. Department of Environmental Engineering, Technical University of Denmark, Lyngby, Denmark.
Marshall, K.C. (ed) (1984): Microbial Adhesion and Aggregation, Report of the Dahlem Workshop, Berlin 1984, Springer, Berlin. (Life sciences research report 31) 433 pp.

Secondary settling (here Tianjin, Peoples Republic of China) is not dealt with in this book. An exellent treatise is given in /25/.

6 Treatment Plants for Nitrification

By Jes la Cour Jansen, Poul Harremoës and Mogens Henze

Ordinary, municipal wastewater contains 20-50 mg/l of nitrogen compounds, most in the form of ammonium (NH_4^+ and NH_3) and organic compounds which are easily converted into ammonium in a treatment plant. There are three reasons why the presence of these compounds is not wanted in the treated wastewater, and why attempts are made to remove or convert them before their discharge into the recipient:

- On biological oxidation (nitrification), ammonium will consume a considerable amount of oxygen. In ordinary municipal wastewater, the oxygen requirement for this oxidation will be approx. 40% of the total oxygen requirement. Hence biological treatment without nitrification will not in many cases, be sufficient to prevent oxygen depletion in the recipient.
- Ammonia (NH_3) is a strong fish poison. Fortunately the content of ammonia in the wastewater is low as the chemical equilibrium $NH_4^+ \rightleftarrows NH_3 + H^+$ at normal pH is shifted towards the left. Only at a pH of more than 8.5, ammonia begins to form an essential part of the total content of ammonium compounds. Especially in streams, the pH can vary considerably. In Denmark it is required that the ammonia content of the stream is less than 0.025 mg/l if the stream is to be graded a fishing stream. This drastic requirement of the ammonia content means that in most cases it is necessary to oxidize the ammonium components before discharge.
- Along with phosphorus, nitrogen constitutes the most important nutrient for plant growth in the recipients. Nitrification in itself does not mean that the nutrient content will change as the plants absorb nitrogen in the form of nitrate or ammonia. However, nitrification is necessary if nitrogen is to be removed from the wastewater by denitrification since biological denitrification, where nitrate is converted into free nitrogen, requires that the ammonium compounds are oxidized into nitrate before they are reduced to gaseous nitrogen. Chapter 7 gives a more explicit explanation of the linking of nitrification and denitrification in treatment plants.

As a consequence of the above effects of ammonium discharge, nitrification (in addition to the removal of organic matter) is also required to an increasing extent when wastewater is discharged into streams, lakes and rivers.

6.1. Mass balances, nitrifying plants

The nitrification process in biological treatment plants occurs either concurrently with a conversion of organic matter or as a separate process. In both cases the same micro-organisms are responsible for the oxidation of ammonium into nitrate. Hence the two processes have much in common.

6.1.1. Separate nitrifying plants

The plant may be an activated sludge treatment plant or a biofilm plant (trickling filter, aerated filter, rotating discs, etc.). Figure 6.1 shows a schematic representation of these two plant types.

The mass balance of a nitrifying plant can be established when the kinetics of the processes taking place in the applied model are known. A model can be based on a process matrix, as shown in Table 6.1. Notice that the yield constant of the entire nitrification process, $Y_{max,A}$, is expressed per amount of nitrate nitrogen produced and *not* per amount of ammonium nitrogen converted. The unit may, for example, be kg COD (B)/kg NO_3^- – N. $f_{XB,N}$ is the content of nitrogen in the biomass, kg N/kg COD(B). As to the change of alkalinity in Table 6.1, assimilation of nitrogen in the nitrifying bacteria has been neglected. The factor 1/7 of the stoichiometric coefficient expresses that two equivalents of alkalinity per mole (= 14 g) of converted nitrogen are removed which corresponds to 2/14 or 1/7 eqv/g N. Nitrification produces COD from the production of biomass from oxidation of ammonium to nitrate (and that is why the mass balance of COD does not fit in the matrix).

The COD mass balance of nitrifying sludge, $X_{B,A}$, is visualized in the following (for symbols and layout see figure 6.1).

$$Q_1 \cdot X_{B,A,1} + r_{V,XB} \cdot V_2 - b_A \cdot X_{B,A,2} \cdot V_2 = Q_3 \cdot X_{B,A,3} + Q_5 \cdot X_{B,A,5} \qquad (6.1)$$

b_A is the decay constant (decay of nitrifying bacteria).

Treatment Plants for Nitrification

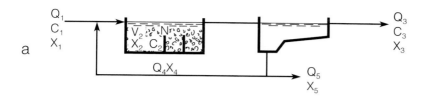

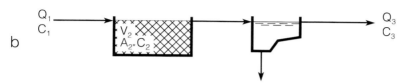

Figure 6.1. Schematic representations of nitrifying plants.
a. Nitrifying activated sludge plant
b. Nitrifying biofilm plant

Component→ Process↓	S_{NH4}	S_{NO3}	S_{O2}	$X_{B,A}$	$X_{S,N}$	S_{ALK}	Reaction rate $r_{v,...}$
1. Aerobic growth, nitrifying bacteria	$-\dfrac{1}{Y_{max,A}}$ $-f_{XB,N}$	$\dfrac{1}{Y_{max,A}}$	$-\dfrac{4.57-Y_{max,A}}{Y_{max,A}}$	1	$f_{XB,N}$	$-\dfrac{1}{7Y_{max,A}}$ $-\dfrac{f_{XB,N}}{14}$	$\mu_{max,A} \dfrac{S_{NH4}}{S_{NH4}+K_{S,NH4,A}} \cdot \dfrac{S_{O2}}{K_{S,O2,A}+S_{O2}} \cdot X_{B,A}$
2. Decay, nitrifying bacteria				-1	$-f_{XB,N}$		$b_A \cdot X_{B,A}$
3. Hydrolysis of organic matter	1				-1		$k_{h,A} \cdot X_{ND}$
Unit	kg N/m³		kg COD/m³		kg N/m³	eqv/m³	
	Ammonium	Nitrate	Oxygen	Nitrifying biomass	Suspended org. nitrogen	Alkalinity	

Table 6.1. Process matrix, separate nitrifying plant.

From Table 6.1 we know that

$$r_{V,XB} = \mu_{max,A} \cdot \frac{S_{NH4}}{S_{NH4,A}+K_{S,NH4,A}} \cdot \frac{S_{O2}}{K_{S,O2,A}+S_{O2}} \cdot X_{B,A} = \mu_{obs,A} \cdot X_{B,A}$$

Activated sludge treatment plants

The mass balance (6.1) can be simplified according to the following hypothesis/substitution:

$X_{B,A,1} = 0$ (there are very few nitrifying bacteria in ordinary wastewater may be 0.1-1 g/m³).

Likewise the following applies

$Q_3 \cdot X_{B,A,3} \ll Q_5 \cdot X_{B,A,5}$

(the major part of the nitrifying bacteria removed from the plant are withdrawn together with the excess sludge)

$\mu_{obs,A} - b_A = \mu_{obs,A,net}$

Hence the simplified mass balance will be:

$$\mu_{obs,A,net} \cdot X_{B,A,2} \cdot V_2 = Q_5 \cdot X_{B,A,5} \tag{6.2}$$

By entering the sludge age, Expression (4.15), is found:

$$\theta_{X,A} = \frac{1}{\mu_{obs,A,net}} \tag{6.3}$$

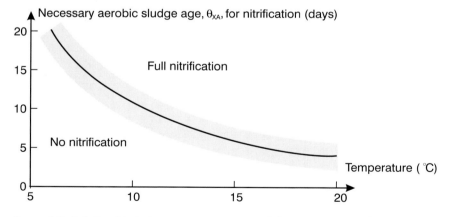

Figure 6.2. Relationship between temperature and the necessary aerobic sludge age for the achievement of nitrification in activated sludge treatment plants. (Operating conditions: Oxygen concentration 2 g/m³). For higher DO concentration the curve will be lowered.

Knowing the net growth rate of the nitrifying bacteria, $\mu_{obs,A,net}$, it is possible to find the necessary sludge age, $\theta_{X,A}$, from Expression (6.3), or the necessary volume of the nitrification tank by means of Expression (6.2).

The sludge age is the easiest parameter to use, as it is very easily calculated for a specific plant seeing that it is assumed to be the same for all the sludge components. Hence it can be calculated based on an arbitrary measuring unit, SS, VSS, COD, etc. and consequently no knowledge of the concentration of nitrifying biomass, $X_{B,A}$, in the plant is required. The sludge age to be calculated is the aerobic sludge age because it was assumed, when writing the mass balance in Expression (6.1), that removals, growth and decay only take place in the aeration tank (with volume V_2). Figure 6.2 shows the necessary aerobic sludge age for the maintenance of the nitrification under normal operating conditions in activated sludge treatment plants.

Expression (6.3) can be used to find the ammonium concentration of the aeration tank, $S_{NH4,2}$, when the sludge age is known:

$$\mu_{obs,A,net} = \frac{1}{\theta_{X,A}}$$

$$\mu_{max,A} \cdot \frac{S_{NH4,2}}{S_{NH4,2} + K_{S,NH4,A}} \cdot \frac{S_{O2,2}}{K_{S,O2,A} + S_{O2,2}} - b_A = \frac{1}{\theta_{X,A}} \quad (6.4)$$

$\mu_{max,A}$, $K_{S,NH4}$ (= K_S), $K_{S,O2,A}$ and b_A can be found in Table 3.10 (the column for ammonium oxidizers or for the entire process), $\theta_{X,A}$ can be found in figure 6.2, after which $S_{NH4,2}$ can be determined.

Example 6.1

The critical point of a nitrifying activated sludge treatment plant is the low temperatures in winter. In the case of an ideally mixed tank we know that the aerobic sludge age is 25 days and that the oxygen concentration is 1.6 g/m³.

What will the effluent concentration of ammonium be at a temperature of 8°C?

The following is known/assumed (see values in Tables 3.10 and 3.6):

$\mu_{max,A} = 0.7 \text{ d}^{-1}$ (20°C)

$\kappa = 0.09° \text{ C}^{-1}$

$K_{S,NH4,A} = 0.5 \text{ g N/m}^3$

$K_{S,O2,A} = 0.8 \text{ g O}_2/\text{m}^3$

$b_A = 0.05 \text{ d}^{-1}$

$\mu_{max,A}$ (8°C) calculated by Expression (3.16).

$\mu_{max,A}$ (8°C) = $\mu_{max,A}$ (20°C) $\cdot$ exp ($\kappa \cdot$ (T–20)) = 0.7 $\cdot$ exp (0.09 $\cdot$ (8–20)) = 0.24 d^{-1}

Likewise b_A (8°C) = 0.02 d^{-1} is found.

When substituting known values into Expression (6.4), the ammonium concentration of the aeration tank, $S_{NH4,2}$, is found:

$$\mu_{max,A} \frac{S_{NH4,2}}{S_{NH4,2} + K_{S,NH4,A}} \frac{S_{O2,2}}{K_{S,O2,A} + S_{O2,2}} - b_A = \frac{1}{\theta_{X,A}} \quad (6.4)$$

$$0.24 \frac{S_{NH4,2}}{S_{NH4,2} + 0.5} \frac{1.6}{0.8 + 1.6} - 0.02 = \frac{1}{25}$$

hence:
$S_{NH4,2} \sim 0.3 \text{ g } NH_4^+ - N/m^3$

Biofilm plants

Nitrification consists of the removal of soluble materials: NH_4^+ and O_2. Therefore, the condition of the biofilm theory, as described in previous sections, has been satisfied, and the theory has shown its practical applicability. Figure 6.3 shows the observed nitrification rate per surface area as a function of the concentration in the bulk water.

The nitrogen mass balance of a nitrifying filter (figure 6.1b) is as follows, see Expression (5.3):

$$Q_1 \cdot C_{N,1} - r_{A,NH4} \cdot A_{2*} = Q_3 \cdot C_{N,3} \quad (6.5)$$

For a separate nitrifying biofilm plant the nitrification rate is limited either by the oxygen concentration or by the ammonium concentration because, in practice, the biofilm can be assumed to be only partially utilized (partially penetrated). The criterion for the limitation is found when calculating Expression (5.32) for ammonium and oxygen.

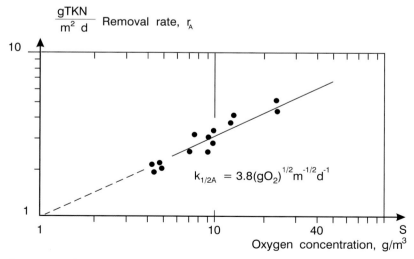

Figure 6.3. Observed nitrification rate per surface area of biofilm as a function of the oxygen concentration in bulk water /1/.

Figure 6.4 shows the penetration conditions for NH_4^+ and O_2 in the two cases of limitation, see Expression (5.32):

$$\frac{S_{NH_4}}{S_{O2}} \gtreqless \frac{D_{O2}}{D_{NH_4}} \cdot v_{O2,NH_4} = \frac{2.1 \cdot 10^{-4}}{1.7 \cdot 10^{-4}} \cdot \frac{1}{4.25} = 0.3 \frac{g\,NH_4^+ - N/m^3}{g\,O_2/m^3}$$

For $S_{NH4} > 0.3 \cdot S_{O2}$, oxygen is the limiting factor, and the reaction rate in the tank is:

$$r_{A,NH4} = (k_{½A,O2} / v_{NH4,O2}) \cdot S_{O2,2}^{½} \qquad (6.6)$$

where $k_{1/2A,O2}$ is the rate constant of oxygen,

$v_{NH4,O2}$ the stoichiometric coefficient between oxygen and ammonium
($4.25\,g\,O_2/g\,NH_4^+ - N$),

$S_{O2,2}$ the oxygen concentration in the tank.

For $S_{NH4} < 0.3 \cdot S_{O2}$, ammonium is the potential limiting factor, and the reaction rate in the tank is:

$$r_{A,NH4} = k_{1/2A,NH4} \cdot S_{NH4,2}^{½} \qquad (6.7)$$

where $k_{1/2A,NH4}$ is the rate constant of ammonium,

$S_{NH4,2}$ is the ammonium concentration in the tank.

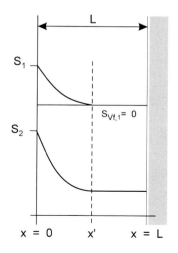
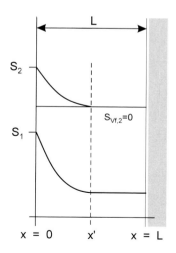

Figure 6.4. Penetration conditions for two substances (f.ex. NH_4^+ and O_2 for nitrification or NO_3^- and CH_3OH for denitrification). To the left, substance 1 is limiting due to partial penetration. To the right substance 2 is limiting.

It depends on the local conditions which of the two expressions for $r_{A,N}$ is to be used. A nitrifying filter will often change from oxygen limited nitrification in the inlet to ammonium limited nitrification later on. In such a case it is possible to make two separate calculations on the filter, see example 6.2.

Example 6.2

Rotating disc filters for the nitrification of landfill leachate has a total filter area of 40,000 m^2. The filter is divided in 2 stages, each ideally mixed.

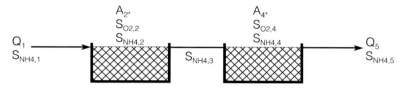

Wastewater characteristics:

$Q_1 = 500 \ m^3/d$

$S_{NH4,1} = 100 \ g \ NH_4^+ - N/m^3$

The oxygen concentration in the two ideally mixed halves of the filter is:

$S_{O2,2} = 5 \ g \ O_2/m^3$

$S_{O2,4} = 7 \ g \ O_2/m^3$

Each of the 2 filters has a disc area of 20,000 m^2.

Reaction specific parameters:

$k_{1/2A,O2} = 4 \ (g \ O_2)^{1/2} \ m^{-1/2} \ d^{-1}$

$v_{NH4,O2} = 4.6 \ g \ O_2/g \ NH_4^+ - N$

$k_{1/2A,NH4} = 0.7 \ (g \ NH_4^+ - N)^{1/2} \ m^{-1/2} \ d^{-1}$

A mass balance for the first plant stage can be expressed as follows:

$Q_1 \cdot S_{NH4,1} - A_{2*} \cdot r_{A,NH4} = Q_3 \cdot S_{NH4,3}$

On the assumption that the nitrification is oxygen limited (to be controlled after calculating), we know from Expression (6.6) that:

$r_{A,NH4} = (k_{1/2A,O2} / v_{NH4,O2}) \cdot S_{O2,2}^{1/2}$, which, if substituted, gives:

$Q_1 \cdot S_{NH4,1} - A_{2*}(k_{1/2A,O2} / v_{NH4,O2}) \cdot S_{O2,2}^{1/2} = Q_3 \cdot S_{NH4,3}$

If we substitute the known values, it is found that:

$500 \cdot 100 - 20{,}000(4 / 4.6) \cdot 5^{1/2} = 500 \cdot S_{NH4,3}$

$S_{NH4,3} = 22 \ g \ NH_4^+ - N/m^3$

(Control of oxygen limitation:

? $S_{NH4,3} > 0.3 \cdot S_{O2,2}$

$0.3 \cdot S_{O2,2} = 0.3 \cdot 5 = 1.5$

$S_{NH4,3} = 22$

that is,

$S_{NH4,3} > 0.3 \cdot S_{O2,2}$

Oxygen limitation **is** ascertained.)

As to the other stage of the filter, ammonium is supposed to be limiting. The mass balance will then be (symbols from figure 6.1b):

$$Q_3 \cdot S_{NH4,3} - A_4 \cdot k_{1/2A,NH} \cdot S_{NH4,4}^{1/2} = Q_5 \cdot S_{NH4,5}$$

The ideal mix can be expressed by the equation $S_{NH4,4} = S_{NH4,5}$

By substitution it is found that:

$$500 \cdot 22 - 20{,}000 \cdot 0.7 \cdot S_{NH4,5}^{1/2} = 500 \cdot S_{NH4,5}$$

from this is found that

$S_{NH4,5} \sim 0.6 \text{ g/m}^3$

(Controlling whether ammonium is limiting or not:

$S_{O2,4} \cdot 0.3 = 2.1$

$S_{NH4,4} = 0.6$

$S_{NH4,4} < 0.3 \cdot S_{O2,4}$

that is, ammonium **is** limiting).

The outlet from the plant will then be approx. $0.6 \text{ g NH}_4^+ - \text{N/m}^3$

pH-inhibition

Nitrification is an alkalinity-consuming process. If the alkalinity of the water is too small, it can be exhausted, resulting in a substantial reduction of the pH, which will inhibit the nitrification process. This happens at pH = 5.8 and limits what can be nitrified without the addition of alkalinity, for example lime.

This is futher complicated by treatment in biofilm processes because the diffusional limitation in the biofilm must be taken into consideration. Nitrification consumes HCO_3^- and produces CO_2 resulting in gradients for diffusion of HCO_3^- into the biofilm. The magnitude of this diffusion will decrease in the interior of the biofilm. Conversely, CO_2 must diffuse out of the biofilm, for which reason higher concentrations are achieved in the biofilm than in the bulk water. The phenomenon is illustrated in figure 6.5. This diffusion (in and out) can be calculated just as for any other substrates and products /3/. The result of such a calculation is shown in figure 6.6. At certain molar ratios there will be a change in the process limitation. The criterion for a diffusion limiting substrate, see Expression (5.32), for O_2 and NH_4^+ is a ratio of 1.4 expressed in moles. Corresponding criteria apply to the ratio between alkalinity and O_2 and alkalinity and NH_4^+, where the ratios are 2.4 and 3.4, respectively, as shown in the figure. If the molar ratio is below 2.4, the pH in the biofilm will fall drastically as shown in the figure.

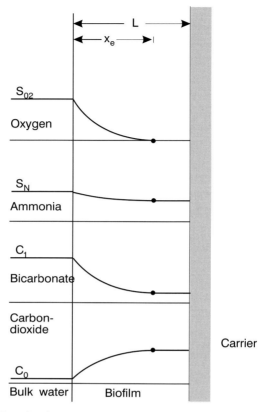

Figure 6.5. Concentration distribution of four components which diffuse into and out of a nitrifying biofilm /3/.

Example 6.3

Wastewater having an alkalinity of 2 eqv/m^3 is to be nitrified. What is the highest ammonium concentration acceptable in the influent?

According to Expression (3.28) approx. 2 moles of HCO_3^- are consumed for each mole of ammonium. In the influent of the nitrifying plant only
1 mol NH_4^+ = 1 mol N/m^3 = 14 g NH_4^+ – N/m^3 can be accepted.

The wastewater is treated in an ideally mixed tank, where an ammonium concentration of max. 2 g NH_4^+ – N/m^3 is required in the bulk water.
The oxygen concentration in the bulk water is 4 g/m^3. Oxygen is the rate limiting substrate because the molar ratio is $\frac{4}{32} / \frac{2}{14} = 0.88 < 1.4$.

Due to the diffusional limitation, an alkalinity above
$2.4 \cdot \frac{4}{32} = 0.30$ eqv/m^3 is required.

Hence only 2.0 – 0.3 = 1.7 eqv/m^3, corresponding to
12 g NH_4^+ – N/m^3, may be removed - unless lime is added.

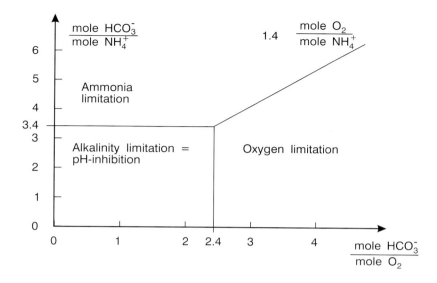

Figure 6.6. The ratio between the components consumed in a biofilm determines which component will be rate limiting /3/.

6.1.2. Combined removal of organic matter and ammonium

Activated sludge
The mass balance of nitrifying sludge is equal to Expression (6.1). Expression (6.4) can be used for calculating the ammonium concentration in a nitrifying activated sludge treatment plant.

Biofilms
As in the case of an activated sludge treatment plant, the question will be whether the heterotrophic bacteria through their growth will prevent the nitrifying bacteria from growing sufficiently. The micro-organisms present in a filter are deposited in a layer on the carrier. Organic matter, ammonium and oxygen are applied through the surface of the biofilm. In practice, oxygen will normally limit the conversion of both organic matter and ammonium. As the nitrifying bacteria grow slowly, they will be ousted and hence eliminated from the plant if the thickness growth of the biofilm, which is primarily conditioned by the growth of the heterotrophic bacteria, is faster than that of the nitrifying bacteria.

If the nitrifying bacteria dispose of too little oxygen, they will be ousted more easily. Therefore the criterion for nitrification will be that oxygen penetrates further into the biofilm than organic matter does, that is, organic matter limits the heterotrophic conversion.

From Expression (5.32) we know that organic matter is limiting if the following inequalities are true:

$$S_{BOD,2} < S_{O2,2} \cdot D_{O2,2} \cdot v_{O2,BOD}/D_{BOD,2} \tag{6.8}$$

Substituting by values from Table 5.2 the inequality will be

$$S_{BOD,2} < 5 \cdot S_{O2,2}$$

where BOD implies dissolved BOD.

Figure 6.7 shows an experimental demonstration of Expression (6.8). It will be seen that partial nitrification may well take place in biofilters.

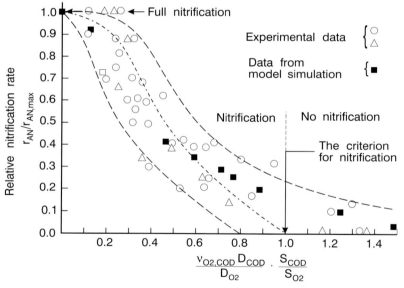

Figure 6.7. Experimental demonstration of the negative influence of organic matter on the nitrification rate in a biofilter /1/,/2/.

6.2. Types of plants for nitrification

Nearly all wastewater containing ammonium also contains organic compounds, which means that nitrification normally takes place concurrently with oxidation of organic matter. Hence the sludge carries out both processes (combined sludge) even though the conversions are produced by very different micro-organisms in

TREATMENT PLANTS FOR NITRIFICATION

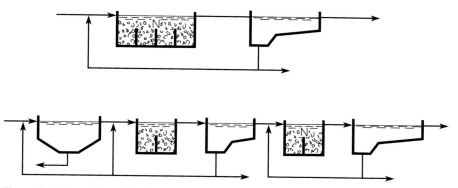

Figure 6.8. Examples of the layout of single and two sludge activated sludge treatment plants for nitrification.

the sludge. (Few industrial plants are supplied with wastewater in which the only oxygen-consuming substance is ammonium in which case the sludge will consist of nitrifying bacteria only (separate sludge)).

Apart from this distinction between plants with combined sludge and separate sludge, the plants can basically be classified as single sludge treatment plants, in which nitrification occurs in the entire biological reactor, and two sludge treatment plants, in which the overall removal of organic matter takes place in the first plant stage without nitrification, whereas nitrification is specially optimized in the second stage.

Figure 6.8 shows examples of the design of single and two sludge activated sludge treatment plants. Figure 6.9 shows some principal types for nitrification in biofilters. These plant types should be regarded as two sludge treatment plants, even though their functional design may suggest that they were single sludge treatment plants. This is due to the fact that the biological tanks shown in figure 6.9 may be considered to have plug flow, whereas the necessary conditions for the presence of nitrifying bacteria are only found in the lower part of the filters.

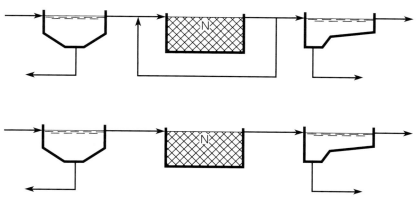

Figure 6.9. Examples of biofilters for nitrification.

221

6.2.1. Nitrification plants with separate sludge

The operation of activated sludge treatment plants for the nitrification of wastewater of which the only oxygen-consuming substance is ammonium is very difficult in practice. Normally the settling behaviour of the sludge is bad, and due to the very small sludge production during the process it is difficult to maintain a reasonable sludge mass in the plant. In practice, plants for this type of wastewater are nearly always designed as biofilters.

6.2.2. Single sludge nitrification plants

This is the most common plant type for nitrification. The entire sludge mass has the same process behaviour, but nevertheless there are various, modified plant types.

The most simple plant design is shown in the photo in figure 6.10. Notice the large aeration tank (compared with the settling tank - the square tank to the left in the photo) which is necessary to maintain a long sludge age. Various modifications

Figure 6.10. Typical Danish design of traditional single sludge system for nitrification of municipal wastewater. The Mårslet wastewater treatment plant (Denmark).

of this plant design are found of which a few types are used in pratice, whereas others are only of theoretical interest.

The contact stabilization plant shown in figure 6.11 is based on the fact that the solids content can be kept high in the stabilization tank, whereas the sludge content in the contact tank corresponds to the content in a plant of an ordinary design. The necessary sludge age (and sludge mass) can therefore be achieved in a plant volume somewhat smaller than that of a traditional plant. The nitrification occurs in both tanks, but only in the stabilization tank there is a sufficiently long hydraulic retention time for the nitrification as a whole to be completed.

The major part of the organic matter is removed in the contact tank, and nitrification will occur to the extent possible within the allowable time. It is characteristic of this plant type that the nitrification can be stable without being complete. The stabilization tank ensures the presence of nitrifying bacteria, but the extent of nitrification depends on the time of contact between wastewater and sludge in the contact tank.

In the plant type with alternating operation as shown in figure 6.12 the tanks are used alternately as aeration tank and settling tank as appears from the applied operational cycle shown in figure 6.13. This plant type is widely used in Germany, the Netherlands and Denmark due to the low construction costs. When considering the possibilities of nitrification in such a plant it should be realized that the aerobic sludge age is much lower than the total sludge age as a substantial part of the sludge is settled on the bottom of the tank that functions as a settling tank. Based on the shown operational cycle, the aerobic sludge age will only be 3/8 of the total sludge age.

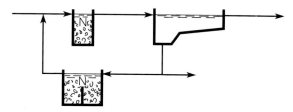

Figure 6.11. Contact stabilization plant for (partial) nitrification.

Example 6.4

Find the necessary tank volume for a plant as shown in figure 6.12 having an operational cycle as shown in figure 6.13.

The wastewater volume is 1,700 m³/d with 0.25 kg BOD/m³. The yield constant is 0.6 kg SS/kg BOD. The plant is to observe an effluent criteria of 0.02 kg BOD/m³ (20 mg BOD/l). The lowest plant temperature is estimated to 8°C.

The sludge production is calculated from Expression (4.13)

$F_{SP} = Y_{obs} \cdot (C_1 - C_3) \cdot Q_1 = 0.6 \cdot (0.25 - 0.02) \cdot 1{,}700 = 235$ kg SS/d

Types of plants for nitrification

Figure 6.12. Typical design of single sludge system with alternating operation for nitrification of municipal wastewater. The Bording wastewater treatment plant (Denmark).

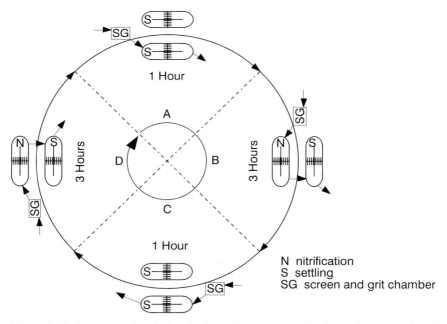

Figure 6.13. Operational cycle for single sludge system with alternating operation for nitrification. In phases A and C the former nitrification tank is prepared to become a settling tank. While this is carried out, raw wastewater is directed into one end of the functioning settling tank (the wastewater will not have time to run out, before the switch to phase B or D occurs).

On the basis of Table 4.5, the sludge concentration in the aeration tanks is estimated at 5 kg SS/m³. From figure 6.2 the necessary aerobic sludge age can be estimated at 14 days. With an operational cycle as shown in figure 6.13, where the aerobic sludge age is 3/8 of the total sludge age (only aeration and nitrification in 3 out of each 8 hours), the necessary total sludge age will be:

$$\theta_X = 14 \cdot \frac{8}{3} = 37 \text{ d}$$

The total volume of the two aeration tanks can be calculated from Expression (4.14)

$$\theta_X = \frac{V_2 \cdot X_2}{F_{SP}}$$

$$V_2 = \frac{\theta_X \cdot F_{SP}}{X_2} = \frac{37 \cdot 235}{5} = 1,740 \text{ m}^3$$

Hence the total volume of the two tanks shall be 1,740 m³ which gives a hydraulic retention time of $\theta = V/Q = 1,740/1,700 \sim 24$ hours.

The example might also have been calculated by another unit for the sludge concentration. That would require that the yield constant and the sludge concentration, X_2, were known in the other unit (for example COD, VSS).

The plant type with step-feed, shown in figure 6.14, is used for tanks with plug flow, where the step-feed of the raw wastewater ensures an even oxygen consumption in the longitudinal direction and a reasonable dilution of the raw wastewater so that load impacts, for example by ammonium, will not lead to inhibition due to oxygen limitations.

6.2.3. Nitrification in two sludge treatment systems

A characteristic of nitrification in a two sludge treatment system, shown in figure 6.15, is that the principal removal of organic matter takes place during the first stage of the treatment plant where the process conditions do not permit nitrification.

The second stage of the plant is specially adapted to nitrification. The necessary sludge age for the achievement of nitrification can be established in a relatively small volume, as the overall part of the sludge production, which is primarily caused by the removal of organic matter, occurs in stage 1. When calculating the sludge age of the sludge in stage 2 (see Expression (4.14)), the denominator (the sludge production) will therefore be strongly reduced. Figure 6.16 shows the reactor volume of the two stages as a function of the COD/TN-ratio of the influent into the nitrification stage. It will be seen that the minimum total volume will be

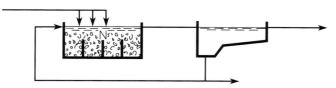

Figure 6.14. Plant for nitrification with "step-feed" inlet of raw wastewater.

TYPES OF PLANTS FOR NITRIFICATION

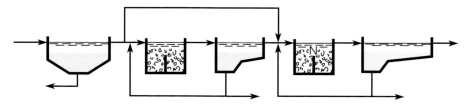

Figure 6.15. Two sludge system for nitrification.

achieved when the overall part of the organic matter is removed in stage 1. This plant type is often used when extending existing plants in which nitrification is required. The oldest parts of the plant, including the sludge treatment parts, often with anaerobic digestion, can continue their usual operation, while the new part of the plant will be based on nitrification and aerobic sludge stabilization. In practice, the sludge properties in the second stage of the plant often turn out to be poor due to the low sludge production by nitrification. However, it will often be possible to solve the problem by applying a certain (small) amount of raw wastewater into the nitrification stage.

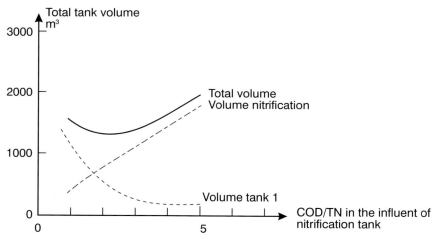

Figure 6.16. The total tank volume and its distribution on the two tanks of a two sludge system, as a function of the composition of the wastewater. Raw wastewater to tank 1 is assumed to contain 1,000 kg COD and 100 kg TN/d.

6.2.4. Nitrification plants with separate sludge in filters

Biological wastewater treatment plants solely intended for oxidation of ammonium are always designed as filters, as the adhesion of the bacteria is so good that a sufficient biomass can be maintained in such a plant in spite of the small sludge production. Normally no settling is needed after the filter due to the small sludge production.

6.2.5. Two sludge nitrification plants in filters

Biofilters in which both organic material and ammonium are to be oxidised should be regarded as two sludge treatment systems. The tanks resemble tanks with plug flow, and as suggested in connection with the criterion for nitrification, Expression (6.8), there will be no nitrification at the inlet of the plant, as the S_{BOD}/S_{O2} ratio is high, but the decreasing concentration of organic matter through the tank means that nitrification will be possible after a certain point in the tank.

At small load, the traditional rock trickling filter as shown in figure 6.17 can be brought to nitrify. The degree of nitrification is often limited and is clearly affected by the temperature dependency of the biological processes as shown in figure 6.18, where annual variations in the nitrification at the Lundtofte treatment plant is shown. German literature /6/ recommends the following to achieve nitrification when treating municipal wastewater (definitions, Chapter 5, Section 5.7):

Rock trickling filters:

B_V = 200 g BOD/(m³·d)
$B_{A,V}$ = 0.4 – 0.8 m/h
R ≤ 1

Figure 6.17. Traditional rock trickling plant where nitrification can occur in periods of low load and high temperature. The Hammel wastewater treatment plant, 1977 (Denmark).

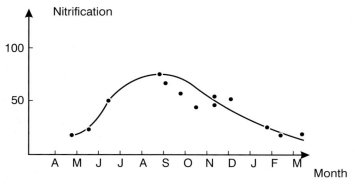

Figure 6.18. Typical variation pattern for nitrification in traditional, low-rate biofilters, the Lundtofte wastewater treatment plant (Denmark).

Trickling filters with plastics media:

B_A = 0.2 - 0.4 kg BOD/(m³ · d)
$B_{A,V}$ = 0.6 - 1.5 m/h   (100 < ω < 200 m² / m³)
R ≤ 1

Rotating disc filters:

B_A = 4 - 5 g BOD / (m² · d)

The German recommendations are normally based on wastewater temperatures above 12°C. At the loads mentioned, effluent concentrations of C_{BOD} = 5 ± 40% and C_{NH4} = 1.5 ± 50% can be achieved based on statistical distributions that are logarithmic.

Through the development of new types of filter media, full nitrification can be achieved in traditional trickling filters, but is often not economically competitive.

Biofilters for nitrification are sometimes designed as rotating disc plants, as shown in figure 6.19. The plants often consist of 4 ideally mix tanks in series, which is a good approximation to a plant with plug flow. The organic matter is removed at the inlet of the plant while ammonium is oxidized near the outlet.

Biofilters for nitrification can also be designed as submerged filters with aeration or pure oxygen supply.

Figure 6.19. Rotating disc filter for nitrification of municipal wastewater.

6.2.6. Combined biofilters and activated sludge treatment plants for nitrification

The design of nitrification plants in two sludge treatment plants can of course be realized with one stage as a biofilter and another as an activated sludge plant. Such a plant type, shown in figure 6.20, was popular in the 60s, but it was ousted by plants with activated slugde only because the initial costs of these are lower. However, this plant type (filter plus activated sludge) has gained renewed interest in recent years, especially for the treatment of concentrated industrial wastewater. This trend is especially due to the development of new plastics media in replacement of the traditional stones as carrier for the bacterial growth.

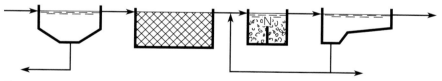

Figure 6.20. Two sludge treatment plant where the first stage is a biofilter, and the second stage, where the nitrification takes place, is an activated sludge treatment plant.

6.3. Design of nitrifying plants

Literature offers countless proposals for the design of nitrifying plants, of which very few - and the most simple at that - have been used in practice. The reason is especially that nitrification is seldom the only plant process on which the design should be based. The design must be determined taking a large number of different aspects into consideration. Plant designs will therefore vary greatly, for which reason it will be difficult to lay down uniform criteria for the design of the plants.

6.3.1. Design of activated sludge treatment plants for nitrification

The design of activated sludge treatment plants for nitrification varies somewhat from the design of plants for the removal of organic matter. Usually it is a question of full or no nitrification as the environment will either permit the nitrifying bacteria to be present, which results in full nitrification, or it will prevent their presence. The design therefore aims at ensuring the presence of the bacteria and not, as is the case by the removal of organic matter, at achieving a required degree of treatment efficiency or effluent quality. Consequently the design criteria will be very simple.

In the following, three levels for the design of activated sludge treatment plants for nitrification will be discussed. Furthermore some principles for optimizing the operation of nitrification for plants with periodic problems regarding the achievement of stable nitrification conditions are discussed.

Design based on sludge load

Design based on sludge load is the design most frequently used. It is characteristic of this design that it does not in any way involve the ammonium content of the wastewater or the content of nitrifying bacteria in the sludge.

When designing based on the sludge load:

$$B_{X,BOD} = Q_1 \cdot C_{BOD,1}/(V_2 \cdot X_2) \qquad (4.11)$$

the volume of the aeration tank, V_2, can be found from the expression:

$$V_2 = Q_1 \cdot C_{BOD,1}/(X_2 \cdot B_{X,BOD}) \qquad (4.21)$$

If a sufficiently small sludge load for other reasons is required (0.15 kg BOD/(kg SS · d)), it can be seen from Table 6.2 that nitrification is expected to be achieved. Many plants have been designed based on sludge stabilization, that is, with a considerably lower load than required with regard to nitrification. In such plants, nitrification is normally achieved without problems.

Sludge load, kg BOD(S)/(kg SS(B) · d)	0.05	0.15	0.3	0.6
Effluent concentration of BOD, g/m^3	5-10	10-20	15-25	20-40
Nitrification	yes	yes	no	no
Aerobic sludge stabilization	yes	no	no	no

Table 6.2. Design of activated sludge treatment plants based on sludge load at 10°C.

Design based on aerobic sludge age

A somewhat more advanced design can be made based on the aerobic sludge age. In this case attention is focused on the living conditions of the nitrifying bacteria in

Treatment Plants for Nitrification

the plant; but it is still the content of organic matter in the wastewater and the total mass of sludge which are used for the design.

Design based on the aerobic sludge age is calculated from Expression (4.14):

$$\theta_{X,\text{aerobic}} = M_X/F_{SP} = V_2 \cdot X_2/F_{SP}$$

$$V_2 = \theta_{X,\text{aerobic}} \cdot F_{SP}/X_2 \tag{6.9}$$

When the sludge content and the form of operation of the aeration tank have been established, the necessary tank volume can be found based on the requirements for the aerobic sludge age. Figure 6.2 shows the curve normally used for the determination of $\theta_{X,\text{aerobic}}$.

Example 6.4

Design of nitrification plants. Figure 6.21 shows the layout and operational cycle of two alternative designs of nitrification plants. An estimate of the necessary plant volume for the two plants is required, based on a criterion of full nitrification down to a temperature of 10°C, but there is no criterion of sludge stabilization.

The BOD load of the plant corresponds to 20,000 PE and arises out of a normal, urban community.

According to figure 6.2, the necessary aerobic sludge age is found to be 11 days for both solutions. Since it is a matter of ordinary, municipal wastewater, no extra safety factor is applied as load variations and limiting substances are not expected to present special problems. In the case of a plant design with separate, secondary settling tank (to the right in figure 6.21) all the sludge in the aeration tanks remains aerobic. With a sludge content of 3 kg

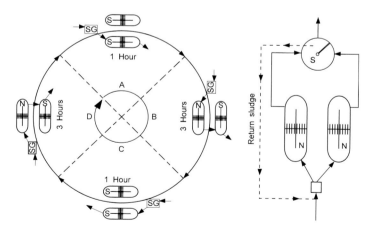

SG screen and grit chamber
N   nitrification
S   settling

Figure 6.21. Layout and operational cycles of two plant types for nitrification of municipal wastewater.

VSS/m³ and a yield constant of 0.6 kg VSS/kg BOD, which is typical for such a plant, the necessary volume is found from Expression (6.9):

$$V_2 = \theta_{X,\text{aerobic}} \cdot F_{SP}/X_2 = \theta_{X,\text{aerobic}} \cdot Y_{obs} \cdot (Q_1 \cdot C_1 - Q_1 \cdot C_3)/X_2$$

$$V_2 = 11 \cdot 0.6 \cdot (1{,}200 - 60)/3 = 2{,}510 \text{ m}^3$$

as $Q_1 \cdot C_1 = 1{,}200$ kg BOD/d is the BOD load (20,000 PE of 60 g/PE · d), and $Q_1 \cdot C_3$ is the effluent quantity of BOD. This is estimated to result in a 95% BOD treatment, that is, $Q_1 \cdot C_3 = 0.05 \cdot 1{,}200 = 60$ kg BOD/d.

In the case of a plant design without a secondary settling tank, only an average of 3/8 of the aeration tank is aerobic. With an average sludge content such as above, the necessary volume which is to be aerobic will again be 2,510 m³, and the total volume of the plant will be $(8/3) \cdot 2{,}510 = 6{,}690$ m³.

Computer-designed plants

There are computer programs of varying complexity which can be used for the design of nitrifying plants. This allows for daily variations to be included in the design. Strong variations in influent nitrogen may cause breakthrough of ammonium in the effluent. An example of the importance of including load variations in a design can be seen in figure 6.22.

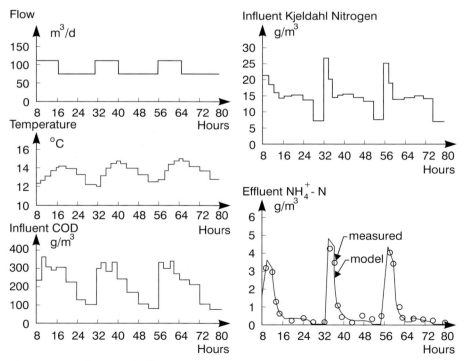

Figure 6.22. Influent and effluent variations, nitrifying plant /7/.

The results from the shown test and modelling show that in this case the plant is overloaded with nitrogen for which reason there will be a breakthrough of ammonium every morning. This may be caused by a combination of high nitrogen concentrations in the influent as well as by pulsating internal recycles with high ammonium contents. This is typical for supernatant from digester or supernatant from sludge concentration or sludge dewatering. These recycles are often let in as pulsating influents in the morning or at noon, that is, at hours when the plant already works at high rate. Through computations, the optimal operation regarding the handling of these internal recycles can be found. More on this topic in chapter 11.

6.3.2. Optimizing operation of nitrifying plants

Many treatment plants have difficulty in maintaining the nitrification, usually in the cold season. Figure 6.23 shows typical results from such a plant.

The aerobic sludge age is too short. In practice the possibilities of increasing the nitrification rate are many. This may be done in several ways. The reaction rate constant can be increased by increasing the temperature, for example by covering

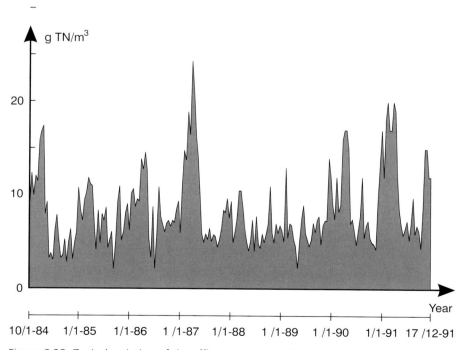

Figure 6.23. Typical variation of the effluent concentration of total nitrogen in a treatment plant with nitrification under stress during the cold season. The peaks during spring are caused by high ammonia concentrations in the effluent. The Søholt wastewater treatment plant, Silkeborg (Denmark).

the tanks. Possible limiting substances can be removed from the wastewater, and finally, in regions with soft water, where nitrification may result in a reduction of the pH, optimal pH can be ensured by lime addition. The substrate dependency of nitrification (see Table 6.1) means that increased contents of ammonium and oxygen will increase the removal rate. But, increased ammonium concentration has little influence as the level will soon reach a height where the removal will be independent of ammonium. However, by increasing the oxygen content there are good possibilities of increasing the nitrification rate. Figure 6.24 shows an example of the oxygen dependency of the nitrification rate. It will be seen that an increase of the oxygen content can cause a significant increase of the rate.

The quantity of nitrifying bacteria cannot be increased more than the wastewater allows for, but in the case of poor nitrification an increase of the sludge content of the plant, or an operational reorganization which increases the aerobic sludge age, will make it easier to maintain the maximum quantity.

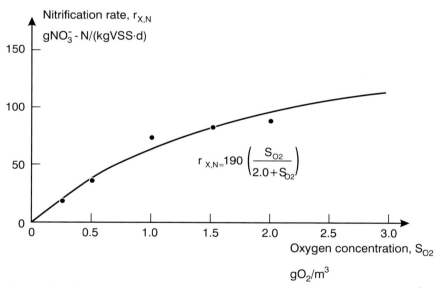

Figure 6.24. The oxygen influence on the nitrification rate. Here $K_{S,O2} = 2$ g O_2/m^3. From /4/.

Design of biofilters for nitrification

There are no well-established rules for the design of biofilters for nitrification. The development making nitrification in biofilters attractive is of a recent date following the development of different stone and plastics media for the filters. Furthermore submerged filters have been developed where air is injected into the filter media on which the biofilm is placed.

In the following, two levels of design are discussed. The first corresponds to the principle most frequently used in practice, while the other which especially used for rotating disc filters solely designed for the nitrification part of the plant, is given as an example of the use of the biofilm kinetic concepts described in Chapter 5.

Design at level I
This design method is based on the removal of ammonium per m² filter per day on an average all over the filter. Hence from Expression (6.5):

$$Q_1 (C_{N,1} - C_{N,3}) = r_{A,NH4} \cdot A_2.$$

The zero order removal rates, $r_{A,NH4}$, used must be found by tests or be procured from plants of a similar design and with the same type of wastewater.

Table 6.3 shows typical removal rates for rotating disc plants for the treatment of different types of industrial wastewater. The magnitude of the ammonium removal found from the table can also be used for the design of plants for domestic wastewater - for the sections of the plant where the main part of the organic matter has been removed. See the criterion for nitrification in biofilters in Expression (6.8).

In Germany, a load of 4-5 BOD/(m² · d) is recommended for the treatment of municipal wastewater with full nitrification.

Type of wastewater	Nitrification rate $r_{A,NH4}$ (g N/(m² · d))	
	min.	max.
Tannery	2.35	2.61
Fertilizer industry	2.36	2.67
Leachate	2.42	2.66
Municipal wastewater, 1	2.03	2.56
Municipal wastewater, 2	1.69	1.82
Municipal wastewater, 3	2.20	2.56
Municipal wastewater, Büsnau	4.61	4.84
Combined sewer overflow	1.53	1.97
Gas works	2.25	2.36
Tar production	0.04	0.12

Table 6.3. Mean removal rates for nitrifying biofilm plants at 20°C /9/.

It will often be necessary to correct the removal rate for temperature variations.

Design at level II
Below a method for the design of rotating disc filters with nitrification is described. The method is based on the main principles of the removal of soluble materials as described in Chapter 5. Figure 6.25 illustrates the main principles of the method and the simplifications made to achieve sufficiently clear criteria.

DESIGN OF NITRIFYING PLANTS

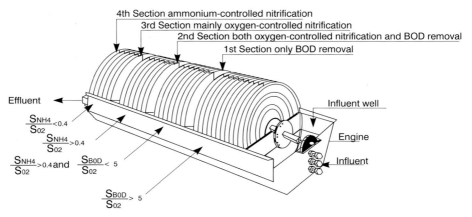

Figure 6.25. Illustration of the main principles for the design of rotating disc plants for nitrification.

There is no nitrification in the first stage of the plant. The content of organic matter is so high that the nitrifying bacteria will not have sufficient time to grow, see the criterion for the achievement of nitrification.

In the second stage of the plant the removal of organic matter as well as nitrification will take place. The ammonium content is high, and the nitrification is limited by the oxygen content and because the proportion of nitrifying bacteria has been "diluted" due to the heterotrophic bacteria which are also present.

In the third stage of the plant the organic matter has by and large disappeared so that this stage functions as a purely nitrifying stage where the removal is limited by oxygen.

In the fourth stage of the plant the ammonium content is so low that it is no longer the oxygen that limits the removal, but the ammonium content.

Hence the design of rotating disc plants is complicated as it is necessary to control not only the removal of ammonium, but also the removal of organic matter and the supply and removal of oxygen. The above division of the plant corresponds to any possible situation in normal plants. In practice, the individual stages of such a plant will be regulated by one of the three limiting phenomena, but all phenomena need not necessarily occur in the individual plant. A practical design computation may start from the influent raw wastewater, compute the plant step by step, or start from the effluent criteria and calculate the treatment step by step upstream the plant. In both cases a number of computations will be necessary to ensure an optimal design.

After finishing the design based on the simplified biofilm kinetics, the plant can be recalculated with more detailed models. This can only be done by computer. Table 11.3 shows the total process matrix used including heterotrophic removal as well as denitrification (see Chapter 7) and autotrophic removal. Tables 11.4 and

11.5 show the used parameters. Table 6.4 shows the computed results for a rotating disc plant in four sections with a total load of 5 g BOD $m^{-2}d^{-1}$ from primary settled wastewater. Ammonium is reduced from 25.0 to 0.6 $gNH_4^+ - N/m^3$; while nitrate is increased from 1.0 to 19.8 $NO_3^- - N/m^3$.

Influent	m^3/d :	4000				
Recycle	m^3/d :	0				
Tank no.		Influent	1	2	3	4
Volume	m^3		100	100	100	100
Surface area	m^2		30000	30000	30000	30000
$K_L a$ value	d^{-1}		600	600	600	600
Oxygen		2.0	6.3	7.4	7.7	9.3
COD degrad.	g/m^3	150.0	34.7	7.0	2.1	1.5
$NH_4^+ - N$	g/m^3	25.0	20.1	11.8	2.7	0.6
$NO_2^- - N$	g/m^3	0.5	0.5	0.5	0.3	0.2
$NO_3^- - N$	g/m^3	1.0	0.6	7.9	17.2	19.8
HCO_3^-	mol/m^3	6.0	5.7	4.6	3.3	2.9
COD inert	g/m^3	30.0	32.9	34.3	35.3	36.1
Heterotrophic biomass	g/m^3	120.0	167.6	175.1	170.6	163.0
Nitrosomonas	g/m^3	0.0	0.1	0.1	0.1	0.1
Nitrobacter	g/m^3	0.0	0.0	0.0	0.0	0.0

Table 6.4. Result of computation of a rotating disc plant in four sections, which have a total load of 5 g BOD/($m^2 \cdot$ d) /4/.

References

/1/ Gönenç, E. and Harremoës, P. (1990): Nitrification in Rotating Disc Systems. II. Criteria for simultaneous mineralization and nitrification. *Water Res.*, **24**, 499-505.

/2/ Gönenç, E. (1982): Nitrification on Rotating Biological Contactors. Department of Environmental Engineering, Technical University of Denmark, Lyngby, Denmark.

/3/ Szwerinski, H., Arvin, E. and Harremoës, P. (1986): pH-Decrease in Nitrifying Biofilms. *Water Res.*, **20**, 971-976.

/4/ Gujer, W. and Boller, M. (1990): A Mathematical Model for Rotating Contactors. *Water Sci. Technol.*, **22**, (1/2), 53-73.

/5/ Boller, M. and Gujer, W. (1986): Nitrification of Tertiary Trickling Filters followed by Deep-bed Filters. *Water Res.*, **20**, 1363-1373.

/6/ Abwassertechnische Vereinigung (1997): Biologische und weitergehende Abwasserreinigung (Biological and advanced wastewater treatment). 4th ed., Ernst & Sohn, Berlin.

/7/ Gujer, W. (1985): Ein dynamisches Modell für die Simulation von komplexen Belebtschlammverfahren (A dynamic model for the simulation of complex processes of activated sludge). EAWAG, Dübendorf, Switzerland.

/8/ EPA (1975): Process Design Manual for Nitrogen Control. U.S. Environmental Protection Agency, Washington, D.C.

/9/ Harremoës, P., Sekoulov, I. and Bonomo, L. (1981): Design of Fixed Film Nitrification and Denitrification Units based on Laboratory and Pilot Scale Results. In: EAS-ISWA '81. 5th European Sewage and Refuse Symposium, Munich 22-26 June 1981, pp. 423-451. GFA e.V., St. Augustin.

Nitrification tank with activated sludge. Excessive aeration produces waste of energy as well as aerosols. Northern Treament Works, Johannesburg (South Africa).

7 Treatment Plants for Denitrification

By Jes la Cour Jansen, Poul Harremoës and Mogens Henze

In many countries it is today recognized that it is necessary to remove nitrogen from wastewater. The discharge of nitrogen and phosphorus results in eutrophication of the recipients and makes them unsuitable for fishing and bathing. At the same time the discharge of nitrogen into fresh recipients means that these become unsuitable as raw water supply for drinking water. In countries where the main part of the drinking water comes from the groundwater, the latter problem is not yet urgent. However, it is a crucial problem in countries where surface water is used as the dominant resource of drinking water, such as in Central Europe, North America and South Africa.

Nitrogen can be removed by physical, chemical and biological treatment methods such as

- ion exchange (drinking water)
- reverse osmosis (drinking water)
- ammonia stripping (wastewater)
- chemical denitrification
- chemical precipitation, $MgNH_4PO_4$
- biological denitrification (drinking water and wastewater)
- assimilation

For wastewater, biological denitrification is the dominant process type.

Denitrification is the process by which bacteria convert nitrate into free nitrogen. As the nitrogen components in ordinary municipal wastewater is present in the form of reduced nitrogen, these components must be nitrified (see Chapter 6) before the denitrification can be established. Therefore, the denitrification process can seldom be looked at separately, but must be seen in relation to the preceding nitrification. By denitrification, nitrate oxidizes an organic compound, that is, nitrate practically replaces oxygen in the degradation of organic matter. This means that the denitrification process should also be viewed as a part of the removal of organic matter. This relationship is stressed by the fact that many bacteria are able to immediately replace oxygen by nitrate as an oxidant in the absence of oxygen.

This relationship between denitrification, nitrification and the removal of organic matter has had an important impact on the development of processes for the removal of nitrogen. The plant layouts which are typically used in full-scale plants have gained their success by appropriately combining the processes.

Denitrification in connection with the treatment of wastewater is a rather new process. Only 30 years ago, the first process designs for efficient nitrogen removal were described in the literature /1/. Later the development has accelerated, but even today few have been in operation for more than 10 years. Consequently process know-how is still to some extent based on experiments in laboratories or on a semi-technical scale. Hence the available know-how about denitrification in biofilters is mainly based on such pilot plants.

7.1. Mass balances, denitrifying treatment plants

As discussed in the introduction, the denitrification process is always related to the oxidation of organic matter (or hydrogen or sulphur). The process can be carried out in a separate process exclusively developed for denitrification, or in a combined process with nitrification and denitrification. The microbiological denitrification process is basically the same, irrespective of the design of the treatment plant. The differences in plant designs may be the consumption of organic matter (the required C/N-ratio) and the size of the fraction of denitrifying organisms in the biomass.

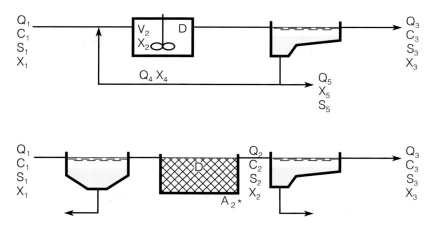

Figure 7.1. Schematic representations for denitrifying plants, separate denitrification.
Top: Denitrification in activated sludge
Bottom: Denitrification in biofilters

Treatment Plants for Denitrification

Component → Process ↓	S_S	X_S	$X_{B,H}$	X_I	S_{NO3}	S_{NH4}	$S_{S,N}$	$X_{S,N}$	S_{ALK}	Reaction rate r_v,...
1. Anoxic growth of heterotrophs	$-\dfrac{1}{Y_{max,H}}$		1		$-\dfrac{1-Y_{max,H}}{2{,}86 \cdot Y_{max,H}}$	$-f_{XB,N}$			$\dfrac{1-Y_{max,H}}{14 \cdot 2{,}86 \cdot Y_{max,H}}$ $-f\, XB,N/14$	$\mu_{max,H}\left(\dfrac{S_S}{K_S+S_S}\right)\left(\dfrac{K_{S,O2,H}}{K_{S,O2,H}+S_{O2}}\right)$ $\left(\dfrac{S_{NO3}}{K_{S,NO3}+S_{NO3}}\right)\cdot \eta_g \cdot X_{B,H}$
2. Decay of heterotrophs		$1-f_{XB,XI}$	-1	$f_{XB,XI}$				$f_{XB,N}$		$b_H \cdot X_{B,H}$
3. Amonification of dissolved org. nitrogen						1	-1		$\dfrac{1}{14}$	$k_a \cdot S_{S,N} \cdot X_{B,H}$
4. Hydrolysis of organic matter	1	-1								$k_h \cdot X_S$
5. Hydrolysis of particulate nitrogen							1	-1		$k_h \cdot X_{S,N}$
Unit	Kg COD/m³				Kg N/m³				ekv./m³	
	Easily degradable org. matter	Slowly degr. org. matter	Heterotrophic biomass	Inert suspended org. matter	Nitrate	Ammonium	Dissolved org. nitrogen	Suspended org. nitrogen	Alkalinity	

Table 7.1. Process matrix for separate denitrification.

7.1.1. Separate denitrifying plant

The process can be carried out in activated sludge plants or in biofilters - see the schematic representation in figure 7.1.

A mass balance for a denitrifying plant can be written when the processes and their kinetics are known. In Table 7.1, a process matrix is shown for a separate denitrifying plant. η_g is the fraction of heterotrophic bacteria which can denitrify.

Activated sludge plant
A nitrate balance for the plant in figure 7.1a looks as follows:

$$Q_1 \cdot S_{NO3,1} - r_{V,S} \cdot V_2 = Q_3 \cdot S_{NO3,3} + Q_5 \cdot S_{NO3,5} \tag{7.1}$$

from which the nitrate concentration in the effluent, $S_{NO3,3}$, is found:

$$S_{NO3,3} = \frac{Q_1 \cdot S_{NO3,1} - r_{V,S} \cdot V_2 - Q_5 \cdot S_{NO3,5}}{Q_3}$$

Assuming that

$Q_3 = Q_1$

$Q_5 \cdot S_{NO3,5} = 0$

the following simplified expression is found:

$$S_{NO3,3} = S_{NO3,1} - r_{V,S} \cdot \theta_2 \tag{7.2}$$

where the hydraulic retention time $\theta_2 = \dfrac{V_2}{Q_1}$

The reaction rate expression, $r_{V,S}$, can be found from:

$r_{V,S} = \nu_{X,S} \cdot r_{V,\ldots}$

where $r_{V,\ldots}$ is found in the last column at the right hand side of Table 7.1, and the stoichiometric coefficient for the biological process, $\nu_{XB,NO3}$, is found in the S_{NO3} column:

$$r_{V,S} = \frac{(1-Y_{max,H}) \mu_{max,H}}{2.86 \cdot Y_{max,H}} \cdot \frac{S_S}{K_S + S_S} \cdot \frac{K_{S,O2,H}}{K_{S,O2,H} + S_{O2}} \cdot \frac{S_{NO3}}{K_{S,NO3} + S_{NO3}} \cdot \eta_g \cdot X_{B,H} \tag{7.3}$$

By combining Expressions (7.2) and (7.3), the nitrate concentration in the effluent, $S_{NO3,3}$, can be found for an ideally mixed tank.

Often the denitrification process can be approximated by a zero order process with respect to organic matter, oxygen and nitrate, that is,

$$r_{V,S} = \frac{\mu_{max,H}}{Y_{max,H}} \cdot \eta_g \cdot X_{B,H} \tag{7.4}$$

Example 7.1

The wastewater from a German chemical company has a high content of nitrate and a high content of organic matter which can be utilized for a direct (separate) denitrification.

The following is known about influent and effluent:

Element	In	Out	Unit
COD	600	140	g COD/m³
$NO_3^- - N$	50	2	g NO_3^- – N/m³

The wastewater flow is

$Q_1 = 2,700$ m³/d

The tank volume for denitrification is

$V_2 = 400$ m³

Find the nitrate removal rate, $r_{V,S}$, when it is assumed to be constant throughout the plant.

Expression (7.2) is used:

$$r_{V,S} = \frac{S_{NO3,1} - S_{NO3,3}}{\theta_2}$$

The hydraulic retention time, θ_2, is

$\theta_2 = V_2 / Q_1 = (400 \text{ m}^3)/(2,700 \text{ m}^3/\text{d}) = 0.15$ d

substituted together with $S_{NO3,1}$ and $S_{NO3,3}$ we have

$$r_{V,S} = \frac{(50 - 2) \text{ g } NO_3^- - N / m^3}{0.15 \text{ d}} = 320 \text{ g } NO_3^- - N / (m^3 \cdot d)$$

Biofilters

A nitrate balance for the plant in figure 7.1b looks as follows:

$$Q_1 \cdot S_{NO3,1} - r_{A,NO3} \cdot A_{2*} = Q_3 \cdot S_{NO3,3} \tag{7.5}$$

where $r_{A,NO3}$ is the nitrate removal rate of the biofilm per unit area as carrier material.

For a denitrifying biofilter, either the organic matter or the nitrate is potentially limiting for the removal. What is limiting is determined by the ratio between the concentrations and the diffusion coefficient for the materials in the biofilm, see Expression (5.32). If the organic matter is easily degradable (that is, if it can diffuse into the biofilm) and the process in the biofilm is a zero order, the following applies:

The organic matter is limiting for the reaction rate when

$$S_{COD,2} < S_{NO3,2} \cdot D_{NO3,2} \cdot \nu_{NO3,COD}/D_{COD,2} \tag{7.6a}$$

where $\nu_{NO3,COD}$ is the stoichiometric coefficient for the biological process between the organic matter (here measured as COD) and nitrate (typically 5.5-6 kg COD/kg $NO_3^- - N$),

$D_{COD,2}$ is the diffusion coefficient for organic matter in the biofilm,

$D_{NO3,2}$ is the diffusion coefficient for nitrate in the biofilm.

Nitrate is limiting when

$$S_{NO3,2} < S_{COD,2} \cdot D_{COD,2} \cdot \nu_{COD,NO3}/D_{NO3,2} \tag{7.6b}$$

If the organic matter is limiting, and if the biofilm is fully penetrated ($\beta > 1$), see Expression (5.13), the reaction rate is:

$$r_{A,COD} = k_{0Vf,COD} \cdot L \tag{7.7a}$$

The reaction rate for a partially penetrated biofilm ($\beta < 1$) is:

$$r_{A,COD} = (2 \cdot D_{COD} \cdot k_{0Vf,COD})^{1/2} \cdot S_{COD,2}^{1/2} \tag{7.7b}$$

If nitrate is limiting and the process is a zero order reaction in the biofilm, we have the following expression for the reaction rate for a fully penetrated biofilm ($\beta > 1$):

$$r_{A,NO3} = k_{0Vf,NO3} \cdot L \tag{7.7c}$$

and for partial penetration ($\beta < 1$) applies:

$$r_{A,NO3} = (2 \cdot D_{NO3} \cdot k_{0Vf,NO3})^{1/2} \cdot S_{NO3,2}^{1/2} \tag{7.7d}$$

Notice that k_{0Vf} has a unit corresponding to the limiting substance; that is, in Expression (7.7c) the unit is for example kg $NO_3^- - N/$ (m^3 **biofilm** $\cdot$ d).

An estimate of the quantity $D_{NO3} \cdot \nu_{NO3,COD}/D_{COD}$ for different organic materials and for $k_{½A,NO3}$, D, $\nu_{NO3,COD}$ and $k_{0Vf,NO3}$ is found in Table 7.2.

	$k_{1/2A,NO3}$	D	$v_{NO3,COD}$	$D_{NO3} v_{NO3,COD}/D_{COD}$	$k_{0f,NO3}$
Acetic acid	1-3	0.3-0.7	5-6	5-20	10-150
Methanol	1.4-3.7	0.8-4	3.5-4.1	1.8-2.3	20-200
Glucose	1-3	0.1-0.7	5-6	5-20	10-150
Unspec. COD	1-3	0.3-0.6	4-5	5-20	10-150
	$g N^{1/2} \cdot m^{-1/2} d^{-1}$	$10^{-4} m^2/d$	$kg COD/kg NO_3^- - N$	$kg COD/kg NO_3^- - N$	$kg NO_3^- - N/(m^3 \cdot d)$

Table 7.2. Kinetic constants in a denitrifying biofilm reactor (20°C). The diffusion coefficient for nitrate is $(0.5-1.0) \cdot 10^{-4} m^2/d$ /2/,/3/.

Figure 7.2 shows simultaneous measurements of $NO_3^- - N$ and CH_3OH in a plug flow filter, measured in the effluent from the filter. The connected measuring points show the distribution down through the filter for the different values of S_M/S_N. The slope of the curves in mean indicates the stoichiometric coefficient: 3.1 kg CH_3OH/kg $NO_3^- - N$.

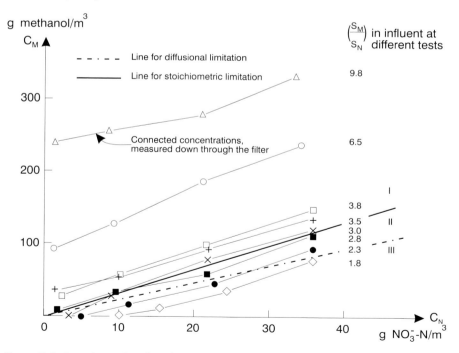

Figure 7.2. Data for a plug flow filter, measured in the effluent from the filter /4/.

An S_M/S_N ratio in the influent greater than 3.1 kg $CH_3OH/NO_3^- -N$ kg (shown as zone I) gives a nitrate limited reaction rate throughout the filter. For S_M/S_N in the influent less than 2.0, methanol will be diffusionally limiting throughout the filter (zone III). Between these cases (zone II) there will be a change in the diffusional limitation from nitrate to methanol.

To achieve a full treatment for nitrate down to 0 g $NO_3^- -N/m^3$ in the effluent, the stoichiometric ratio (3.1) must be complied with at any point of the filter. Therefore nitrate will in such a case be limiting for the reaction rate as discussed above.

Figure 7.3 shows results from a filter test with the removal of nitrate from wastewater. The filter is a downflow filter with a layer thickness of 3.0 m consisting of 3-5 mm gravel. A varying volume of water enters the filter with an influent concentration of $S_{NO3,1} = 23$ g $NO_3^- -N/m^3$. Sampling is carried out for different positions down through the filter. The figure to the right shows the nitrate removal per unit volume and time, $r_{V,NO3}$, as a function of the concentration in the observed sections:

$$r_{V,NO3} = k_{\frac{1}{2}V,NO3} \cdot S_{NO3}^{\frac{1}{2}}$$

$$k_{\frac{1}{2}V,NO3} = 51 \text{ g m}^{-3} \text{h}^{-1}$$

hence a parabola with its top point at the zero point.

The figure to the left shows the measured profiles down through the filter which gives a straight line for the square root of the concentration plotted as a function of the retention time derived from Expressions (5.37) and (5.42):

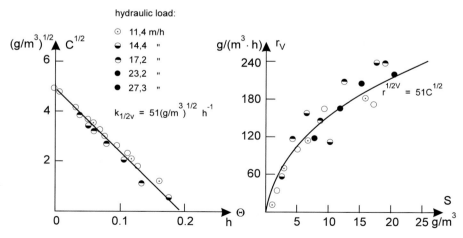

Figure 7.3. Example of a half order reaction. The filter consists of 3-5 mm gravel covered by biomass which reduces nitrate to atmospheric nitrogen. Temperature 17°C. /5/.

$$S_{NO3,1}^{\frac{1}{2}} - S_{NO3}^{\frac{1}{2}} = \frac{1}{2} k_{\frac{1}{2}V,NO3}\, \theta$$

$$k_{\frac{1}{2}V,NO3} = 2\, \frac{S_{NO3,1}^{\frac{1}{2}} - S_{NO3}^{\frac{1}{2}}}{\theta}$$

It clearly appears from the figure that it is a half order reaction.

The points fall on the same line for different hydraulic loads. This proves that it is not hydraulic film diffusion; that is, the diffusion to the biofilm is not limiting for the process in comparison with the diffusion in the biofilm. More detailed information about the experiment is found in /4/ and /5/.

Example 7.2

In the biofilter, from which the data shown in figure 7.3 are derived, the thickness of the biofilm is estimated to L = 200 μm. The specific surface has been found to be ω = 1000 m^2/m^3. The diffusion coefficient for nitrate is set at 0.7 · 10^{-4} m^2/d. How big should the nitrate concentration be to obtain full penetration?

$$k_{\frac{1}{2}V,NO3} = 51\, (g/m^3)^{\frac{1}{2}}/h$$
$$= 1.2 \cdot 10^3\, g\, NO_3^- - N^{\frac{1}{2}} m^{-\frac{3}{2}} d^{-1}$$

$$k_{\frac{1}{2}A,NO3} = k_{\frac{1}{2}V,NO3} / \omega$$
$$= 1.2\, g\, NO_3^- - N^{\frac{1}{2}} m^{-\frac{1}{2}} d^{-1}$$

This is less than stated in Table 7.2 because only a certain part of the biofilm is exposed to free-flowing water.

$$k_{\frac{1}{2}A,NO3} = (2 D_{NO3} \cdot k_{0Vf,NO3})^{\frac{1}{2}}$$

$$k_{0Vf,NO3} = \frac{(k_{\frac{1}{2}A,NO3})^2}{2 D_{NO3}}$$

$$= \frac{1.2^2}{2 \cdot 0.7 \cdot 10^{-4}} = 10\, kg\, NO_3^- - N / (m^3 \cdot d)$$

$$\beta^2 = 1 = \frac{2 D_{NO3} \cdot S_{NO3}}{k_{0Vf,NO3} \cdot L^2}$$

$$S_{NO3} = \frac{k_{0Vf,NO3} \cdot L^2}{2 D_{NO3}} = \frac{10000 \cdot (2 \cdot 10^{-4})^2}{2 \cdot 0.7 \cdot 10^{-4}}\, \frac{gNm^{-3}d^{-1}m^2}{m^2 d^{-1}}$$

$$S_{NO3} = 2.9\, g\, NO_3^- - N / m^3$$

7.1.2. Combined nitrification and denitrification

This type of plant is normally found in activated sludge plants, see the schematic representation in figure 7.4. The mass balances for the plant can be written when a process matrix is known. An example is shown in Table 11.1.

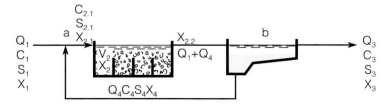

Figure 7.4. Schematic representation of a combined nitrification and denitrification, activated sludge plant.

Among the biomass balances, especially the balance for nitrifying bacteria is interesting. This balance can be used to examine whether a washout of the nitrifying bacteria takes place. The mass balance for the nitrifying biomass fully corresponds to the balance set out in Expression (6.1). By using the symbols, defined in figure 7.4, the balance is:

$$Q_1 \cdot X_{B,A,1} + r_{V,XB} \cdot V_3 - b_A \cdot X_{B,A,3} \cdot V_3 = Q_4 \cdot X_{B,A,4} + Q_6 \cdot X_{B,A,6} \quad (7.8)$$

Similar to Expression (6.1), Expression (7.8) can be converted into:

$$\theta_{X,A} = 1/\mu_{obs,A,net} \quad (7.9)$$

Remember that only tank V_3 is included in the calculation of the aerobic sludge age.

Denitrification may occur in biofilm even if there are aerobic conditions in the water flowing through the filter. This is the case where nitrate enters deeper into the biofilm than oxygen, for example due to a high concentration of nitrate and a low concentration of oxygen. Denitrification may occur in the innermost part of the biofilm if organic matter is present. The denitrification rate is, however, reduced due to the longer diffusion path.

Figure 7.5 shows concentration profiles in the biofilm. For the three zones, it applies:

aerobic zone
$0 < x < x_0$
$$\frac{\partial^2 S_{NO3}}{\partial x^2} = 0 \quad , \quad \frac{\partial^2 S_{O2}}{\partial x^2} = \frac{k_{0Vf,O2}}{D_{O2}} \quad , \quad \frac{\partial^2 S_{Me}}{\partial x^2} = \frac{k_{0Vf,Me}}{D_{Me}} \quad (7.10)$$

anoxic zone
$x_0 < x < x_N$
$$\frac{\partial^2 S_{NO3}}{\partial x^2} = \frac{k_{0Vf,NO3}}{D_{NO3}} \quad , \quad \frac{\partial^2 S_{O2}}{\partial x^2} = 0 \quad , \quad \frac{\partial^2 S_{Me}}{\partial x^2} = \frac{k_{0Vf,Me}}{D_{Me}} \quad (7.11)$$

anaerobic zone
$x_N < x < L$
$$S_{NO3} = 0 \quad , \quad \frac{\partial^2 S_{O2}}{\partial x^2} = 0 \quad , \quad \frac{\partial^2 S_{Me}}{\partial x^2} = \frac{k_{0Vf,Me}}{D_{Me}} = \text{constant} \quad (7.12)$$

In the aerobic zone the concentration profile for nitrate is straight. In the anoxic zone, the concentration profile is a parabola with top point in x_N.

Treatment Plants for Denitrification

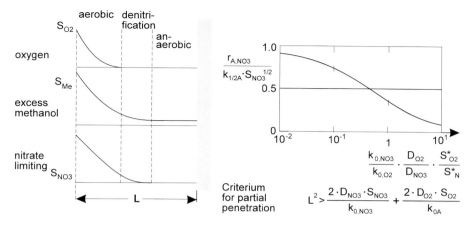

Figure 7.5. The figure shows how denitrification may occur in spite of aerobic conditions in the free water. Nitrate diffuses through the aerobic part of the biofilm and is denitrified in the anoxic zone. To the right the reduced denitrification rate is shown compared with denitrification without oxygen.

The figure shows the reduced denitrification rate, $r_{A,NO3}$, in relation to the denitrification rate if no oxygen was present: $k_{1/2A,NO3} \cdot S_{NO3}^{1/2}$, as a function of the ratio between the concentration of oxygen and nitrate in the wastewater.

The penetration of nitrate can be expressed by the penetration of oxygen and nitrate:

$$\beta_{NO3,red}^2 = \beta_{O2}^2 + \beta_{NO3}^2$$

where β_{O2} is the degree of penetration of oxygen

β_{NO3} is the degree of penetration of nitrate

$\beta_{NO3,red}$ is the degree of penetration of nitrate when oxygen is present

$$\beta_{NO3,red}^2 = \frac{2 D_{O2} S_{O2}}{k_{0Vf,O2} \cdot L^2} = \frac{2 D_{NO3} \cdot S_{NO3}}{k_{0Vf,NO3} \cdot L^2}$$

Example 7.3

In the effluent from an ideally mixed filter, the following concentrations are measured:

Oxygen S_{O2} =   2 g O_2 / m³
Methanol S_{Me} =   20 g CH_3OH / m³
Nitrate S_{NO3} =   5 g $NO_3^- - N$ / m³

What is the removal of methanol and nitrate per m² of filter medium?

For the processes, the following constants are chosen (Tables 5.1, 5.2 and 7.2 which are calculated in g CH_3OH, not as g COD).

249

Methanol and oxygen:

$k_{0Vf,Me} = 90 \text{ kg } CH_3OH \text{ m}^3/d$
$D_{Me} = 1.0 \cdot 10^{-4} \text{ m}^2/d$
$D_{O2} = 1.7 \cdot 10^{-4} \text{ m}^2/d$
$v_{Me,O2} = 1.1 \text{ g } O_2/\text{g } CH_3OH$

Methanol and nitrate:

$k_{0f,NO3} = 30 \text{ kg } NO_3^- - N \text{ m}^3/d$
$k_{\frac{1}{2}A,NO3} = 3 \text{ g } NO_3^- - N^{\frac{1}{2}} \text{ m}^{-\frac{1}{2}} d^{-1}$
$D_{NO3} = 0.7 \cdot 10^{-4} \text{ m}^2/d$
$v_{NO3,Me} = 3 \text{ g } CH_3OH/\text{g } NO_3 - N$

$$\frac{S_{O2}}{S_{Me}} = \frac{2}{20} < \frac{D_{Me}}{D_{O2}} \cdot v_{Me,O2} = \frac{1.0}{1.7} \cdot 1.1 = 0.65$$

The oxygen is limiting for the removal.

$$\begin{aligned} r_{A,Me} &= (2D_{O2}k_{0Vf,Me}/v_{Me,O2})^{\frac{1}{2}} \cdot S_{O2}^{\frac{1}{2}} \\ &= (2 \cdot 1.7 \cdot 10^{-4} \cdot 90 \cdot 10^3/1.1)^{\frac{1}{2}} \cdot 2^{\frac{1}{2}} \\ &= 7.5 \text{g } CH_3OH \text{ m}^2/d \end{aligned}$$

The oxygen penetration of the biofilm is:

$$\begin{aligned} \beta L &= \left(\frac{2D_{O2} \cdot S_{O2}}{k_{0Vf,O2}}\right)^{\frac{1}{2}} \\ &= \left(\frac{2 \cdot 1.7 \cdot 10^{-4} \cdot 2}{90 \cdot 10^3 \cdot 1.1}\right)^{\frac{1}{2}} \\ &= 83 \mu m \end{aligned}$$

For nitrate removal applies:

$$\begin{aligned} r_{A,NO3,max} &= k_{\frac{1}{2}A,NO3} \cdot S_{NO3}^{\frac{1}{2}} \\ &= 3 \cdot 5^{\frac{1}{2}} = 7 \text{ g } NO_3^- - N/(m^2 \cdot d) \end{aligned}$$

$$\frac{k_{0Vf,NO3} \cdot D_{O2} \cdot S_{O2}}{k_{0Vf,O2} \cdot D_{NO3} \cdot S_{NO3}} = \frac{30 \cdot 1.7 \cdot 10^{-4} \cdot 2}{90 \cdot 1.1 \cdot 0.7 \cdot 10^{-4} \cdot 5} = 0.29$$

It appears from figure 7.5 that the denitrification rate is approx. 60 per cent of maximum.

$r_{A,NO3} = 0.6 \cdot 7 = 4.2 \text{ g } NO_3^- - N \text{ m}^{-2} d^{-1}$

In spite of the aerobic conditions in the water, the denitrification rate is considerable.

The full penetration of oxygen and nitrate is

$$\begin{aligned} \beta L &= \left(\frac{2D_{O2} \cdot S_{O2}}{k_{0Vf,O2}} + \frac{2D_{NO3} \cdot S_{NO3}}{k_{0Vf,NO3}}\right)^{\frac{1}{2}} \\ &= \left(\frac{2 \cdot 1.7 \cdot 10^{-4} \cdot 2}{90 \cdot 10^3 \cdot 1.1} + \frac{2 \cdot 0.7 \cdot 10^{-4} \cdot 5}{30 \cdot 10^3}\right)^{\frac{1}{2}} \\ &= 174 \mu m \end{aligned}$$

The enthusiastic reader is welcome to solve the equations as regards the diffusion of methanol to find out whether the concentration of methanol is big enough to penetrate so deeply.

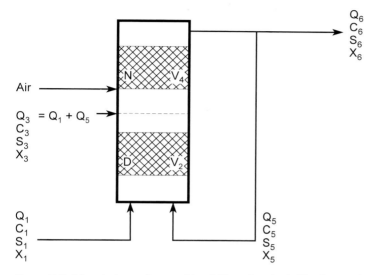

Figure 7.6. Mass balance for combined filters for denitrification and nitrification.

Figure 7.6 shows a type of plant based on the combination of two biofilters (BAF) of which the upper one is arerated. This type is found in full scale /6/. The denitrification tank has plug flow with longitudinal diffusion, Peclet number, Pe_L, approx. 50. Due to the aeration, the nitrification tank has a higher longitudinal diffusion.

7.2. Types of plants for denitrification

A discussion of the various plant alternatives for denitrification as suggested in the literature will be very long. In a systematic survey /7/, three main types are described, of which there are countless variants. But since then, the process development within the field of biofilters has added a number of new major types as well as a multitude of new variants.

For activated sludge plants, the decisive distinction is whether the process is only set up for denitrification (separate sludge), or whether the denitrification process is integrated in the overall process of the plant, in which case the sludge, apart from the denitrification, also participates in other processes (combined sludge).

Figure 7.7 shows typical plant designs based on the two principles.

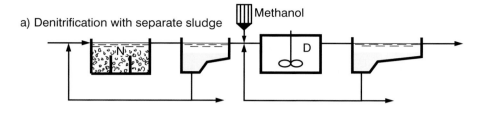

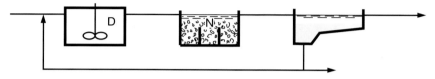

Figure 7.7. Design of a denitrification plant with separate and combined activated sludge, respectively.

The other distinction between plant types is the use of organic matter from wastewater or from external carbon sources (for example methanol, starch or acetic acid).

For biofilters, the development of plant types for denitrification has taken place in connection with a radical renewal of the whole filter technology by the application of submerged filters. The reason is that the special process conditions, which must be made for the denitrification to occur, may be established in the new filter types whereas the conventional trickling filters are difficult to adapt to the process.

Figure 7.8 shows the principle for the different filter technologies. It should be noted that denitrification in biofilters normally takes place with a separate sludge system.

Figure 7.8 only shows only the design of the reactors for denitrification. The separation of the treated water and surplus sludge from the process takes place in conventional secondary settling tanks for the rotating discs whereas for the submerged filters and for the fluidized filters the top can be designed to function as a separation unit. Surplus sludge can be withdrawn directly from the filter. As yet, experiences with the operation of denitrifying filters are few.

In the following, a number of plant types for denitrification are described. For activated sludge plants, the types are described which are typically used in full-scale plants supplemented by a few which are suitable to illustrate important characteristics of the process. For biofilters, of which very few types are used commercially, examples are given of the designs which are considered to be the most attractive.

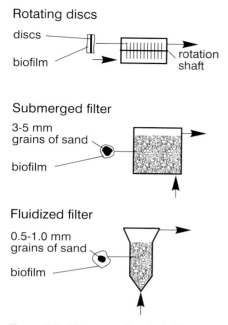

Figure 7.8. Filter types for denitrification.

7.2.1. Denitrification plants with separate sludge

Plants with separate sludge have typically been used as full-scale plants, especially in the USA. The process design, where the plant for denitrification forms an independent treatment stage, was used as an extension of plants with nitrification and removal of organic matter where further requirements are made for the removal of nitrogen. Figure 7.7a shows the design of such a post denitrification plant.

The denitrification plant is coupled to the effluent of the existing plant, but as organic matter is needed for denitrification, it must be added, such as in the form of methanol or other organic material.

Plants with separate sludge for denitrification are also known from industrial plants where nitrate is present in the raw wastewater. In such cases the plant may be designed in a single treatment stage as shown in figure 7.9. The organic matter for the denitrification can be added (for example in the form of methanol), or it may already be present in the wastewater.

TYPES OF PLANTS FOR DENITRIFICATION

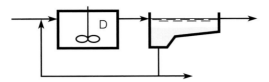

Figure 7.9. Industrial plant for the treatment of wastewater containing nitrate. Separate denitrification.

7.2.2. Denitrification plants with combined sludge

Plants with combined sludge are the dominant full-scale plant type for denitrification (figure 7.7b). The main reason is that the process makes it possible to use organic matter in the raw wastewater for the denitrification. This eliminates the addition of organic matter (for example methanol or acetic acid) and at the same time saving on oxygen is obtained, as part of the organic matter of the wastewater is oxidized by nitrate instead of by oxygen. Figure 7.10 shows the flow scheme of a post denitrification plant with the content of the organic matter in the sludge itself as an energy source for the denitrification process.

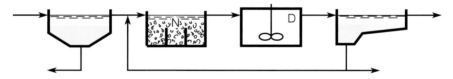

Figure 7.10. Conventional post-denitrification plant with sludge as organic matter.

The plant type is merely of an historical interest as it was used for the first experiments with denitrification of wastewater /1/. The sludge of the plant carries out removal of organic matter and nitrification in the first stage whereas in the second stage it carries out denitrification. The organic matter for denitrification comes from the degradation processes in the sludge which means that the process is slow.

There is a possibility of obtaining a faster process by adding a carbon source directly to the denitrification process as shown in figure 7.11. The carbon source may be external (methanol, acetic acid, etc.) or internal (wastewater, supernatant, hydrolyzate, etc.). Carbon sources containing greater amounts of reduced nitrogen (especially ammonium), will require a recycle of water back to the nitrification process to have the ammonium oxidized into nitrate. If the extent of the necessary recycle increases considerably, a pre-denitrification process will be used instead as shown in figure 7.12. By this process, a better, direct utilization of the carbon sources of the raw wastewater for the denitrification is obtained. The recycle ratio depends on the required treatment efficiency, but will normally be 2-5 (including the recycle which is achieved through the return sludge pumping).

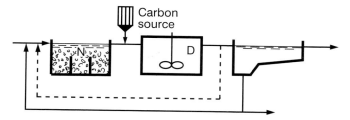

Figure 7.11. Post-denitrification with addition of carbon source to the denitrification. If the carbon source contains ammonium, it is necessary to recycle (as shown with the dotted line).

The recycle plant according to the principle shown in figure 7.12 can be designed in many ways. The recycle may be an external recycle or it may take place internally in the combined nitrification-denitrification tank. The latter technology has especially been developed in South Africa where it is used in many plants.

In many plant designs there is not one special tank for nitrification and another one for denitrification. In certain periods, a tank, or part of a tank, is used for nitrification and in other periods for denitrification. In terms of the process, it may be pre-denitrification (as shown in figure 7.12), or a mix of pre-denitrification and post denitrification.

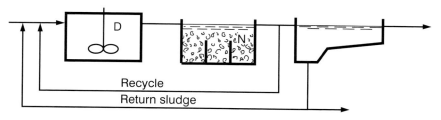

Figure 7.12. Pre-denitrification with recycle.

The alternating process (figure 7.13) is a pre-denitrification where the recycle has been replaced by a switch between nitrification and denitrification in the tank(s). The plant facilitates control and process optimization. The system was developed in Denmark (Biodenitro), and it is dominant among Danish full-scale plants for the removal of nitrogen.

Figure 7.14 is an aerial view of the Frederikssund central treatment plant (Denmark) which has applied alternating operation since 1978. Notice that on the plant there are a total of 4 tanks which function in pairs, 2 by 2. In this plant operated as shown in figure 7.13, one cycle lasts for 4 hours.

Figure 7.15 shows an alternating type of plant with nitrification and denitrification in one tank. In terms of time, the individual processes occur separately in a certain cycle. This type of plant (an alternating plant) can be designed in

Types of plants for denitrification

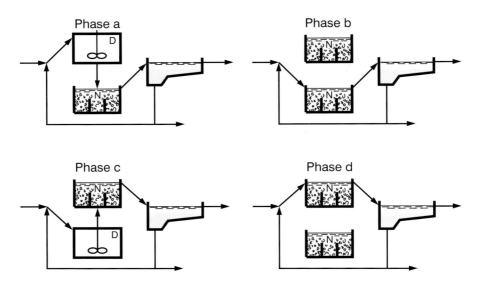

Figure 7.13. Denitrification with alternating operation.

many ways. If it is provided with a separate settling tank, the wastewater can be discharged continuously. If there is a possibility of collecting the wastewater ahead of the process tank, it may conveniently take place in phases C and D so that wastewater only enters in phases A and B. Maximum use of the organic matter of

Figure 7.14. The Frederikssund central treatment plant (Denmark).

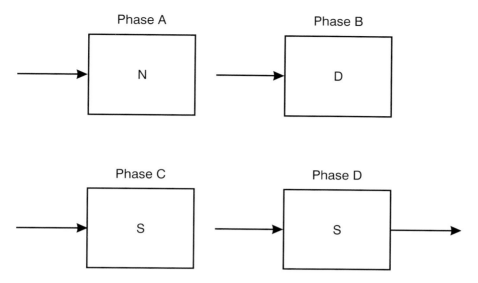

Figure 7.15. Alternating nitrification and denitrification in one tank (SBR-plant or oxidation ditch plant). If there is a separate settling tank, phases C and D can be left out, and effluent from the tank is then also possible in phases A and B.
D: Denitrification
N: Nitrification
S: Settling

the raw wastewater for denitrification is attained by only letting in wastewater in phase B.

Another form of an alternating process is shown in figure 7.16. The supply of oxygen is designed in such a way that zones with oxygen and zones without oxygen occur throughout the tank. The processes occur alternately when the sludge with the water is carried from zone to zone through the plant. The system is for example used on the large Vienna-Blumental treatment plant in Vienna, Austria. The organic matter for the denitrification comes from the wastewater as well as from the degradation of the sludge.

A treatment plant, as shown in figure 7.16, requires a very precise control to function. During changes in the loading or changes in the process rates, it may be

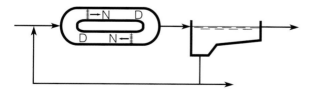

Figure 7.16. Simultaneous denitrification plant with varying aerobic and anoxic conditions.

difficult to control the supply of oxygen so as to achieve an optimum distribution between zones with nitrification and zones with denitrification.

The switching between nitrification and denitrification are so frequent in this type of plant (1-2 minutes) that it more resembles a simultaneous nitrification-denitrification. In this type of plant as well as in other plants with combined sludge, a denitrification will take place simultaneously with the nitrification. The reason is that in the innermost part of the flocs or the biofilm there will be oxygen-free zones where nitrate is present which will result in denitrification. There may also be zones in the nitrification tank where the aeration is not efficient. This will also facilitate "simultaneous" denitrification.

7.2.3. Biofilters for denitrification

All biofilters for denitrification are submerged, which means they are water-filled. Where rotating discs or packed filter media (plastics discs or loose carriers) are used, the sludge production is removed from the treated water by settling as shown in figure 7.17. For filters with backwash (gravel, Leca and the like) the plant design is shown in figure 7.18, whereas for fluidized filters the design is shown in figure 7.19. Fluidized filters can either use sand or plastics as carrier or function with spherical bacterial flocs without carrier (granules). Full-scale experiences with denitrifying filters are few. Often an external carbon source is added in the form of methanol, acetic acid or industrial wastewater.

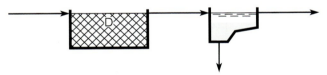

Figure 7.17. Denitrifying filter (packed plastics filters or rotating discs) with continuous sludge separation.

The combination with nitrification will normally be performed in a pre-denitrification process, that is, with a nitrifying filter after the denitrifying filter and recycle of the nitrified water. As the nitrifying filter should not be loaded with too large amounts of organic matter, it is necessary that the effluent from the denitrifying filter only has a minor content of suspended solids. The requirement of recycle means that denitrifying filters without settling of the wastewater flow, that is, filters with backwash (figure 7.18 top) or fluidized filters (figure 7.18 bottom) are superior. For the other types, the recycle will mean that the settling tanks will be very large (and hence very expensive).

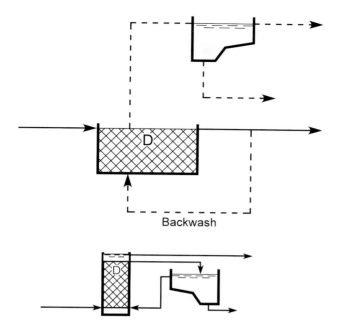

Figure 7.18. Denitrifying filters. Top with sand or porous materials resembling Leca as filter medium. Bottom fluidized with a carrier of sand

7.3. Design of denitrifying plants

For the design of denitrifying plants, a number of special process conditions should be taken into consideration. In this section the following is discussed:

- C/N ratio
- Oxygen/stirring
- Simultaneous nitrification-denitrification
- Nitrogen gas in settling tanks
- Oxygen recovery by denitrification with raw wastewater carbon
- Alkalinity
- Removal rate

7.3.1. C/N ratio
By denitrification, organic matter and nitrogen are removed. As a certain amount of organic matter should be used for the denitrification process, the C/N ratio is important for the success of the process.

In connection with an evaluation of a denitrification plant, the COD/N - or the BOD/N - ratio in the **influent** is often calculated for the part of the nitrogen to be removed which is unambiguous for separate denitrification plants. For plants with combined sludge, the C/N ratio should be calculated for the influent into the **denitrification tank** in the case of pre-denitrification - see figure 7.12. In the case of post denitrification with combined sludge, it is the influent into the **nitrification tank** which is used for the C/N calculation.

The C/N ratio in the influent can be used for a rough evaluation of process possibilities and to find out if it is necessary to add an external carbon source.

The C/N ratio in the influent, $(C/N)_1$, is calculated using symbols from figure 7.4, and hence

$$(COD/TN)_1 = C_{COD,1}/C_{TN,1} \tag{7.13}$$

or as

$$(BOD/TN)_1 = C_{BOD,1}/C_{TN,1} \tag{7.14}$$

A more detailed analysis of the consumption of organic matter and the removal of nitrogen will give a better understanding in cases where the composition of the wastewater deviates from the ordinary one, or where a new type of plant is being considered.

In a denitrification plant, organic matter will be used for three purposes:

1. Conversion of nitrate (or nitrite) into nitrogen gas.
2. Sludge production.
3. Respiration with oxygen.

If the quantity of these three consumptions is known, the necessary C/N ratio for the plant concerned can be calculated. A surplus of organic matter in relation to nitrogen can only be used for respiration with oxygen or for sludge production, and it may leave the treatment plant with the treated water and hence, in respect of organic matter, reduce the quality of the effluent.

Based on the equations of reaction, the realized C/N ratio can be calculated for a current process, that is, the amount of organic matter which has been removed and the amount of nitrogen which has been removed. The three 3 elements mentioned above contribute to the C/N ratio in the following way as organic matter is here calculated in COD which gives an exact calculation of the C/N ratio in the form of the COD/N ratio.

1. $NO_3^- + 1.25\ "COD" \rightarrow 1/2\ N_2 + CO_2$
   COD-consumption = 1.25 moles/mole $NO_3^- - N$
   = 2.86 kg COD/kg $NO_3^- - N$
   The 1.25 "COD" represents an amount of organic matter which would consume 1.25 moles of oxygen = 40 g oxygen by a COD analysis, whereas NO_3^- represents 1 mole of nitrogen = 14 g nitrogen.
   If the denitrified amount of nitrate is $\Delta NO_3^- - N$, the COD-consumption is $2.86 \cdot \Delta NO_3^- - N$

2. Sludge production = F_{SP}. If it is measured in COD-units we have the COD-consumption directly.
COD-consumption = $F_{SP,COD}$
3. Oxygen respiration. If the oxygen consumption is F_{O2}, the COD-consumption = F_{O2}.

The total COD-consumption per volume of wastewater is then:

$2.86 \cdot \Delta NO_3^- - N + F_{SP,COD} + F_{O2}$

The total nitrogen removal is denitrified nitrate ($\Delta NO_3^- - N$) plus assimilated nitrogen (in the sludge production):

$\Delta NO_3^- - N + F_{SP,COD} \cdot f_{B,N}$

Hence the C/N ratio with the unit kg COD/kg N, realized by the current process, is

$C/N = (2.86 \cdot \Delta NO_3^- - N + F_{SP,COD} + F_{O2})/(\Delta NO_3^- - N + F_{SP,COD} \cdot f_{B,N})$ (7.15)

The C/N ratio can also be determined on the basis of an equation of reaction.
Note that what is included in the C/N ratio, as defined in Expression (7.15), is not the total COD and total N in the influent, hence it is not $(COD/N)_1$; for both COD and nitrogen it applies that the amounts in the effluent should be deducted.
For a low-loaded nitrification-denitrification plant, the observed yield constant will for example be = 0.4 kg COD(B)/kg COD(S). A corresponding equation of reaction with organic matter, $C_{18}H_{19}O_9N$, and biomass, $C_5H_7NO_2$, will look as follows:

$C_{18}H_{19}O_9N + 8.4\ NO_3^- + 8\ H^+ + 0.4\ NH_4^+ \rightarrow$
$1.4\ C_5H_7NO_2 + 11\ CO_2 + 4.2\ N_2 + 9.4\ H_2O$ (7.16)

Here the COD-consumption ($C_{18}H_{19}O_9N$ has a COD-content of 560 g O_2/mol):

$(560/(8.4 \cdot 14)) = 4.8$ kg COD/kg $NO_3^- - N$ or

$560/(8.4 \cdot 14 + 0.4 \cdot 14) = 4.5$ kg COD/kg TN

The realized C/N-ratio may in this case also be expressed as:

$C/N = (4.5\ \text{kg COD} \cdot \Delta N + F_{O2})/(\Delta N)$,

where ΔN is denitrified + assimilated nitrogen.
The part of the C/N ratio which is due to the oxidation of organic matter with oxygen, F_{O2}, is a "waste" of organic matter. Besides, this is the consumption which depends on the design and operation of the plant whereas the other two contributions to the COD-consumption are given through the stoichiometry in the denitrification process and the effluent requirement for nitrogen.

An efficiency factor, $f_{C/N}$, can be defined as:

$$f_{C/N} = (2.86 \cdot \Delta NO_3^- - N + F_{SP})/(2.86 \cdot \Delta NO_3^- - N + F_{SP} + F_{O2}) \tag{7.17}$$

If no organic matter is "lost" for oxygen respiration, that is, $F_{O2} = 0$, then $f_{C/N} = 1$.

For $f_{C/N} = 1$, an optimum C/N ratio is obtained which is the lowest possible which can be obtained. In practice, it will therefore always be necessary with a higher C/N ratio.

The C/N ratio is in practice found from the following expression:

$$(C/N)_{practice} = (C/N)_{optimum} \cdot \frac{1}{f_{C/N}} \tag{7.18}$$

If the (C/N) ratio in raw wastewater corresponding to the $(C/N)_{practice}$ is calculated, we have the following expression:

$$(COD/TN)_{practice,1} =$$

$$(COD/TN)_{optimum} \cdot \frac{1}{f_{C/N}} \cdot \frac{C_{COD,1}}{C_{COD,1} - C_{COD,3}} \cdot \frac{C_{TN,1} - C_{TN,3}}{C_{TN,1}} \tag{7.19}$$

or

$$(COD/TN)_{practice,1} = (COD/TN)_{optimum} \cdot \frac{1}{f_{C/N}} \cdot \frac{E_{TN}}{E_{COD}} \tag{7.20}$$

A corresponding expression for BOD/N is:

$$(BOD/TN)_{practice,1} = (BOD/TN)_{optimum} \cdot \frac{1}{f_{C/N}} \cdot \frac{E_{TN}}{E_{BOD}} \tag{7.21}$$

where E is the treatment efficiency.

The efficiency factor for organic matter, $f_{C/N}$, depends on the design of the plant and the process control. For separate denitrification plants, $f_{C/N}$ comes close to 1 whereas it is considerably lower for combined nitrification-denitrification plants.

For domestic wastewater and municipal wastewater, the C/N ratio in raw wastewater will normally be sufficiently high to ensure a reasonably high denitrification rate, provided that the utilization factor is kept sufficiently high. In Table 7.3, the optimum C/N ratio is stated for different types of organic matter, and in Table 7.4, the efficiency factor for a number of plant designs is stated. If the design and the operation of the plant are inadequate, the efficiency factor may be lower than stated in the table.

Organic matter	(C/N)$_{optimum}$	Unit
Organic matter in wastewater	3-3.5	kg BOD/kg N
	4-5	kg COD/kg N
Organic matter in sludge	1.5-2.5	kg BOD/kg N
	2.9-3.2	kg COD/kg N
Methanol	2.3-2.7	kg MeOH/kg N
	3.5-4.1	kg COD/kg N
	1.0-1.2	mole MeOH/mole N
Acetic acid	2.9-3.5	kg HAc/kg N
	3.1-3.7	kg COD/kg N
	0.9-1.1	mole HAc/mole N

Table 7.3. Optimum (C/N) ratio for different types of organic matter to be used for denitrification.

Plant type	Fig	$f_{C/N}$
Activated sludge:		
Separate	7.9	0.9-1.0
Post denitrification	7.10	0.2-0.5
Post denitrification (with extra carbon source)	7.11	0.8-0.9
Recycle	7.12	0.4-0.6
Alternating	7.13	0.4-0.6
Alternating, one-tank	7.15	0.3-0.6
Simultaneous	7.16	0.3-0.5
Filters:		
Submerged filters/rotating discs with recycle	–	0.4-0.7
Submerged filters/rotating discs without recycle	7.17	0.9-1.0
Filter with backwash without recycle	7.18	0.5-0.8
Fluidized filter	7.18	0.4-0.7

Table 7.4. The efficiency factor, $f_{C/N}$, for organic matter for different plant designs for denitrification. Data from /15/.

If the C/N ratio in the raw wastewater is too low, it means that the denitrification process will be either only partial, or it will occur at a very reduced rate which also, in most cases, means a reduced treatment efficiency for nitrate. A low C/N ratio will increase the effluent of intermediate products from the denitrification, for example of dinitrogen oxide, N_2O /14/.

Example 7.4

What is the necessary C/N ratio, $(BOD/N)_{practice,1}$, for denitrification in a recycle plant where organic matter from wastewater is the carbon source?

The treatment efficiency in the plant is 80 per cent for total nitrogen and 95 per cent for BOD.

The C/N ratio in practice can be found by using Expression (7.21)

$$(BOD/TN)_{practice,1} = (BOD/TN)_{optimum} \cdot \frac{1}{f_{C/N}} \cdot \frac{E_{TN}}{E_{BOD}} \qquad (7.21)$$

From Table 7.3 it is estimated that $(C/N)_{optimum}$ = 3.2 kg BOD/kg TN, and from Table 7.4 we have $f_{C/N}$ = 0.4-0.6.

This means that if we "substitute" the range in the $f_{C/N}$ interval, we find

$$(BOD/TN) = 3.2 \cdot \frac{1}{0.6} \cdot \frac{0.8}{0.95} - 3.2 \cdot \frac{1}{0.4} \cdot \frac{0.8}{0.95} = 4.5 - 6.7 \text{ kg BOD/kg TN}.$$

The higher end of the interval is high for municipal wastewater, see Table 1.12. Therefore we must be careful not to squander the organic matter in the treatment process.

7.3.2. Oxygen/stirring

For processes with activated sludge, stirring is required to ensure the suspension of the sludge (prevent settling) and to ensure contact between the wastewater and the sludge. Stirring must take place without unnecessary air/oxygen supply as this would inhibit the denitrification process by reducing the denitrification rate and by reducing the efficiency factor, $f_{C/N}$, for the organic matter. The result of the inhibition of the denitrification process is that some of the denitrifying bacteria respire with oxygen instead of nitrate. Respiration of 32 g oxygen (= 1 mole of O_2) corresponds to a removal of 4 electron equivalents (as oxygen is reduced from oxidation step 0 to –2). That is, 8 g of O_2 ~ 1 electron equivalent. Correspondingly, 14 g of nitrate nitrogen (= 1 mole of N) represent 5 electron equivalents (reduction of nitrogen from oxidation step +5 to 0), that is, 2.8 g $NO_3^- - N$ ~ 1 electron equivalent.

That means that the nitrate respiration is reduced by

$$\frac{2.8 \text{ g } NO_3^- - N / 1 \text{ eqv}}{8 \text{ g } O_2 / 1 \text{ eqv}} = 0.35 \text{ g } NO_3^- - N / \text{g } O_2 \qquad (7.22)$$

Inhibition with oxygen may also occur by the supply of oxidized wastewater to the denitrification tank, for example in a recycle plant.

Example 7.5

In a denitrification tank with paddle stirring, a K_La-value of 3 d^{-1} at 8°C is measured in the wastewater. The oxygen concentration is 0 g/m^3 and the sludge concentration 4 kg VSS/m^3. Calculate how much the denitrification rate is reduced at 8°C.

The oxygen flux from the atmosphere to the tank is $\frac{A}{V} \cdot N = K_L a (C_m - C)$ according to the theory for aeration.

The saturation concentration of oxygen, C_m, at 8°C is found to be 11.8 g/m^3. Hence the oxygen flux is:

$= 3\ d^{-1}\ (11.8 - 0)\ g\ O_2/m^3 = 35.4\ g\ O_2/(m^3 \cdot d)$

If the bacteria which utilized the supplied oxygen had denitrified instead, they would have denitrified:

$35.4\ g\ O_2/(m^3 \cdot d) \cdot 0.35\ g\ NO_3^- - N/g\ O_2 = 12.4\ g\ NO_3^- - N/(m^3 \cdot d)$

If it is assumed that the denitrification rate without the influence of oxygen, $r_{X,S}$, corresponds to the rate stated in figure 3.13 for raw wastewater, approx. 12 g NO_3^- – N/(kg VSS · d), a specific volumetric denitrification rate is found:
$r_{V,S} = r_{X,S} \cdot X = 12 \cdot 4 = 48\ g\ NO_3^- - N/(m^3 \cdot d)$

Hence the current denitrification rate is:

$r_{V,S}$ (current) $= r_{V,S} - 12 = 48 - 12 = 36\ g\ NO_3^- - N/(m^3 \cdot d)$

That is, the denitrification rate is in this case reduced by 12/48 = 25 per cent. At a higher temperature, the reduction is lower.

7.3.3. Simultaneous nitrification-denitrification.

The oxygen concentration affects the nitrification as well as the denitrification rate. This means that at low to moderate oxygen concentrations, both processes can run simultaneous. It has to all times been a temptation to designers of treatment processes to design simultaneous processes. Often the rationale has been that if nitrification and denitrification happens at the same time in the same tank, it should be possible to save half the total tank volume. This is not so, the reason being that in these simultaneous nitrification-denitrification processes, both processes runs with reduced speed. At best both might run simultaneously at half speed. This is illustrated in figure7.19. It is seen from the figure that the favourable DO region for simultaneous nitrogen removal is from 0-1 ppm. There is still some denitrification at higher DO, but not of any significance. Floc structure and -size has impact on the rate of the processes, as it influence the effect of diffusion limitation. This means that high turbulence will decrease simultaneous denitrification (due to small flocs or smaller zones without oxygen) and increase nitrification, also due to smaller flocs and less diffusion limitation.

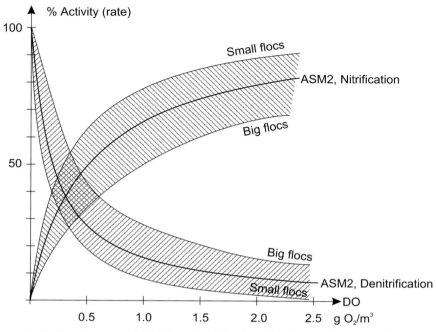

Figure 7.19. Simultaneous nitrification-denitrification. The figure is based on the commonly used Monod half-saturation constants for nitrification and denitrification /21/.

7.3.4. Nitrogen gas in settling tanks and biofilters

If denitrification occurs in the settling tank, it may cause problems with rising sludge as a result of nitrogen gas bubbles. It is difficult exactly to say when it becomes a problem as the water in a settling tank can be super-saturated with nitrogen without nitrogen bubbles being formed.

The increased pressure at the bottom of the settling tank will increase the saturation concentration for nitrogen and will subsequently counteract the tendency of super-saturation.

Oxygen in the sludge in the settling tank will reduce the denitrification (see Example 7.5).

With a saturation concentration of nitrogen at an atmospheric pressure of about 20 g N_2/m^3, an increased hydraulic head of 2 m means that the saturation concentration increases to approx. 24 g N_2/m^3. A denitrification of 4 g $NO_3^- - N/m^3$, while the water and the sludge are in the settling tank, will therefore not worsen the situation.

For a normally well-functioning denitrification process and a well-functioning settling tank, the amount of nitrate, which can be denitrified, will not cause problems. If, however, the sludge retention time in the tank is high, and the denitrification process is only partially functioning, or does not function at all, problems with rising sludge may occur. Experiences indicate that 5-10 g $NO_3^- - N/m^3$ at the inlet to a settling tank, in combination with a temperature of 20°C or higher, will cause problems /20/.

In biofilters, the production of nitrogen gas bubbles may cause problems inside the biofilm and in the water volume of the reactor itself.

Figure 7.20 shows a submerged filter with profiles for the nitrate concentration, varying from 30 to 3 g $NO_3^- - N/m^3$. In the influent, the nitrogen concentration is in equilibrium with the atmosphere, 80 per cent N_2 corresponds to approx. 20 g N_2/m^3. Compared with pure nitrogen, the saturation is then 25 g N_2/m^3. In the figure, a concentration of N_2 is shown which corresponds to the nitrate removed by denitrification. It is seen that just ½ m down the filter there is super-saturation. In practice, there are nitrogen gas bubbles in all denitrification filters. By fine-grained filter media this will block the flow of water and necessitate frequent backwash to remove the bubbles.

Figure 7.21 shows the concentration profile in a denitrifying biofilm. Here, too, super-saturation occurs. This may lead to the formation of bubbles in the boundary layer between the carrier and the biofilm (figure 7.22). In practice, this is one of the mechanisms to slough off biofilm.

For the nitrogen production in the biofilm we have

$$\frac{d^2 S_{N2}}{dx^2} = -\frac{k_{0Vf, N2}}{D_{N2}} \qquad (7.23)$$

where $k_{0Vf,N2}$ is the denitrification rate per volume of biofilm, expressed as N_2-production

D_{N2} is the diffusion coefficient for nitrogen.

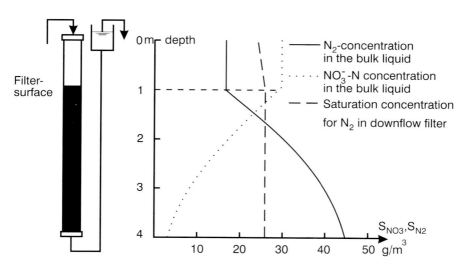

Figure 7.20. Concentration profiles for nitrate and nitrogen in a submerged filter /8/.

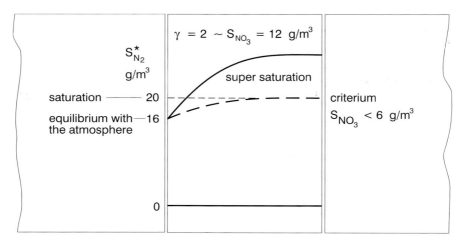

Figure 7.21. Concentration distribution of dissolved nitrogen in a denitrifying biofilm. In the interface, there is equilibrium with the atmosphere. For nitrate concentrations in the free water of more than 6 g NO_3^--N/m^3, super-saturation occurs /8/.

Figure 7.22. Bubbles in the interface between carrier and biofilm due to super-saturation with free nitrogen in the biofilm /8/.

When the equation is solved, the following expression for the nitrogen concentration behind the anoxic zone in the biofilm /8/ is found to be

$$S_{N2,L} = S_{N2,x=0} + \frac{D_{NO3}}{D_{N2}} \cdot S_{NO3,x=0} \qquad (7.24)$$

Example 7.6

How high can the nitrate concentration be to cause a noticeable supersaturation and bubble formation inside the biofilm?

At the surface of the biofilm, the nitrogen concentration is at least 16 g N_2/m^3 (equilibrium with the atmosphere at 15°C). Inside the biofilm, the concentration may increase to 20 g N_2/m^3 before bubble formation. This is equal to a supplement of 4 g N_2/m^3.

$D_{NO3} = 0.7 \cdot 10^{-4}$ m²/d

$D_{N2} = 1.1 \cdot 10^{-4}$ m²/d

$S_{NO3,max} = 4 \cdot \frac{1.1}{0.7} = 6$ g $NO_3^- - N/m^3$

Any higher concentration in the water outside the biofilm will cause super-saturation and bubble formation. It is a phenomenon which must be taken into account in any denitrification filter.

7.3.5. Oxygen consumption

If we leave out of consideration the unwanted oxygen consumption, as discussed in Section 7.3.2, the denitrification process represents in itself a negative oxygen consumption as nitrate replaces oxygen by oxidation of organic matter. It should be remembered, however, that a larger amount of oxygen is used for the nitrate production than the amount which is "saved" by the denitrification.

If we look at nitrification and denitrification as an overall process:

$NH_4^+ \rightarrow NO_3^-$

$NO_3^- \rightarrow 0.5\,N_2$

$\overline{NH_4^+ \rightarrow 0.5\,N_2}$

we can find the oxygen consumption for the overall process:

$$NH_4^+ + 0.75\,O_2 + OH^- \rightarrow 0.5\,N_2 + 2.5\,H_2O \qquad (7.25)$$

The "oxygen saving" by denitrification, viewed separately, corresponds to 1.25 mol O_2/mol $NO_3^- - N$, or 2.86 g O_2/g $NO_3^- - N$ (as the oxidation step for nitrogen is decreased by 5, which is equivalent to 1.25 moles of O_2, which is also decreased by (4 · 1.25 =) 5 in the overall oxidation step).

Table 7.5 lists the oxygen and alkalinity consumption by nitrification, denitrification and combined nitrification-denitrification.

Process	Oxygen consumption			Alkalinity consumption	
	$\dfrac{mol\ O_2}{mol\ N}$	$\dfrac{g\ O_2}{g\ N}$	$\dfrac{g\ O_2}{g\ N}$ *)	$\dfrac{eqv.\ alk.}{mol\ N}$	$\dfrac{eqv.\ alk.}{mol\ N}$ *)
Nitrification	2.0	4.57	4.3	2.0	1.9
Denitrification	−1.25	−2.86	−2.4	−1.0	−0.8
Nitrification + denitrification	0.75	1.71	1.9	1.0	1.1

*)Value in practice, inclusive of nitrogen in the sludge production.

Table 7.5. Oxygen consumption and alkalinity consumption by nitrification and denitrification.

7.3.6. Alkalinity

By denitrification, alkalinity is produced, see Expression (3.30). Table 7.5 lists the changes.

As discussed in Chapter 6, nitrification, which is alkalinity consuming, may cause a lowering of the pH-value in cases where the consumption is higher than the alkalinity in the water. This may partially be compensated for by combination with denitrification because the alkalinity consumption by the overall nitrification-denitrification process is halved.

In biofilms, special conditions apply because the pH in the film may be very different from the pH in the bulk water. The alkalinity producing denitrification process will increase the pH in the biofilm as illustrated in figure 7.23. At worst, the pH can be increased to 9.7 (methanol), 9.9 (acetic acid) and 10.5 (methane).

The increased pH may result in precipitations in the biofilm. Under special circumstances it may lead to a large accumulation of an inorganic mass on the carrier. That may lead to clogging and to increase in weight which may be mechanically destructive for rotating disc plants.

7.3.7. Design of activated sludge plants with denitrification

The basic design parameter is the sludge specific denitrification rate, $r_{X,S}$ (g $NO_3^- - N/(kg\ VSS \cdot h)$). This rate can be estimated from figure 3.13.

Combined with the determined sludge concentration, X_2, the specific volumetric removal rate can be calculated:

$$r_{V,S} = r_{X,S} \cdot X_2 \qquad (7.26)$$

$r_{X,S}$ is estimated for the lowest temperature at which the process is to function.

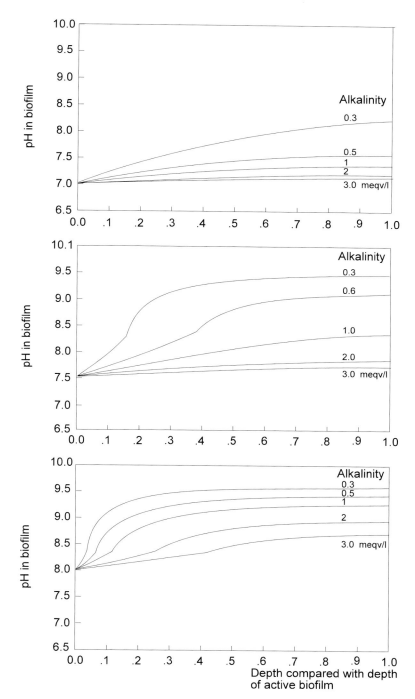

Figure 7.23. An example of calculated pH profiles in a denitrifying biofilm for different pH values (7.0-8.0) and alkalinity (0.3-3.0 meqv/l) in the bulk water /10/.

By means of the nitrogen balance over the plant, with the use of symbols in figure 7.1a, the amount of nitrogen to be denitrified, M_N, can be calculated:

$$M_N = Q_1 \cdot C_1 - Q_3 \cdot C_3 - Q_5 \cdot C_5 \tag{7.27}$$

$Q_1 \cdot C_1$ are given assumptions for the composition of the raw wastewater,
C_3 is the acceptable effluent concentration,
$Q_5 \cdot C_5$ is the nitrogen in the excess sludge production ($= f_{X,N} \cdot F_{ESP}$).

When M_N is known, it can be checked whether the C/N ratio is big enough, or whether it is necessary to reduce the estimated denitrification rate, $r_{X,S}$, due to a too low C/N ratio, or possibly add an external carbon source. Subsequently the necessary tank volume for denitrification, V_2, may be determined on the basis of:

$$M_N = V_2 \cdot r_{V,S}$$
$$V_2 = M_N / r_{V,S} \tag{7.28}$$

Example 7.7

Design of plant for denitrification with activated sludge.

How big should the denitrification tank be when a volume of wastewater of 5,000 m³/d is to be denitrified?

The amount of nitrogen to be denitrified, M_N, is found from a nitrogen balance, see figure 7.1a.

$$M_N = Q_1 \cdot C_1 - Q_3 \cdot C_3 - Q_5 \cdot C_5 \tag{7.18}$$

The following is known:

C_1 = 50 g TN/m³

C_3 = 7 g TN/m³

Furthermore, Q_1 = 5,000 m³/d is known here. It is reasonable to assume that $Q_3 \sim Q_1$, that is, Q_3 = 5,000 m³/d.

Nitrogen in the sludge production, $f_{X,N}$, is estimated at 0.05 kg N/kg SS. The sludge production is assumed to be 700 kg SS/d. This means that $Q_5 \cdot C_5$ is 700 kg SS/d · 0.05 kg N/kg SS = 35 kg N/d.

By substitution we find:

M_N = 5,000 m³/d · 50 g N/m³ - 5,000 m³/d · 7 g N/m³ - 35 kg N/d = 180 kg N/d

The lowest temperature in the denitrification tank is estimated to be 8°C. From figure 3.13 we find a denitrification rate, $r_{X,S}$, of approx. 0.5 g N/(kg VSS · h) (for raw wastewater at 8°C). The sludge concentration in the denitrification tank, X_2, is set at 4 kg VSS/m³. By substitution in Expression (7.26) we find:

$r_{V,S}$ = (0.5 g N/(kg VSS · h)) · 4 kg VSS/m³ = 2 g N/(m³ · h) = 0.048 kg N/(m³ · d)

The tank volume, V_2, is found from Expression (7.28):

V_2 = (180 kg N/d)/(0.048 kg N/(m³ · d)) = 3,750 m³

7.3.8. Model based process design

Combined nitrification-denitrification plants with several tanks and recycles, or with alternating operation, are suitable for an analysis of the operation by means of computer models. The interaction between the three different processes, each of which has opposite interests, is complicated and the design is difficult to carry out, without the extra computational capacity supplied by a computer model.

The present computer models cannot be used for the design of the plants, but they can be used to analyze the effect of the diurnal variations, rain, temperature changes, or inhibition on the chosen design. As a result of a computer analysis, it may also be necessary to make changes in the dimensions of the plant, or in its mode of operation.

The models are also suitable for optimization of the operation of existing plants. The effect of changes in the oxygen concentration, recycle, phase length or other controls can be tested and optimized before they are taken into use on the plant itself. In the long term, further developed models can be directly incorporated in the control of the plants.

Figure 7.24 shows a comparison between a model calculation and a direct measurement in a tank on an alternating plant with nitrogen removal. In order to obtain the shown compliance, a comprehensive analysis and calibration are required, as further illustrated by Example 7.8 /16/.

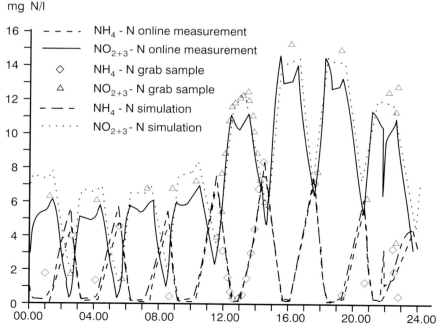

Figure 7.24. Alternating nitrification-denitrification. NH_4^+ and NO_3^- measured online, with spot samples and calculated with model /16/.

Design of denitrifying plants

Example 7.8.

Model calculation by means of a computer model /19/.

The work comprises 3 main activities:

- Characterization of the wastewater
- Characterization of activated sludge
- Determination of process constants through tests

The influent to the plant is wastewater which is preprecipitated with polymerized aluminium chloride. The preprecipitated sludge is hydrolyzed and the dissolved hydrolyzate is added to the denitrification phase to increase both the rate and capacity of the denitrification. The design of the plant is shown in figure 7.25.

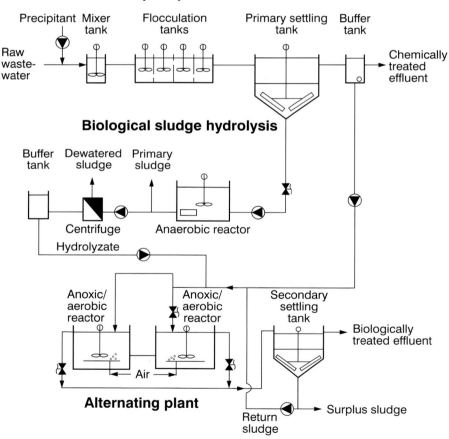

Figure 7.25. Sketch of the test plant /17/. The model calculations only concern the alternating plant consisting of the two anoxic/aerobic tanks.

Characterization of wastewater. Diurnal variations in the influent to the alternating tanks have been measured. The COD, Kjeldahl nitrogen and ammonium as well as SS and VSS have been measured. Figure 7.26 shows the diurnal variations for total and dissolved COD.

The low concentration levels occur because the preprecipitation reduces the COD by approx. 65 per cent. This also explains the very low content of suspended COD.

The detailed composition of COD in the influent is determined on the basis of the oxygen uptake rate (OUR). The dissolved COD is characterized by means of three fractions: inert, S_I, very easily degradable matter, S_{HAc}, and easily degradable matter, S_S. Based on the effluent from the process, dissolved inert COD can be estimated at approx. 10 per cent of the dissolved COD in the influent. The other two fractions are determined by a respiration test.

When determining the fractions in suspended matter, the yield constant observed by the process is used. The model calculations of this yield constant can be adjusted by making changes in the inert fraction, X_I. In this case it is thus found that the inert particulate COD accounts for 25 per cent of the particulate COD in the influent. The residue of the particulate COD of the influent is assumed to consist of slowly degradable COD and biomass, that is, X_S, $X_{B,H}$ and $X_{B,A}$.

The COD fractions vary in the course of the day, and in Table 7.6 we see the result of the wastewater characterization as an independent characterization of 8-hour flow-proportional samples has been carried out.

Date	Hour	S_{HAc}	S_S	S_I	$X_S + X_{B,H} + X_{B,A}$	X_I
		(% of COD soluble)			(% of COD particulate)	
18/12	09-17	20	70	10	75	25
18/12	17-01	30	60	10	75	25
19/12	01-09	20	70	10	75	25
19/12	09-17	10	80	10	75	25
19/12	17-01	30	60	10	75	25
20/12	01-09	20	70	10	75	25

Table 7.6. Applied fractionation of COD for the influent to the alternating process /16/.

Characterization of the sludge. Based on respiration experiments with oxygen (OUR), nitrate (NUR) and ammonium oxidation (AUR) /18/, the composition of sludge is determined to:

$X_{B,H}$ = 0.80 kg/m^3
$X_{B,A}$ = 0.20 kg/m^3
X_I = 2.80 kg/m^3
η_g = 0.5

Determination of process constants. The saturation constants for oxygen have by batch experiments been determined for both nitrifying and denitrifying bacteria, and the following values have been found:

$K_{S,O2,A}$ = 0.5 g/m^3
$K_{S,O2(NO3)}$ = 0.023 g/m^3

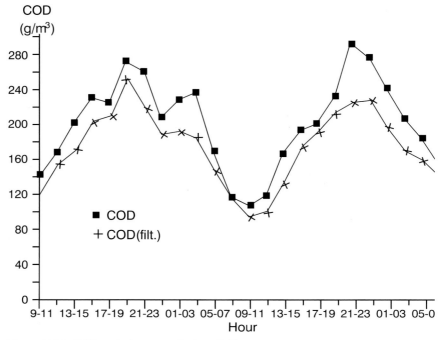

Figure 7.26. COD-variations over 2 days /16/.

By model calculations, four constants were changed in relation to the model default values. These are (the default values are stated in brackets):

b_H	=	1.5 d^{-1}	(0.62 d^{-1})
$K_{S,COD}$	=	2 g COD / m^3	(2.5 g COD / m^3)
$K_{S,O2,H}$	=	0.5 g O_2 / m^3	(0.2 g O_2 / m^3)
$K_{S,O2,A}$	=	0.5 g O_2 / m^3	(0.4 g O_2 / m^3)

The process explanation of these changes is as follows:

- The higher value for decay, b_H, is assumed to be due to the special characteristics of the influent water in the form of a low content of suspended solids. In this way another composition of the sludge is developed which may have another decay constant than for a plant into which raw wastewater, or primary settled wastewater is discharged. The decay constant also includes phenomena such as grazing (predation) which will also be influenced by the sludge conditions in the plant.
Note the interaction between decay and maximum growth rate. In some cases we will get about the same result from the computer simulation, whether the decay is increased, or the maximum growth rate is reduced.

- The somewhat lower value for the saturation constant is due to the very easily degradable substrate which is included in the process and in the model. It has been found from experience that the more easily degradable a substrate is, the lower saturation constant should be used.

- The saturation constant for oxygen by denitrification (and removal by oxygen) is found to be 0.5 g/m³ which is higher than the normal standard value in the model used. The difference between the value determined by the batch experiment and in the pilot plant itself is assumed to be due to the small flocs which are found in the batch experiment. The smaller the flocs, the smaller saturation constant for oxygen, see the biofilm kinetics.

By using the procedure outlined above, we can in this case obtain results from computer-simulation as the one shown in figure 7.24. The major part of the work is to characterize the influent water. Normally it is only necessary to change very few of the process constants as it is also seen in this case. Once the constants have been found for a given plant, experiences seem to indicate that they are rather constant with time. One of the reasons is that in spite of the fact that the wastewater changes its concentration in the course of weeks and months, the percentage distribution of the individual substance fractions is assumed to be rather constant. For the process constants it applies that some of them are influenced by the design of the plant and its operation (for example floc sizes and sludge composition).

7.3.9. Design of biofilters for denitrification

The design is carried out on the same principle as for nitrifying filters, Chapter 6, Section 6.3.3. The design is carried out by means of a mass balance by finding the necessary surface area of the carrier, A_{2^*}.

Based on the concentrations of organic matter and nitrate through the filter (it **may** be ideally mixed, but normally it looks more like plug flow), it is determined (by using Expression (7.6a) or (7.6b)) which component is limiting.

Next the reaction kinetics are determined by using Expression (5.13), and finally the reaction rate expression is determined by Expression (7.7a), (7.7b), (7.7c), or (7.7d). From Expression (7.5), the necessary area of the carrier, A_{2^*}, is found.

Example 7.9

A plug flow filter for the denitrification of 1,800 m³/d municipal wastewater has to be designed. The water is treated in a nitrifying activated sludge plant after which it is denitrified with acetic acid, HAc, as the carbon source.

The influent concentration to the filter is $S_{NO3,1} = 23$ g $NO_3^- - N/m^3$.
The effluent concentration, $S_{NO3,3}$, must be 3 g $NO_3^- - N/m^3$.

An acetic acid dose is chosen. From Table 7.2, the stoichiometrically needed dosage is found (inclusive of assimilation) to be 5.5 g COD/g $NO_3^- - N$.

We choose to add more acetic acid than this ratio, say 6.5 g COD/g $NO_3^- - N$. That is,

$$S_{COD,1,min} = (S_{NO3,1} - S_{NO3,3}) \nu = (23 - 3) \frac{g\, NO_3^- - N}{m^3} \cdot 6.5 \frac{g\, COD}{g\, NO_3^- - N}$$

$$= 130 \frac{g\, COD}{m^3}$$

which corresponds to

$$130 \text{ g COD} / m^3 \cdot v_{COD,HAc} = 130 \frac{\text{g COD}}{m^3} \cdot 0.937 \frac{\text{g HAc}}{\text{g COD}} = 122 \frac{\text{g HAc}}{m^3}$$

By means of Expression (7.6b) it is considered whether nitrate is limiting in the influent and effluent.

From Table 5.1 we find $\frac{D_{NO3}}{D_{COD}} = 1.6$.

$$\frac{S_{COD,1}}{S_{NO3,1}} = \frac{130 \text{ g COD}/m^3}{23 \text{ g NO}_3^- - N/m^3} = 5.6 \frac{\text{g COD}}{\text{g NO}_3^- - N} < \frac{D_{NO3}}{D_{COD}} \cdot v_{NO3,COD}$$

$$= 1.6 \cdot 5.5 = 8.8 \frac{\text{g COD}}{\text{g NO}_3^- - N}$$

That is, acetic acid is limiting and nitrate is not limiting.

In the effluent, the acetic acid content (and the nitrate content) is reduced.

From the ratio $v_{NO3,COD} = 5.5$ g COD/g NO_3^- – N from Table 7.2 and a mass balance for COD, the effluent concentration is determined:

in + removed = out

$$S_{COD,1} - (S_{NO3,3} - S_{NO3,1}) \; v_{NO3,COD} = S_{COD,3}$$

$$S_{COD,3} = 130 \frac{\text{g COD}}{m^3} - (23 - 3) \frac{\text{g NO}_3^- - N}{m^3} \cdot 5.5 = 20 \text{ g} \frac{COD}{m^3}$$

For the effluent it is considered whether nitrate is limiting

$$\frac{S_{COD,3}}{S_{NO3,3}} = \frac{20}{3} = 6.7 \frac{\text{g COD}}{\text{g NO}_3^- - N} < \frac{D_{NO3}}{D_{COD}} \cdot v_{NO3,COD} = 8.8 \frac{\text{g COD}}{\text{g NO}_3^- - N}$$

Just as for the influent, nitrate is not limiting and the acetic acid is limiting.

By using Expression (5.13), β is determined in the filter, first in the influent, as $k_{0Vf,COD}$ is measured at

260 kg COD/($m^3 \cdot$ d)/(6.5 kg COD/kg NO_3^- - N) = 40 kg NO_3^- - N/($m^3 \cdot$ d)

L is set at 2 mm and D_{COD} at $0.5 \cdot 10^{-4}$ m²/d.

That is,

$$\beta = \left(\frac{2 \cdot D_{COD} \cdot S_{COD,1}}{k_{0Vf,COD} \cdot L^2}\right)^{1/2} = \left(\frac{2 \cdot 0.5 \cdot 10^{-4} \text{ m}^2/\text{d} \cdot 130 \text{ g COD}/m^3}{260 \cdot 10^3 \text{ g COD}/(m^3 \cdot d) \cdot (2 \cdot 10^{-3} m)^2}\right)^{1/2}$$

$$= 0.013 < 1$$

that is, the film is only partially penetrated, and hence it is a half order reaction.

Correspondingly it is found for the effluent: β = 0.002 <1.

The whole filter therefore has half order kinetics with respect to the really limiting component: acetic acid.

The mass balance for the removal of acetic acid or nitrate can now be set up.
For acetic acid we have

$$QS + r_A \cdot \omega \cdot A \cdot dy = Q\left(S + \frac{\delta S}{\delta y} \cdot dy\right)$$

$$r_{A,COD} = k_{1/2A,COD} \cdot S_{COD}^{1/2}$$

According to Expression (5.37) and (5.42), the solution is:

$$S_{COD,3}^{1/2} = S_{COD,1}^{1/2} - 0{,}5 \cdot k_{1/2A,COD} \frac{A_{2*}}{Q_1}.$$

As

$$k_{1/2A,COD} = \left(2 \cdot D_{COD,2} \cdot k_{0Vf,COD}\right)^{1/2}$$

A_{2*} can be found:

$$A_{2*} = \frac{(S_{COD,3}^{1/2} - S_{COD,1}^{1/2}) \cdot Q_1}{0.5 \cdot (2 \cdot D_{COD,2} \cdot k_{0vf,COD})^{1/2}}$$

$$= \frac{(130^{1/2} - 20^{1/2})\ (g\ COD/m^3)^{1/2}}{0.5\ (2 \cdot 0.5 \cdot 10^{-4}\ m^2/d \cdot 260 \cdot 10^3\ g\ COD/(m^3 \cdot d))^{1/2}} = 4{,}900\ m^2$$

If the specific volumetric area in the filter, ω, is 100 m²/m³, the volume of the filter, V_2, is
$V_2 = A_{2*}/\omega = 4{,}900\ m^2/100\ m^{-1} = 49\ m^3$

(corresponding to a hydraulic retention time for an empty filter of approx. 40 min.).

Well, that was a long example. What about its credibility? We will answer that question by quoting the German philosopher Georg Wilhelm Friedrich Hegel: "If reality does not want to abide by the idea - it's too bad for reality".

7.4. Redox-zones in the biomass

In treatment plants where biomass is present in the form of biofilms or flocs, the removal may be limited by diffusion. This will usually be dominant in biofilters, but it may also play a role in activated sludge plants. In both cases processes may occur which cannot immediately be explained without understanding the possible limitation of the diffusion in respect of the reaction. It is important for the understanding of the observed phenomena that the diffusion does not just lead to limitations in the reaction rate; it is equally important to understand that the diffusional limitation may lead to layers of redox- zones in the biomass. The section on simultaneous aerobic removal of organic matter and denitrification in the rear of the biofilm is a good example; but it is just an example of something of a more general nature.

Figure 7.27 is an example of layers of redox-zones as they may occur in a biofilm. In each layer is indicated whether the substance is removed or produced. In zones where the substance is removed, the curvature is negative. In zones with

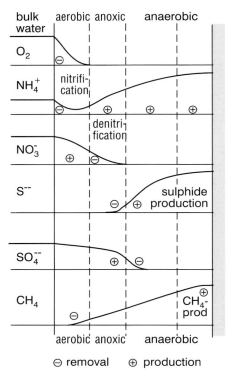

Figure 7.27. Illustration of layers of redox zones in a biofilm.

production the curvature is positive, like the second derivative of the concentration profile (Section 5.1).

The outer layer, exposed to the bulk, oxygen-rich water, is aerobic due to the diffusion of oxygen into the biofilm. It is the penetration depth of the oxygen which limits the aerobic removal in the reactor. In this zone the substances from the wastewater, which can be degraded aerobically, are oxidized (especially diffusible organic matter and ammonium), as described in Chapters 5 and 6. It should be realised that it is not just the substances coming from the outside which may be oxidized. A diffusion of reduced substances from the anaerobic conversion of organic matter in the internal layers may diffuse to the outer layers of the biofilm to be oxidized there. In the figure, this is illustrated with the profile for ammonia.

If nitrate is present in the water and/or if nitrate is produced by nitrification in the aerobic zone, nitrate will diffuse into the oxygen-free zone where denitrification will occur (the anoxic zone). The denitrification requires the presence of a reductant which has so far been described solely as originating from the diffusion (into) from the water. This reductant may also originate from the internal, reduced zones in the biofilm. This may cause phenomena which will influence the net removal.

In the internal, reduced zones, all relevant processes may take place under anaerobic conditions. These may comprise the well-known redox processes for an electron potential less than $p(\varepsilon)° = 5.8$, see for example /9/. The most important processes in connection with water treatment are:

- Reduction of iron to ferro components which are soluble and hence may diffuse into the outer zones.
- Reduction of organic matter to fatty acids. Several of the intermediate products as well as the fatty acids themselves may diffuse to the outer zones.
- Reduction of sulphate to sulphide, as illustrated in the figure.

In connection with anaerobic degradation of organic matter, a release of ammonium takes place which diffuses into the aerobic zones.

The reduction of sulphate in the biofilm or the oxidation of sulphide diffusing out from anaerobic degradation of the organic matter is of special potential importance. The produced sulphide can be precipitated with metals, for example iron, but it is otherwise accessible for diffusion into the outer oxidizing zones.

The ultimate anaerobic reduction of organic matter is the conversion into methane and carbon dioxide. The methane may diffuse out, to be oxidized in the outer zones. Due to the low solubility in water, an accumulation of methane may occur in the internal zone of the biofilm to such an extent that the solubility is exceeded. This may lead to bubble formation which may result in sloughing off of the biofilm from the support material.

All these processes are acidity or alkalinity producing and will to some extent influence the pH in the biofilm which is again important to the redox conditions. In this way the processes interact in a complicated way, determined by the diffusion conditions, the resulting redox conditions, the resulting pH conditions, the growth and degradation rates in each zone, etc. The history of the growth of the biofilm and the mechanical influence of hydrodynamic conditions are important factors for the sloughing off and new growth.

Relatively little is known about these conditions and their importance to the engineering application of the biofilters in wastewater treatment. All these processes fulfil an important role in the old-fashioned, low-rate trickling filters. In these filters there was sufficient space for local clogging. The biofilm was degraded anaerobically and finally sloughed off to give room for new growth. These filters were problematic due to these conditions and today they are not competitive.

In trickling filters, submerged filters and rotating discs with normal load without nitrification, it is not known to what extent anaerobic conditions play a role for the net removal.

If the load is sufficiently low to allow nitrification, there are three zones: the aerobic, the anoxic and the anaerobic zones. Denitrification occurs in such filters, as already described. The importance of the anaerobic zones is practically unknown.

It is difficult to examine the processes in the interior of such complicated biofilms, solely based on the bulk reaction in the filter. Recently, methods have been developed for the detailed analysis of these processes by measuring inside the biofilms with microsensors /12/. It is today possible to measure the following substances by means of microsensors:

$pH, S^{2-}, O_2, NO_2^-, NO_3^-$, glucose.

The interaction between the processes in the different redox- zones can be simulated with models for diffusion and for the processes. The problem is to know enough about the processes to be able to describe the combination of redox-zones. The research on biofilms has been intensified. A much better understanding of the microbiological aspects of biofilms, the structure of biofilms and the redox-zones in biofilms will emerge within the foreseeable future.

References

/1/ Wuhrmann, K. (1964): Stickstoff- und Phosphorelimination: Ergebnisse von Versuchen im technischen Masstab. (Nitrogen and phosphorus elimination: Results from technical experiments). *Z. Hydrologie*, **XXVI**, 520-533.

/2/ Jansen, J. la Cour (1983): Removal of Soluble Substrates in Fixed Films. Department of Environmental Engineering, Technical University of Denmark, Lyngby, Denmark.

/3/ Jansen, J. la Cour and Harremoës, P. (1984): Removal of Soluble Substrates in Fixed Films. *Water Sci. Technol.* **17**, (2/3), 1-14.

/4/ Harremoës, P. and Riemer, M. (1977): Pilot-Scale Experiment on Down-Flow Filter Denitrification. *Prog. Water Technol.*, **10**, (5/6), 149-165.

/5/ Harremoës, P. and Riemer, M. (1975): Report on Pilot Studies of Denitrifikation in Down-Flow Filter. Department of Environmental Engineering, Technical University of Denmark, Lyngby, Denmark.

/6/ Rogalla, F. and Bourbigot, M.M. (1990): New Developments in Complete Nitrogen Removal with Biological Aerated Filters. *Water Sci. Technol.*, **22**, (1/2), 273-280.

/7/ Henze, M. and Harremoës, P. (1977): Biological Denitrification of Sewage: A Literature Review. *Prog. Water Technol.*, **8**, (4/5), 509-555.

/8/ Harremoës, P, Jansen, J. la Cour and Kristensen, G.H. (1982): Practical Problems Related to Nitrogen Bubble Formation in Fixed Film Reactors. *Prog. Water Technol.*, **12**, (6), 253-269.

/9/ Harremoës, P. Henze, M., Arvin, E. and Dahi, E. (1989): Teoretisk vandhygiejne. (Water Chemistry) Polyteknisk Forlag, Lyngby, Denmark.

/10/ Arvin, E. and Kristensen, G.H. (1982): Effect of Denitrification on the pH in Biofilm. *Water Sci. Technol.*, **14**, (8), 833-848.

/11/ Kinner, N. (1983): A Study of the Microorganisms Inhabiting Rotating Biological Contactor Biofilms during various Operating Conditions, Ph.D. Disertation. University of New Hampshire, Durham, N.H.

/12/ Revsbech, N.P. (1989): Microsensors: Spatial Gradients in Biofilms. In: Characklis, W.G. and Wilderer, P.A. (eds.), Structure and Function of Biofilms, pp. 129-144. John Wiley and Sons, New York, N.Y.

/13/ Dalsgaard, T., Kühl, M., Jensen, K. Jørgensen, B.B. and Revsbech, N.P. (1991): Stofomsætning i biofilm målt med mikrosensorer. (Removal of substances in biofilm measured with microsensors). *Vand og Miljø*, **8**, 471-477.

/14/ Hanaki, K., Hong, Z. and T. Matsuo (1992): Production of Nitrous Oxide Gas during Denitrification of Wastewater. *Water Sci. Technol.*, **26**, (5/6), 1027-1036.

/15/ Henze, M. (1991): Capabilities of Biological Nitrogen Removal Processes from Wastewater. *Water Sci. Technol.*, **23**, (4-6), 669-679.

/16/ Kristensen, G.H., la Cour Jansen, J., Jørgensen, P.E., Christensen, H.W. and Henze, M. (1991): Simuleringssystem til styring af renseanlæg. (Simulation system for the control of treatment plants). Report to Teknologirådet. Vandkvalitetsinstituttet, Hørsholm, Denmark.

/17/ Kristensen, G.H., Jørgensen, P.E., Strube, R. and Henze, M. (1992): Combined Pre-precipitation, Biological Sludge Hydrolysis and Nitrogen Reduction - a Pilot-demonstration of Integrated Nutrient Removal. *Water Sci. Technol.*, **26**, (5/6), 1057-1066.

/18/ Kristensen, G.H., Jørgensen, P.E. and Henze, M. (1992): Characterization of Functional Groups and Substrate in Activated Sludge and Wastewater by AUR, NUR and OUR. *Water Sci. Techol.*, **25**, (6), 43-57.

/19/ EFOR (1991): Edb-model for rensningsanlæg. (Computer model for treatment plants). User manual, version 2.0. Efor aps, Søborg, Denmark.

/20/ Henze, M., Dupont, R., Grau, P. and de la Sota, A. (1993): Rising sludge in secondary settlers due to denitrification. *Water Res.*, **27**, 231-236.

/21/ Henze, M., Gujer, W., Mino,T., Matsuo,T., Wentzel, M.C. and Marais, G.v.R. (1995): Activated Sludge Model No.2. IAWQ, London. (IAWQ Scientific and Technical reports, No.3).

Nitrification and denitrification tank with oxygen meter. Frederikssund treatment plant (Denmark). One of the first Danish plants with alternating operation.

8 Plants for Biological Phosphorus Removal

By Mogens Henze

Phosphorus stimulates plant growth (eutrophication) in streams, lakes, rivers, and in the ocean. A large part of the phosphorus which is discharged into the receiving waters derives from wastewater. Due to increasing eutrophication problems all over the world, the demand for phosphorus removal from wastewater is increasing. Biological phosphorus removal is a process which can be fully or partly used for the solution of the problem. It will sometimes be necessary to combine the process with chemical precipitation and/or filtration to remove some of the phosphorus. Sludge from treatment plants with biological phosphorus removal will be well-suited as fertilizer if it is not overloaded with metals or xenobiotics.

8.1. Mass balances, biological phosphorus removal plants with activated sludge

Figure 8.1 gives a schematic representation of a biological phosphorus removal plant. Two mass balances are of special interest:

a) A mass balance for easily metabolisable organic matter (acetate and propionate) in tank V_2. (This balance can tell how much is available of the organic matter for the biological phosphorus removal which can then be used to calculate the phosphorus removal potential).
b) A phosphorus balance for the whole plant. (On the basis of which the current phosphorus removal can be calculated).

The mass balance for acetate and propionate, here expressed as acetic acid, HAc, looks as follows in tank V_2:

$$\underbrace{Q \cdot S_{HAc,1}}_{\text{in}} + \underbrace{Q_5 \cdot S_{HAc,5}}_{\text{in}} + \underbrace{v_{S,HAc} \cdot k_h \cdot S_2 \cdot V_2}_{\text{hydrolysis / fermentation}} - \underbrace{r_{V,HAc} \cdot V_2}_{\text{bio} - \text{PHAc uptake}}$$

$$= \underbrace{Q_{2.2} \cdot S_{HAc,2.2}}_{\text{out}} \qquad (8.1)$$

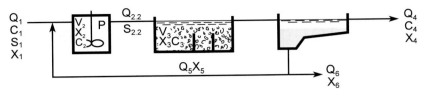

Figure 8.1. Schematic representation, biological phosphorus removal.
V_2 = anaerobic phosphorus tank
V_3 = aerobic/anoxic tank(s)

The process is so far carried out in activated sludge systems. Full-scale biofilm plants have not yet been developed for biological phosphorus removal.

Normally, Expression (8.1) can be simplified by assuming that:

$S_{HAc,5} = 0$ (all easily degradable organic matter has been removed in the treated water/recycled sludge)

Based on Expression (3.37), the storage of PHA or removal of easily degradable organic matter, $r_{V,HAc}$, can be simplified into:

$$r_{V,HAc} = k_P (S_{HAc}/(S_{HAc} + K_{S,HAc})) \tag{8.2}$$

where k_P is 1-2 kg COD(S)/(m³ · d) at 20°C, ($k_p = q_{PHA} \cdot X_{B,PAO}$)
$K_{S,HAc}$ is 2-6 g COD(S)/m³.

A combination of Expressions (8.1) and (8.2), and an assumption that $S_{HAc,5} = 0$, gives for an ideally mixed anaerobic tank V_2:

$$Q_1 \cdot S_{HAc,1} + v_{S,HAc} \cdot k_h \cdot S_2 \cdot V_2 - \frac{(k_P \cdot V_2 \cdot S_{HAc,2})}{(S_{HAc,2} + K_{S,HAc})} = Q_2 \cdot S_{HAc,2} \tag{8.3}$$

$S_{HAc,2}$ can be found from Expression (8.3) if the influent ($Q_1 \cdot S_{HAc,1}$), tank volume (V_2) and the hydrolysis (k_P, $K_{S,HAc}$) are known.

The removal of acetate and propionate by the PAOs under anaerobic conditions is closely related to the final uptake of phosphorus. Using the stoichiometric coefficient from example 3.11, $v_{HAc,P} = 0.05$ g P/g HAc-COD a mass balance for the treatment plant can be established.

A mass balance is established for the plant shown in figure 8.1 with respect to phosphorus:

in	out	surplus sludge structural	surplus sludge PP
$Q_1 \cdot C_{P,1}$ =	$Q_4 \cdot C_{P,4}$ +	$Q_6 \cdot C_{P,6}$ +	$Q_6 \cdot C_{PP,6}$

(8,4)

The Bio-P process is complicated, and the approach here is simplified. Table 8.1 presents a stoichiometric matrix describing the processes for the PAOs. It looks complicated, but this matrix is also a simplification.

Process	S_{O2}	S_A	S_{N2}	S_{NO3}	S_{PO4}	X_I	X_S	X_{PAO}	X_{PP}	X_{PHA}
10 Storage of X_{PHA}		-1			Y_{PO4}				$-Y_{PO4}$	
11 Aerobic storage of X_{PP}	$-Y_{PHA}$				-1				1	$-Y_{PHA}$
12 Anoxic storage of X_{PP}			$-v_{12,NO3}$	$v_{12,NO3}$	-1				1	$-Y_{PHA}$
13 Aerobic growth of X_{PAO}	$v_{13,O2}$				$-i_{PBM}$			1		$-1/Y_H$
14 Anoxic growth of X_{PAO}			$-v_{14,NO3}$	$v_{14,NO3}$	$-i_{PBM}$			1		$-1/Y_H$
15 Lysis of X_{PAO}					$v_{15,PO4}$	f_{XI}	$1-f_{XI}$	-1		
16 Lysis of X_{PP}					1				-1	
17 Lysis of X_{PHA}		1								-1

Table 8.1 The ASM2d stoichiometric matrix for PAOs. For detailed information see /3/.

The phosphorus in the surplus sludge consists of structural phosphorus, present in all biomass, $C_{P,6}$, and polyphosphate, $C_{PP,6}$ present in the PAOs only. The structural phosphate will be similar in concentration to the normal phosphorus content in the excess sludge (1-1.5% P in SS). The polyphosphate concentration in the excess sludge is calculated based on the availability of fatty acids in the anaerobic tank:

$$Q_1 \cdot (S_{HAc,1} + S_{F,HAc,1})/v_{HAc,P} = Q_6 \cdot C_{PP,6} \tag{8.5}$$

where $S_{F,HAc,1}$ is the amount of HAc produced by fermentation in the anaerobic tank

$Q_6 \cdot C_{PP,6}$ is the amount of polyphosphate stored, assuming that all available HAc (in wastewater and produced by fermentation) is taken up as PHB by the PAOs, and that all stored polyphosphate is found in the excess sludge and nothing in the effluent. This is a reasonable assumption in most cases in practice.

Example 8.1

Wastewater, 3,600 m³/d, contains 60 g of easily degradable organic matter (COD) per m³. It is assumed that all is present as HAc in the anaerobic tank. A plant for biological phosphorus removal with an ideally mixed anaerobic tank of 200 m³ has been built. Calculate the effluent concentration of HAc from the anaerobic tank.

The mass balance for HAc is:

$$Q_1 \cdot S_{HAc,1} - \frac{(k_P \cdot V_2 \cdot S_{HAc,2})}{(S_{HAc,2} + K_{S,HAc})} = Q_2 \cdot S_{HAc,2} \tag{8.3}$$

k_P is estimated to 1.46 kg COD(S)/(m^3 · d)

$K_{S,HAc}$ is estimated to 3 g COD/m^3

Substitution gives:

3,600 m^3/d · 0.060 kg COD(S)/d − (1.46 kg COD(S)/(m^3 · d)) · $S_{HAc,2}$ · 200 m^3/($S_{HAc,2}$ + 0.003 kg COD(S)/m^3) = 3,600 m^3/d · $S_{HAc,2}$

from which $S_{HAc,2}$ can be found:

$S_{HAc,2}$ = 0.006 kg COD/m^3

that is, the following amount of HAc has been taken up in the anaerobic tank:

$S_{HAc,1} - S_{HAc,2}$ = 0.060 − 0.006 = 0.054 kg COD/m^3

If the stoichiometric coefficient $v_{HAc,PO4}$ = 0.06 kg P/kg COD, the biological phosphorus process has removed

$(S_{HAc,1} - S_{HAc,2}) \cdot v_{HAc,PO4}$ = (0.060 − 0.006) · 0.06 = 0.0032 kg P/m^3
from the wastewater.

(provided that such an amount of dissolved phosphate has been available in the raw wastewater. This will normally be the case in countries without excessive wastewater generation per capita).

8.2. Plant types, biological phosphorus removal

All plant designs for biological phosphorus removal consist of a combination of biological phosphorus removal and an aerobic process (oxidation of organic matter). The reason is that in order to function the process requires that the microorganisms are alternately exposed to aerobic and anaerobic conditions.

Biological phosphorus removal can also be combined with nitrification and denitrification in a combined activated sludge process, but not with nitrification alone in a combined activated sludge process. Details on design of Bio-P plants are not abundant, but take a look in /1,4/.

8.2.1. Biological phosphorus removal with nitrification-denitrification and an internal carbon source

Figure 8.2 shows two typical plant designs. In the anaerobic tank it has been attempted to establish something which resembles plug flow. In one of the examples in the form of a number of ideally mixed tanks in series, in the other example in the form of a long, plug flow tank. Plug flow ensures to a certain degree that there are parts of the anaerobic front tank where no nitrate is present even if nitrate should be present in the return flow from sludge treatment. Any nitrate present will be removed in the first part of the anaerobic tank (hence it is not anaerobic after all)

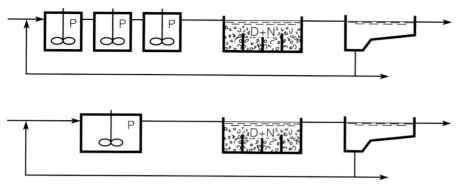

Figure 8.2. Biological phosphorus removal (P) in combination with nitrification and denitrification (D + N). Internal carbon source. The anaerobic tank need not be the first step in the proces. It can be incorporated in the phases during intermittent nitrification-denitrification

which will then function as a denitrification tank. This is inconvenient as far as the process is concerned, as easily degradable organic matter, which could otherwise have been used for the biological phosphorus removal, will disappear.

Figure 8.3. Anaerobic tank for biological phosphorus removal. Odense NØ treatment plant (Denmark).

The easily degradable organic matter in raw wastewater is used directly for the biological phosphorus removal together with substances which might be produced by fermentation in the anaerobic tank.

Figure 8.3 shows an anaerobic tank for biological phosphorus removal.

8.2.2. Biological phosphorus removal with nitrification-denitrification and an external carbon source

If an insufficient amount of easily degradable organic matter is present in the raw wastewater, an external carbon source can be applied which may be acetate or industrial wastewater/waste products from the food processing industry. Figure 8.4 shows the schematic representation of the plant layout.

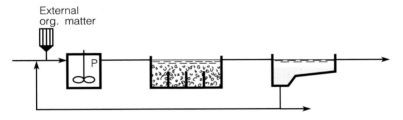

Figure 8.4. Biological phosphorus removal with external input of organic matter.

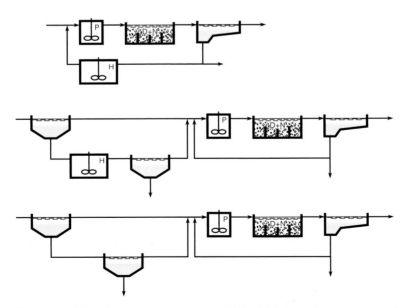

Figure 8.5. Biological phosphorus removal (P) with internally produced easily degradable organic matter (by hydrolysis (H)). D + N are denitrification and nitrification, S is sedimentation.

8.2.3. Biological phosphorus removal with internally produced easily degradable organic matter

If there is a shortage of easily degradable organic matter (acetic acid etc.) in the wastewater, it can be produced internally in the plant by a hydrolysis process. Activated sludge can hydrolyse in an anaerobic tank at a rate allowing for Bio-P activity to occur. Figure 8.5 gives a schematic representation of plant designs. In the lower plant, the hydrolysis takes place in the primary tank and in a concentration tank.

8.2.4. Biological phosphorus removal without nitrification-denitrification

The biological phosphorus removal can take place in a plant design as shown in figure 8.6. It is important here to avoid nitrification in summer, that is, a low aerobic sludge age is a prerequisite.

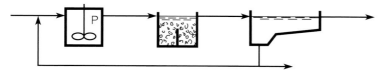

Figure 8.6. Biological phosphorus removal (P) in combination with removal of organic matter. No nitrification-denitrification (A/O or Phoredox).

8.3. Design of biological phosphorus removal

8.3.1. Easily degradable organic matter

The presence of easily degradable organic matter is a prerequisite for biological phosphorus removal. Roughly 1 kg of dissolved phosphorus per 20 kg of easily degradable COD, $S_A + S_F$ may be removed assuming that S_F is fermented in the anaerobic tank. The easily degradable COD must be acetic acid or similarly small organic molecules.

The amount of easily degradable COD available in the anaerobic tank appears from the first three terms of the mass balance, Expression (8.1).

If nitrate is introduced to the anaerobic tank, it will mean that an amount of easily degradable matter is consumed by denitrification. The amount consumed is 4-6 kg COD/kg $NO_3^- - N$.

The amount available of easily degradable organic matter can be estimated on the basis of 3 contributions:

influent: $\quad Q_1 \cdot S_{HAc,1}$

hydrolysis/
fermentation: $\quad r_{V,HAc} \cdot \exp(\kappa(T-20)) \cdot V_2$

denitrification: $\quad -v_{NO3,HAc} \cdot Q_5 \cdot S_{NO3,5}$

Using the following assumptions:

$r_{V,HAc} \quad = 0.25$ kg COD(S)/(m³ · d)

$\kappa \quad = 0.1$ °C^{-1}

$v_{NO3,HAc} \quad = 5$ kg COD/kg $NO_3^- - N$

and that 0.05 kg P/kg COD can be removed. Hence the maximum possible biological phosphorus removal can be calculated:

$$Q_1 (C_{P,1} - C_{P,4}) = 0.5 (Q_1 \cdot S_{HAc,1} + 0.25 \cdot \exp(0.1(T-20)) \cdot V_2 - 5 \cdot Q_5 \cdot S_{NO3,5}) \qquad (8.7)$$

Here it is assumed that all organic matter is removed so that the effluent concentration of easily degradable organic matter, $S_{HAc,2}$, from the anaerobic tank is 0.

Example 8.2

Calculate the maximum possible biological phosphorus removal at 20°C and at 8°C for the plant described in Example 8.1. Additional information is that the nitrate concentration in the return sludge flow, 3,000 m³/d, is 0.003 kg $NO_3^- - N$/m³.

Maximum phosphorus removal corresponds to $S_{HAc,2} = S_{HAc,5} = 0$.

By substitution into Expression (8.7) we find for 20°C

3,600 m³/d $(C_{P,1} - C_{P,4})$ = 0.05 kg P/kg COD(S) (3,600 m³/d · 0.060 kg COD(S)/m³ + 0.25 kg COD(S)/(m³ · d) · exp (0.1 (20 –20)) 200 m³ –
5 kg COD(S)/kg $NO_3^- - N$ · 3,000 m³/d · 0.003 kg $NO_3^- - N$/m³)

$C_{P,1} - C_{P,4}$ = 0.05 (216 + 50 – 45)/3,600 = 0.003 kg P/m³

At 8°C, all terms, with the exception of the hydrolysis, are unchanged. The hydrolysis is 0.25 kg COD(S)/(m³ · d) · exp (0.1(8–20)) · 200 m³ =
15 kg COD(S)/d

$C_{P,1} - C_{P,4}$ = 0.1 (216 + 15 – 45)/3,600 = 0.0025 kg P/m³

Hence a maximum of respectively 3 and 2.5 g P/m³ can be removed by biological phosphorus removal at 20°C and 8°C.

8.3.2. Design of tanks for biological phosphorus removal

The design of the anaerobic tank for biological phosphorus removal is based on the amount of VFA that is needed for the process. In most cases the VFA is the limiting factor with respect to phosphorus removal. The design results in a hydraulic retention time for the anaerobic tank (or the anaerobic fraction of the hydraulic retention time in the aeration tanks). The anaerobic tank can be of the pre-anaerobic type, like in figure 8.2, or an internal-anaerobic tank, like the top one in figure 8.7.

Pre-anaerobic tank

If the anaerobic tank is the first step in the process train, the influent wastewater VFA is directly available. On top of this comes what is produced by fermentation. The hydraulic retention time is normally designed for 1-3 h /4/. The long retention time can give significantly more VFA produced from fermentation, and thus improved Bio-P performance. It is important to control the overall process SRT, as increased SRT will decrease the rate of fermentation.

More advanced design can be made by modelling the process /3/ based on lab – or pilot experiments.

Internal-anaerobic tank

Here all VFA present in the influent will have been removed in the aerobic and anoxic process steps preceding the anaerobic tank. In this case all VFA must be generated by fermentation in the tank. The retention time can be from 1-10 h, depending on the rate of fermentation and the need for VFA for the P-removal.

Modelling will in this case give better information on the possibilities for fermentation in the anaerobic tank. A key factor for significant fermentation in an internal-anaerobic tank is to keep the total SRT as low as possible, but still cater for sufficient aerobic SRT if nitrification is part of the process.

Example 8.3

Find the total effluent concentration of dissolved phosphorus from the plant in Example 8.2 when the influent concentration of total phosphorus is 6 g/m^3 and the effluent concentration of suspended solids is 20 g SS/m^3. It is assumed that there are 5% phosphorus in suspended solids.

At 20°C we have from Example 8.2:

$C_{P,1} - C_{P,4} = 3$ g P/m^3

$C_{P,1} = 6$ g P/m^3, that is,

$C_{P,4} = 3$ g P/m^3

$C_{P,4} = S_{P,4} + f_{X,P} \cdot X_4$ \hfill (8.8)

By combination we find:

$S_{P,4} = C_{P,4} - 0.05 \cdot 20 = 3 - 1 = 2$ g P/m^3

At 8°C we have:

$S_{P,4} = 3.5 - 0.05 \cdot 20 = 2.5$ g P/m^3

If detergents with phosphate are used in the catchment area, the influent concentration might be 10 g P/m^3. In such a case, the effluent concentration of dissolved phosphorus will be 7-7.5 g P/m^3 depending on the temperature.

Calculation of phosphorus content in surplus sludge

By using the phosphorus balance, Expression (8.5), the phosphorus amount in the sludge can be calculated:

$$Q_1 \cdot C_{P,1} = Q_4 \cdot C_{P,4} + Q_6 (C_{P,6} + C_{PP,6}) \tag{8.5}$$

$$Q_1 \cdot C_{P,1} = Q_4 \cdot S_{P,4} + Q_4 \cdot X_{P,4} + Q_6 (C_{P,6} + C_{PP,6})$$

The phosphorus amount in the sludge is:

$$Q_4 \cdot X_{P,4} + Q_6(C_{P,6} + C_{PP,6}) = Q_1 \cdot C_{P,1} - Q_4 \cdot S_{P,4} \tag{8.9}$$

The overall sludge production is found from Expression (4.13) as it can be calculated for a normal aerobic process:

$$F_{SP} = Y_{obs}(C_1 - C_3) \cdot Q_1 \tag{4.13}$$

where C_1 and C_3 are the influent and effluent concentrations of organic matter.

The combination of Expressions (8.8), (8.9) and (4.13) give the phosphorus content in the sludge, $f_{X,P}$:

$$f_{X,P} = (Q_1 \cdot C_{P,1} - Q_4 \cdot C_{P,4})/F_{SP} \tag{8.10}$$

Example 8.4

Calculation of the phosphorus content in the sludge from Example 8.3 at 8°C.

The total phosphorus in the influent is 6 g P/m^3. Total COD, C_1, in the influent is 300 g/m^3 and in the effluent, C_4, 50 g/m^3. The volume of wastewater is 14,000 m^3/d.

From Example 8.3 we have: $C_{P,4} = S_{P,4} + X_{P,4} = 2 + 1 = 3$ g P/m^3.

It is assumed that $Q_4 = Q_1$.

It is also assumed that $Y_{obs} = 0.7$ g SS/g COD.

By substitution into Expression (8.10) we find:

$$f_{X,P} = (Q_1 \cdot C_{P,1} - Q_4 \cdot C_{P,4})/F_{SP} \tag{8.10}$$

$f_{X,P} = (C_{P,1} - C_{P,4})/(Y_{obs} (C_1 - C_3))$

Substituting known values give:

$f_{X,P} = (6 - 3$ g P/m$^3)/(0.7$ g SS/g COD$(300 - 50)$ g COD/m^3

$f_{X,P} = 0.017$ g P/g SS

that is, a phosphorus content of 1.7% of SS.

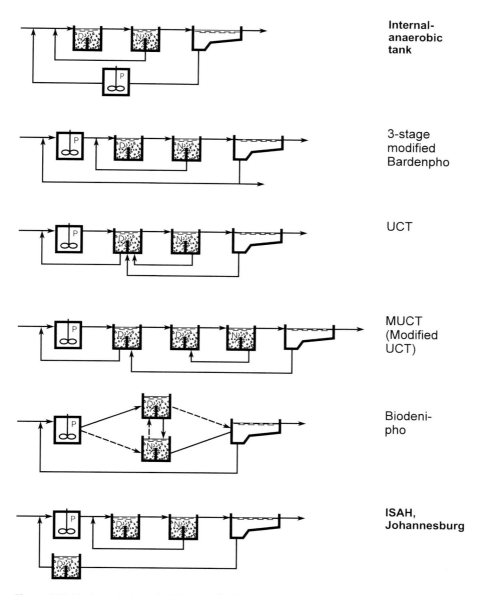

Figure 8.7. Various designs for biological phosphoros removal process lay-out.

Optimization of plant operation, biological phosphorus removal

Models are available for optimization /2,3/. The essential problems are nitrate in the anaerobic tank and a shortage of easily degradable organic matter. Nitrate compete through denitrification for the VFAs, see Chapter 3, Section 3.6. The problem can be solved in two ways:

a. Improvement of the denitrification so that there is very little nitrate in the recycled sludge.
b. Introduction of a denitrification tank (anoxic sludge stabilisation) in the return sludge flow.

The use of a denitrification tank is shown in figure 8.8.

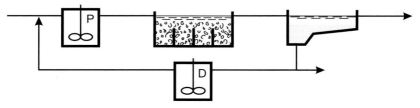

Figure 8.8. Anoxic sludge stabilisation to reduce denitrification carry-over to the pre-anaerobic tank.

Other bacteria, G-bacteria, or GAO (glycogen accumulating bacteria) may take up easily degradable organic matter and compete with the PAO's. The GAO's do not accumulate or release phosphate and do not normally influence the phosphorus removal process.

A shortage of easily degradable matter can be compensated for in several ways:
a. external input of acetic acid,
b. external supply of industrial wastewater,
c. increased hydrolysis/fermentation in the anaerobic tank by increasing the retention time,
d. separate hydrolysis of sludge or wastewater which is subsequently supplied to the anaerobic tank, see figure 8.5 and figure 8.7 top.

The input of oxygen to the anaerobic tank will have the same effect as the input of nitrate, that is, there will be a consumption of easily degradable organic matter by oxidation. The input of oxygen can take place with the flow of recycle sludge (normally it is no problem), or by aeration of the anaerobic tank in connection with the mixing/turbulence which is necessary to keep the biomass suspended.

A third problem for the biological phosphorus removal may be the release of phosphorus from the sludge before it is removed from the plant. This may take place if anaerobic conditions occur in some of the other tanks simultaneously with the presence of easily degradable organic matter. It will be a risk in connection with the input of an external carbon source in excess for the denitrification. When nitrate has been removed, a release of phosphorus from the sludge will take place. The aerobic sludge age should be kept as low af possible. The higher $\theta_{x,aerobic}$, the less efficient the bio-P process will be.

References

/1/ Ramadori, R. (ed.) (1987): Biological Phosphate Removal from Wastewater. Proceedings of an IAWPRC Specialized Conference held in Rome 28-30 September 1987. Pergamon Press, London. (Advances in Water Pollution Control).

/2/ Henze, M., Gujer, W. Mino, T., Matsuo, T., Wentzel, M.C. and G. v. R. Marais (1995): Activated Sludge Model No.2. IAWQ, London. (IAWQ Scientific and Technical Report. No.3).

/3/ Henze, M., Gujer, W., Mino, T., Matsuo, T., Wentzel, M.C., Marais, G.v.R. and van Loosdrecht, M.C.M. (1999): Activated sludge model No. 2d, ASM2d. *Water Sci. Technol.*, **39**, (1), 165-182.

/4/ Scheer, H. (1994): Vermehrte biologische Phosphorelimination – Bemessung und Modelierung in Theorie und Praxis – (Biological phosphorus removal – design and modelling in theory and practice). Universität Hannover, Germany. (Veröff. des Instituts für Siedlungswasserwirtschaft und Abfalltechnik, Heft 88).

9 Hydrolysis/fermentation and Anaerobic Wastewater Treatment

By Mogens Henze

The treatment steps described in this chapter are truly anaerobic, i.e neither oxygen nor nitrate/nitrite is present or fed into the process tanks (except for the minor amounts with the influent to the processes). The anaerobic wastewater treatment can be integrated with aerobic/anoxic processes or can be separate. Hydrolysis is always integrated with other treatment processes.

9.1. Hydrolysis/fermentation

Hydrolysis/fermentation can be used for many purposes in the treatment plant as shown in table 9.1. The processes are in all cases slow compared with aerobic and anoxic processes. In evaluating the potential hydrolysis yield, it is important to take into consideration, the large amounts of for example activated sludge present in a treatment plant. If all the sludge is kept anaerobic for a few hours, a huge amount of VFA will be generated.

Substrate	Purpose	Used for	Yield in practice % of total COD	%/d at 20 °C
Wastewater	$S_F \rightarrow S_{HAc}$	Denitrification, Bio-P	2-6	1-5
Primary sludge	$X \rightarrow S_{HAc}$	Denitrification, Bio-P Biogas production	10-20	1-4
Activated sludge	$X \rightarrow S_{HAc}$	Bio-P	1-3	0.1-1

Table 9.1. Types of hydrolysis/fermentation /13,14,15/

The soluble COD produced during hydrolysis/fermentation can be separated from the sludge by settling or centrifugation. The soluble COD can then be added at a point in the nutrient removal process where there is a carbon source deficit. The experience with hydrolysate tells that it gives very fast reactions rates, often higher than pure acetate carbon source. The reason seems to be the high content of small easily degradable organic molecules of many different kinds. The VFAs accounts for 60-80% of the soluble COD in hydrolysate and with acetate accounting for 60-

80% of the VFAs. The yield as given in table 9.1 can reach the high value for long hydraulic retention time.

Design of hydrolysis/fermentation based on hydraulic retention time

The design is normally based on hydraulic retention time, with 1-3 days as typical values at 20 °C. The longer the hydraulic retention time, the higher potential soluble COD yield. The drawback can be hydrogen sulphide production, which can cause inhibition of the biological processes and which will consume VFA or other organic compounds. A second problem with high hydraulic retention time can be methane production, which will also consume/convert VFA to methane and carbon dioxide. The third problem is the investment cost for the big tank volume.

Design based on anaerobic sludge age, $\theta_{X,anaerobic}$.

The hydrolysis/fermentation process is a first order process with respect to biomass concentration. The impact of this is that the anaerobic sludge age, $\theta_{X,anaerobic}$ is the correct design parameter. $\theta_{X,anaerobic}$ is often 1-3d or 5-15% of the total sludge age, and the control of this to low values can also avoid methane production. A possibility to control hydrogen sulphide production is to recycle sludge over the aerobic/anoxic tanks, thus keeping the anaerobic retention time per passage low. Figure 8.5 and 8.7 shows types of hydrolysis/fermentation processes, which takes place in the hydrolysis tanks (H) and in the anaerobic Bio-P tanks (P).

9.2. Anaerobic wastewater treatment

9.2.1. Introduction

Anaerobic wastewater treatment is not the same as anaerobic sludge treatment. The reason is that a larger part of the organic matter in the wastewater is dissolved in anaerobic wastewater treatment. When dissolved organic matter is to be removed, it means that treatment processes are to be used which ensure a good and a sufficiently long contact between the dissolved substances in the wastewater and the microorganisms carrying out the anaerobic processes. It means that for anaerobic *wastewater* treatment there is a big difference between the hydraulic retention time and the sludge age which is usually not the case in anaerobic sludge treatment.

The first anaerobic treatment plant for industrial wastewater was built in 1929 for the treatment of wastewater from yeast production in Slagelse, Denmark. Even if this plant was in operation for almost 30 years, the development went very slowly. The development of this process did not gather momentum until the Dutch UASB plant type was introduced around 1980 /6/.

Anaerobic treatment processes are cheap in operation and yield a gas production of a considerable value. The processes are especially favourable for concentrated wastewater where an aerobic oxidation of the organic matter would otherwise result in high electricity costs.

It is a further advantage if the wastewater to be treated is warm, as the necessary tank volume is hereby reduced. A method to reduce the tank volume is to utilize processes where it is possible to maintain a high sludge concentration in the tank. The development within anaerobic processes is thus moving towards such processes, for example sludge blanket plants and fluidized filters of which many are already known from other branches of biotechnology and from chemical technology /11/.

Anaerobic wastewater treatment plants are by and large only used for the treatment of industrial wastewater. By percentage, plants for anaerobic industrial wastewater treatment remove a lot of organic matter and almost no nutrients. Anaerobic plants must therefore be considered as pretreatment plants which are to be followed by other treatment processes/plants.

9.2.2. Mass balances, anaerobic plants

A schematic representation for an anaerobic plant is shown in figure 9.1.

A mass balance for an anaerobic treatment plant can be written when we know the processes which are included, and their kinetics. Table 9.2a shows a process matrix for anaerobic processes. Like all other process matrices, it is very simplified when compared to reality. The process matrix in Table 9.2b is even more simplified. It can be used in situations where the methane production can be considered as the rate-limiting process, for example by a constant load. Note that for an anaerobic wastewater treatment process, the methane production is about the only process which contributes to the removal of COD from the water. By removal of suspended solids, for example by settling, centrifugation or membrane separation, parts of the suspended COD can be removed after the anaerobic biological process.

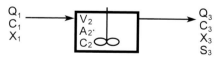

Figure 9.1. Schematic representation, anaerobic wastewater treatment.

Component → Process ↓	$X_{B,S}$	$X_{B,M}$	X_I	X_S	S_s	S_{HAc}	S_{CH4}	S_{ALK}	Reaction rate $r_{v,?}$
1. Hydrolysis of organic matter				-1	1				$k_{hx} \cdot \frac{X_S/X_{B,S}}{K_X + (X_S/X_{B,S})} \cdot X_{B,S}$
2. Acid production	1				$-\frac{1}{Y_{max,S}}$	$\frac{1-Y_{max,S}}{Y_{max,S}}$		$\frac{5,9}{Y_{max,S}}$	$\mu_{max,S} \frac{S_S}{S_S + K_{S,S}} \cdot X_{B,S}$
3. Methane production		1				$-\frac{1}{Y_{max,M}}$	$\frac{1-Y_{max,M}}{Y_{max,M}}$	$\frac{5,8}{Y_{max,M}}$	$\mu_{max,M} \frac{S_{HAc}}{S_{HAc} + K_{S,M}} \cdot X_{B,M}$
4. Decay acid producers	-1		$f_{XB,XI}$	$1-f_{XB,XI}$					$b_{H,S} \cdot X_{B,S}$
5. Decay methane producers		-1	$f_{XB,XI}$	$1-f_{XB,XI}$					$b_{H,M} \cdot X_{B,M}$
Unit	kg COD/m³							eqv/m³	
	Acid producing biomass	Methane producing biomass	Inert suspended org. matter	Slowly degradable org. matter	Easily degradable org. matter	Acetic acid	Methane	Alkalinity	

Table 9.2a. Process matrix, anaerobic process

Component → Process ↓	X_B	C_S	S_{CH4}	Reaction rate $r_{V,X}$
Methane production	1	$-\frac{1}{Y_{max}}$	$\frac{1-Y_{max}}{Y_{max}}$	$\mu_{max} \frac{C_S}{C_S + K_S} \cdot X_B$
	kg COD/m³ Biomass	kg COD/m³ Org. matter	kg COD/m³ Methane	

Table 9.2b. Simplified process matrix, anaerobic process. Y_{max} is the total maximum anaerobic yield constant.

With symbols from figure 9.1, a COD mass balance for the wastewater can be written as follows (by using the simplified process matrix in Table 9.2b):

$$Q_1 \cdot C_1 - r_{V,X} \cdot V_2 = Q_3 \cdot C_3 \tag{9.1}$$

where $r_{V,X} = \frac{1-Y_{max}}{Y_{max}} \cdot \mu_{max} \frac{C_S}{C_S + K_S} \cdot X_B$

The produced amount of methane corresponds to the COD removed from the water and is therefore $r_{V,X} \cdot V_2$.

Suspended process

In a suspended anaerobic process, the sludge age will often be the design basis as the plant is laid out in such a way that the slow-growing methane bacteria can survive.

The observed net growth rate of the methane bacteria, μ_{obs}, can be found from a mass balance based on Table 9.2a to be:

$$\mu_{obs,M,net} \cdot X_{B,M} = \mu_{max,M} \cdot \frac{S_{HAc}}{S_{HAc} + K_{S,M}} \cdot X_{B,M} - b_{H,M} \cdot X_{B,M}$$

Hence a methane mass balance over the plant in figure 9.2 is:

$$Q_1 \cdot X_{B,M,1} + \mu_{obs,M,net} \cdot X_{B,M,2} \cdot V_2 = Q_3 \cdot X_{B,M,3} + Q_5 \cdot X_{B,M,5} \qquad (9.2)$$

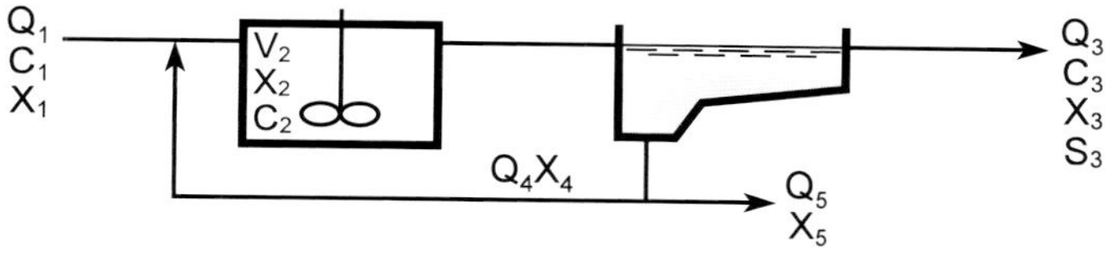

Figure 9.2. Suspended process/sludge blanket process, anaerobic treatment.

The amount of methane bacteria in the influent, $Q_1 \cdot X_{B,M,1}$, can normally be assumed to be zero. By introducing the sludge production $F_{SP} = Q_3 \cdot X_{B,M,3} + Q_5 \cdot X_{B,M,5}$ (see Expression (4.12)), the sludge mass $M_X = X_{B,M,2} \cdot V_2$ (see Expression (4.10)) and the sludge age $\theta_{X,M}$ (see Expression (4.14)), we find:

$$\theta_{X,M} = 1/\mu_{obs,M,net} \qquad (9.3)$$

In figure 9.3, the necessary sludge age, $\theta_{X,M}$, is shown as a function of the temperature.

The volume of the anaerobic tank, V_2, can be found from Expression (9.4):

$$V_2 = \frac{\theta_{X,M} \cdot F_{SP}}{X_{B,M,2}} \qquad (9.4)$$

Example 9.1

An anaerobic contact plant (suspended sludge) is going to treat wastewater from a sugar factory. The net growth rate of the methane producing bacteria at 30°C has been measured at 0.05 d^{-1}. How high is the sludge age going to be?

From Expression (9.3) we find:

$\theta_{X,M} = 1/\mu_{obs,M,net} = 1/0.05$ d^{-1} = 20 d

From figure 9.3 we can estimate the necessary sludge age to 20-28 d. The necessary sludge age is therefore in the lower end of the normal experience range.

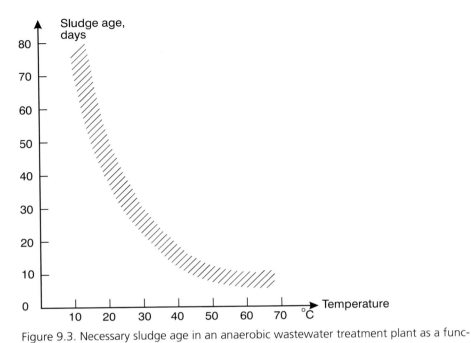

Figure 9.3. Necessary sludge age in an anaerobic wastewater treatment plant as a function of the temperature.

What would the necessary sludge age be estimated at if a thermophilic process at 55°C was to be used?

We see from figure 9.3 that $\theta_{X,M}$ ~ 6-12 d. If it is assumed that the net growth rate is high at 55°C, just like at 30°C, the necessary sludge age can be estimated at 6 d.

Biofilm plants

A COD mass balance for the wastewater looks as follows:

$$Q_1 \cdot C_1 - r_{A,M} \cdot A_{2^*} = Q_3 \cdot C_3 \tag{9.5}$$

The reaction rate, $r_{A,M}$, can often be described with a zero order kinetics, both in the biofilm itself and in the liquid (that is, the biofilm is fully utilized).

$$r_{A,M} = k_{0Vf,M} \cdot L$$

If the volume of the biofilm is $A_{2^*} \cdot L_2$ (L_2 is the thickness), and the concentration of biomass in the biofilm itself is $X_{B,f,2}$, we have

$$r_{A,M} \cdot A_{2^*} = r_{X,M} \cdot X_{B,f,2} \cdot L_2 \cdot A_{2^*} = r_{X,M} \cdot X_{B,2} \cdot V_2$$

Where $X_{B,2}$ is the volumetric concentration of biomass of an anaerobic tank.
Substituted in (9.5) we find

$$Q_1 \cdot C_1 - r_{X,M} \cdot X_{B,f,2} \cdot L_2 \cdot A_{2^*} = Q_3 \cdot C_3 \qquad (9.6)$$

The concentration of biomass, $X_{B,f,2}$, can be high in the biofilm itself (10-50 kg COD/m^3) which means that the concentration compared with the total tank volume, X_2, can also be high.

9.3. Plant types, anaerobic processes

Like for other biological processes, the plant types for anaerobic processes can be divided into plants with suspended sludge and plants with biofilm. As anaerobic treatment plants often treat concentrated wastewater, sometimes with high sludge concentrations, it may be necessary to pretreat the wastewater. The pretreatment may consist in a separation of suspended and dissolved solids.

9.3.1. Pretreatment of wastewater, anaerobic plants

The pretreatment may consist in a phase separation in the form of:

- fine screening
- settling
- flotation
- hyper-filtration

possibly in combination as for example settling and hyper-filtration can be combined. Figure 9.4 shows an example of a plant layout where the fixed and liquid phases are separated, and subsequently, two separate anaerobic plants are used. The sludge can be handled in a traditional digester whereas the dissolved part of the organic matter can be treated in one of the many plant types discussed in the next chapter.

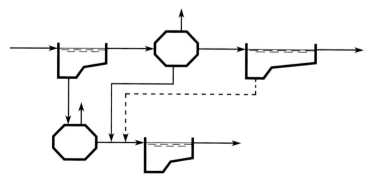

Figure 9.4. Anaerobic treatment plant with two separate anaerobic reactors. Suitable for wastewater with a high content of suspended solids.

Hyper-filtration can be used for concentration of dilute wastewater, for example from ordinary households, or as an actual treatment process for the water phase in highly concentrated wastewater types (industry, agriculture).

It applies for all plants with filters that a pretreatment is necessary to ensure against clogging. The pretreatment can be anything from fine screening to hyper-filtration.

9.3.2. Plants with suspended sludge

The plant type resembles activated sludge plants. The sludge is kept mechanically stirred or in certain cases stirred through the gasification in the process and the pumping in of wastewater.

The traditional plant is shown in figure 9.5 and is called the contact process. The biological process takes place in a tall tank, 5-10 m height, whereas sludge separation takes place by settling, lamella sedimentation, possibly with preceding de-gasification or refrigeration, see figure 9.6.

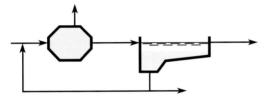

Figure 9.5. Contact process, anaerobic wastewater treatment.

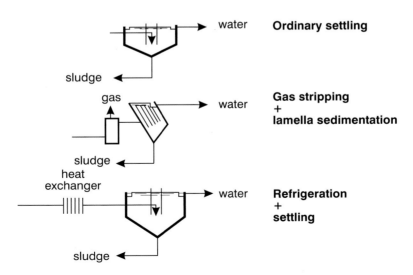

Figure 9.6. Sludge separation techniques, anaerobic wastewater treatment.

Degasification or gas stripping prevents or delays the time for the production of rising sludge as a certain gas production is going to take place after the degasification before the water is sufficiently super saturated for gas bubbles to be formed again. Degasification may take place mechanically/hydraulically, or by means of vacuum. Refrigeration stops the biological process temporarily and consequently prevents the formation of new gas bubbles for a period of time.

The contact process is a historical starting point for anaerobic wastewater treatment. The first plant in the world was built in Slagelse (Denmark) in 1929 for the treatment of wastewater from the Danish Destillers' yeast factory. The contact process is very widespread, especially in connection with the sugar and spirits industries. Figure 9.7 shows the anaerobic contact process at Copenhagen Pectin A/S (Denmark).

A modern variant of the contact process is the sludge blanket plant (UASB) which is shown in figure 9.8. This plant consists of a biological tank with upflow combined with a settling tank.

At a suitable arrangement of the biological tank, and provided that the content of organic matter in the wastewater is easily degradable (preferably dissolved), the biomass can produce spherical granules. These granules are permanent as opposed

Figure 9.7. Anaerobic contact process, Copenhagen Pectin A/S, Ll. Skensved (Denmark). Sludge separation takes place by means of lamella separation. The plant was originally built around 1970 and has since been extended in stages.

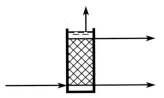

Figure 9.8. Sludge blanket plant, anaerobic wastewater treatment. The top of the plant consists of a complicated separator which separates gas, water and sludge. Suitable for wastewater with a high content of acetic acid, for example hydrolyzed wastewater from sugar factories.

to ordinary flocs which are degraded and regenerated all the time. The spherical granules have a high concentration of biomass and a high density. In this way not just high tank volumetric removal rates are obtained, the granules also remain in the active part of the tank as their settling velocity is greater than the upflow water velocity. The sludge concentration in sludge blanket plants is typically twice as high as in contact process plants, that is 5-15 kg VSS/m^3 (7-20 kg COD/m^3). The sludge blanket plants can have hydraulic problems, due to the distribution of the influent and the separation of water and gas at the top of the tank. This type of plant is today the most widespread among anaerobic treatment plants. An example is shown in figure 9.9.

Figure 9.9. UASB-plant for anaerobic treatment of wastewater from sugar production. CSM (The Netherlands).

9.3.3. Anaerobic filter processes

The design of the different filter plants (biofilm plants) for anaerobic treatment corresponds to what is used for other biological processes. The filter plants normally require that there is only a small amount of suspended solids in the wastewater, or clogging may occur. The small amount of suspended solids in the influent, combined with the small sludge production, means that there are only small amounts of suspended solids in the effluent. Therefore, many anaerobic filter plants have no final settling. Another reason for the lack of final settling is that the plants are often used for pretreatment.

For all anaerobic filter processes it applies that a certain part of the biomass is suspended. It applies especially to the fixed upflow filters /1/ and the rotating discs.

The fixed filter, figure 9.10, is the most commonly used plant type among the anaerobic filter plants. It is built using plastic filter media and is submerged and with upflow.

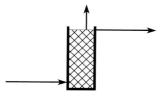

Figure 9.10. Fixed anaerobic filter. Suitable for pretreatment of (warm) wastewater from the food industry. Gas is collected from the top of the filter.

In figure 9.11 three other filter plant layouts are shown. They are not yet very widespread. In the fluidized filter, sand is usually used as a carrier. Both in the fluidized and in the expanded filter, a large recycle is required to keep the filter layer moving. There are some hydraulic problems in connection with fluidized filters, and the problems are reinforced by anaerobic processes producing gas bubbles which can inhibit the separation of water and granules at the top of the filter. The gas bubbles may also cause problems for the biofilm as they can be produced between the film and the carrier and thus causing sloughing off of the biofilm.

In fluidized filters, the filter media are cleaned mechanically for the adhered biomass whereas it is the aim of the function of the expanded filter that the cleaning takes place in the filter itself in that the filter particles rub against each other. In the rotating disc, the sloughing off of the slime is controlled by the rotation rate of the disc.

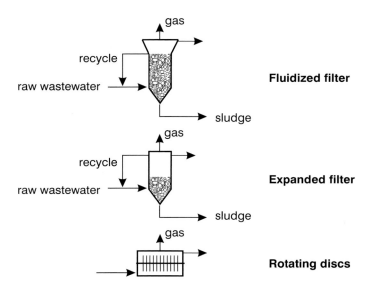

Figure 9.11. Various anaerobic filter processes. Normally used without final settling. The height of the filter layer in the fluidized filter is controlled by withdrawal of filter grains (not shown in the figure) which are returned in the column after the treatment. The treated sludge can be treated separately or be mixed in the treated flow of water.

9.4. Design of anaerobic plants

The design of anaerobic plants is not very advanced. The normal design rules are all based on an extensive removal of organically degradable matter. It is not possible to design for example for 50 per cent removal. The reason is the slow growth of the methane producing bacteria which necessitates such a low load that most of the organic matter is being hydrolyzed and converted into short-chained fatty acids. In order to avoid inhibition of the methane production, just about all the fatty acids must be further converted which explains why it is not possible to carry out a partial treatment.

The purpose of the design is therefore to ensure the survival of the methane producing bacteria (see figure 9.3 and Example 9.1).

The temperature is important. Normally the plants are built for an operational temperature of approx. 35°C (mesophillic). Thermophilic processes (at 50-60°C) are seldom used whereas plants with lower temperatures (15-25°C) are sometimes used.

9.4.1. Design of plants with suspended sludge

These plants are normally designed based on the COD volumetric load, the COD sludge load, or the sludge age.

Design based on the volumetric load

The volumetric load, $B_{V,COD}$, is defined as follows (symbols from figure 9.1):

$$B_{V,COD} = Q_1 \cdot C_{COD,1}/V_2 \qquad (9.7)$$

from which the necessary tank volume, V_2, can be found:

$$V_2 = Q_1 \cdot C_{COD,1}/B_{V,COD} \qquad (9.8)$$

Notice that $C_{COD,1}$ is biodegradable COD.

The volumetric load varies with the type of plant as shown in Table 9.3. The geometric data of the plants and the technical details are given in Table 9.4.

Plant type	Volumetric load kg COD/(m³·d)		
	15-25°C	30-35°C	50-60°C
Contact plant	0.5-2	2-6	3-9
Sludge blanket plant	1-3	3-10	5-15
Fixed filter	1-3	3-10	5-15
Rotating disc	1-3	3-10	5-15
Fluidized filter	1-4	4-12	6-18
Expanded filter	1-4	4-12	6-18

Table 9.3. Volumetric load (biodegradable COD) for anaerobic plants at different temperatures. Treatment efficiency for COD: 80-90 per cent.

Example 9.2

Design of an anaerobic sludge blanket plant by means of the volumetric load.

1,200 m³ of wastewater/d is to be treated. The content of organic matter in the wastewater is divided into fractions, see Expression (2.4):

S_I = 0.1 kg COD/m³

S_S = 2.5 kg COD/m³

X_I = 0.2 kg COD/m³

X_S = 0.2 kg COD/m³

C_{COD} = 3.0 kg COD/m³

The temperature of the plant is set at 35°C. From Table 9.3 we find the volumetric load, $B_{V,COD}$ = 3-10 kg COD/(m³ · d). Here we choose $B_{V,COD}$ = 6 kg COD/(m³ · d).

From Expression (9.8), the volume, V_2, of the anaerobic tank can be determined:

	Unit	Fluidized filter	Expanded filter	Rotating filter	Fixed filter	Contact plant	Sludge blanket plant
Filter medium, type	-	sand activated carbon	sand gravel plastics	plastics	gravel stones plastics	-	-
Filter medium, diameter	mm	0.2-1	0.3-2	1000-3000	-	-	-
Filter medium, disc distance	cm	-	-	10-20	-	-	-
Filter medium, rotating speed	rpm	-	-	2-5	-	-	-
Filter medium, circumferential velocity	m/s	-	-	0.3	-	-	-
Filter medium, degree of submersion	%	100	100	75-100	100	-	-
Porosity, void	%	-	-	-	50-98	-	-
Porosity, with biomass	%	-	-	-	20-90	-	-
Specific surface	m^2/m^3	200-600	200-600	100-200	60-200	-	-
Height of tank	m	4-8	2-3	-	3-6	5-10	2-4
Radius of tank	m	3-4	2-3	-	(5-20)	(5-20)	(5-20)
Upflow water velocity, nominal	m/h	4-8	1-5	-	-	-	0.01-0.15
Recycle ratio	-	5-500	2-50	-	-	0.5-2	-
Sludge concentration, tank	kg SS/m^3	10-30	10-40	5-15	5-15	3-10	5-20
Sludge concentration, tank	kg VSS/m^3	8-25	8-30	4-12	4-12	2-8	3-16
Sludge concentration, tank	kg COD/m^3	10-30	10-40	5-15	5-15	3-10	5-20
Sludge adhered to filter medium	% of total	95-100	70-90	50-80	20-80* 80-90**	0	0
Sludge, suspended	% of total	0-5	10-30	20-50	20-80* 10-20**	100	100
Sludge concentration, effluent	g SS/m^3	20-100	20-100	20-300	20-300	20-100	20-100
Gas flux	Nm^3/(m^2·d)	5-40	5-40	-	5-20	-	5-20
Gas flux, maximum	Nm^3/(m^2·d)	30-40	30-40	-	10-20	-	10-20
Energy, pumping, water	Wh/m^3	15-30	10-20	-	-	-	-
Energy, pumping, raw water	Wh/m^3	75-3000	20-1000	5-10	20-40	10-30	15-30
Energy for rotation	Wh/(m^3 tank)	-	-	20-80	-	-	-
Energy for stirring	Wh/(m^3 tank)	-	-	-	-	5-15	-

* upflow
** downflow

Table 9.4. Technical data for anaerobic plants partly based on /1, 2, 3, 4, 5, 11, 12/.

$$V_2 = Q_1 \cdot C_{COD,1} / B_{V,COD} \tag{9.8}$$

The biodegradable COD concentration is 2.7 kg COD/m^3 as inert dissolved, S_I, and inert suspended COD, X_I, must be deducted.

By substitution we find:

$V_2 = (1{,}200 \text{ m}^3/\text{d}) \cdot (2.7 \text{ kg COD/m}^3)/(6 \text{ kg COD}/(\text{m}^3 \cdot \text{d})) = 540 \text{ m}^3$

Design by means of sludge load

The sludge load, $B_{X,COD}$, is defined as:

$$B_{X,COD} = Q_1 \cdot C_{COD,1}/M_X$$

The sludge mass, M_X, is the sludge mass which is found in the biologically active parts of the plant.

For the plant shown in figure 9.2 we have:

$$M_X = V_2 \cdot X_2$$

which, when substituted and after isolation of the tank volume, V_2, gives:

$$V_2 = Q_1 \cdot C_{COD,1}/(X_2 \cdot B_{X,COD}) \tag{9.9}$$

The sludge load for the different plant types is the same, see Table 9.5. However, the sludge concentration in the anaerobic tank, X_2, varies with the plant type and the actual mode of operation.

Organic matter in wastewater dominated by:	Sludge load kg COD/(kg VSS·d)		
	15-25°C	30-35°C	50-60°C
Acetic acid	1.5-3	5-10	7-15
Dissolved	easily degradable matter	0.7-1.5	3-5
Suspended	slowly degradable matter	0.1-0.3	0.5-1

Table 9.5. Sludge load for anaerobic plants, irrespective of type. Treatment efficiency for COD: 80-90 per cent.

Design by means of sludge age

The purpose of this design is to ensure that the methane producing bacteria can survive in the plant. The sludge age, $\theta_{X,M}$, is given by the following expression:

$$\theta_{X,M} = M_X/F_{SP}$$

where the sludge mass, M_X, is to be considered as being the part of the sludge of the plant which directly carries out an anaerobic removal. In most anaerobic plants it means that it is the sludge which is present in the anaerobic tank.

Expression (9.4) can be used to find the necessary volume, V_2:

$$V_2 = \theta_{X,M} \cdot F_{SP}/X_{B,M,2} \tag{9.4}$$

The sludge age, $\theta_{X,M}$, is found from figure 9.3, the sludge concentration, $X_{B,M,2}$, is a plant parameter which can be chosen/estimated by using Table 9.4 within an interval of 5-15 kg VSS/m³ (7-20 kg COD/m³) whereas the sludge production, F_{SP}, is to be calculated/estimated, see below.

The sludge production can be calculated from the following expression:

F_{SP} = suspended inert organic matter in wastewater + sludge production from the biological process

Using the symbols from figure 9.2 we have:

$$F_{SP} = Q_1 \cdot X_{I,1} + Q_1 \cdot (C_1 - S_{S,3}) \cdot Y_{obs} \quad (9.10)$$

where $S_{S,3}$ is dissolved biodegradable organic matter in the effluent,

C_1 is biodegradable organic matter in the influent.

The overall yield constant, Y_{obs}, varies in the interval 0.05-0.10 kg COD/kg COD, see Chapter 3, Section 3.7.2.

Sludge separation technique	Hydraulic surface load m³ water/(m² tank · h)	Sludge surface load kg SS/(m² tank · h)
Settling, traditional	0.5-1.0	2-4
Lamella sedimentation	1.0-2.0	3-6
Gas/sludge separator, sludge blanket plant	0.5-1.0	-
Fluidized filter, top of tank	0.5-1.0	2-4
Hyper-filtration	0.005-0.040*	-

* m³ water/(m² membrane · h)

Table 9.6. Typical data for sludge separation plants, among others based on /5/,/6/. The figures are based on the average hourly flow rate, $Q_{h,av}$.

Design of sludge separation

Sludge separation is designed on the basis of the hydraulic surface load and the sludge surface load. Table 9.6 lists the loads for a number of separation techniques used in anaerobic plants.

Example 9.3

Design of anaerobic contact plants by using of the sludge age.

The wastewater from Example 9.2 is going to be treated in an anaerobic contact plant. The effluent concentration of dissolved COD is assumed to be 0.3 kg/m³.

The necessary tank volume, V_2, can be found by using Expression (9.4):

$$V_2 = \theta_{X,M} \cdot F_{SP}/X_{B,M,2} \quad (9.4)$$

The sludge age, $\theta_{X,M}$, is estimated at 20 d by means of figure 9.3. The sludge concentration is estimated at 8 kg COD/m³ by means of Table 9.4. From Expression (9.10) we find the sludge production:

$$F_{SP} = Q_1 \cdot X_{I,1} + Q_1 \cdot (C_1 - S_{S,3}) \cdot Y_{obs} \quad (9.10)$$

Here Q_1 = 1,200 m³/d

$X_{I,1} = 0.2$ kg COD/m^3

$C_1 = 2.7$ kg COD/m^3

The effluent concentration of biodegradable organic matter, $S_{S,3}$, can be found from the expression:

$S_{TOT,3} = S_{S,3} + S_{I,3}$

of which $S_{S,3} = S_{TOT,3} - S_{I,3} = 0.3 - 0.1 = 0.2$ kg COD/m^3 as $S_{I,3}$ is assumed to be ~ $S_{I,1}$.

The yield constant, Y_{obs}, is in the interval 0.05-0.10 kg COD/kg COD. It is here estimated at 0.08 kg COD/kg COD.

By substituting the now known/estimated values in Expression (9.10) we find the sludge production:

$F_{SP} = 1{,}200$ m^3/d $\cdot$ 0.2 kg COD/m^3 + 1,200 m^3/d (2.7 kg COD/m^3 - 0.2 kg COD/m^3) $\cdot$ 0.08 kg COD/kg COD = 480 kg COD/d

$\theta_{X,M}$, F_{SP} and $X_{B,M,2}$ are substituted in (9.4) after which the necessary tank volume, V_2, can be found:

$V_2 = (20$ d $\cdot$ 480 kg COD/d$)/(8$ kg COD/m$^3) = 1{,}200$ m^3

(that is, the hydraulic retention time, θ, is 1 d)

If the whole sludge production leaves the plant together with the treated water, the concentration of suspended COD, $X_{COD,3}$, is as follows:

$X_{COD,3} = F_{SP}/Q_1 = (480$ kg COD/d$)/(1{,}200$ m^3/d$) = 0.4$ kg COD/m^3

The total effluent concentration, $C_{COD,3}$, is:

$C_{COD,3} = S_{COD,3} + X_{COD,3} = 0.3 + 0.4 = 0.7$ kg COD/m^3

9.4.2. Design of anaerobic filter plants

Filter plants can be designed on the basis of:

- volumetric load (Expression (9.8))
- sludge load (Expression (9.9))
- sludge specific removal rate, $r_{X,M}$

Design of anaerobic filter area

By means of the mass balance in Expression (9.6), the size of the anaerobic filter can be calculated.

The necessary area of the carrier, A_{2*}, is:

$$A_{2*} = (Q_1 \cdot C_1 - Q_3 \cdot C_3)/(r_{X,M} \cdot X_{2*} \cdot L_2) \tag{9.11}$$

Knowing the sludge specific removal rate, $r_{X,M}$, the sludge concentration in the biofilm, X_{2*}, and the thickness of the biofilm, L_2, the necessary filter area, A_{2*}, can easily be calculated. The only problem is that normally we do not know X_{2*} and L_2.

By using the relationship

$X_{2^*} \cdot L_2 \cdot \omega = X_2 \Rightarrow X_{2^*} \cdot L_2 = X_2/\omega$

expression (9.11) can be transformed to

$$A_{2^*} = \omega \cdot (Q_1 \cdot C_1 - Q_3 \cdot C_3)/(r_{X,M} \cdot X_2) \qquad (9.12)$$

The volume of the anaerobic filter, V_2, is:

We have $A_{2^*} = V_2 \cdot \omega$ which gives

$V_2 \cdot \omega = \omega \, (Q_1 \cdot C_1 - Q_3 \cdot C_3)/r_{X,M} \cdot X_2)$

$$V_2 = (Q_1 \cdot C_1 - Q_3 \cdot C_3)/(r_{X,M} \cdot X_2) \qquad (9.13)$$

The specific filter surface area, ω, and the sludge concentration, X_2, is found in Table 9.4. The sludge specific removal rate, $r_{X,M}$, can be estimated from Table 9.7. Subsequently the area as well as the volume of the filter can be estimated.

The organic matter removed in the filter, $Q_1 \cdot C_1 - Q_3 \cdot C_3$, can approximatly be expressed as $Q_1 \cdot (C_1 - S_{S,3})$,

where C_1 is the biodegradable substance in the influent,
 $S_{S,3}$ is the dissolved biodegradable substance in the effluent.

Hereby the suspended organic substance in the effluent, X_3, is understood to be the yield of the biological removal which is not quite wrong, especially for plants without separate withdrawal of excess sludge.

Hence expression (9.13) can be written:

$$V_2 = Q_1 \cdot (C_1 - S_{S,3})/(r_{X,M} \cdot X_2) \qquad (9.14)$$

It clearly appears from the above design example that the area in anaerobic filters is of less importance. The filter medium functions as the carrier for the sludge (biomass), and diffusional limitations in the film hardly ever occur. It means that what is decisive for the anaerobic filters is the total sludge concentration which can be obtained in the filter whereas the area of the filter medium is of no special importance.

Organic matter in wastewater dominated by:	Removal rate $r_{X,M}$, kg COD(S)/(kg COD(B)·d)		
	15-25°C	30-35°C	50-60°C
Acetic acid	0.7-1.5	2.5-5	3.5-7.5
Dissolved, easily degradable matter	0.3-0.7	1.2-2.5	1.5-3
Suspended slowly degradable matter	0.05-0.15	0.25-0.5	0.5-1.0

Table 9.7. Sludge specific removal rate, $r_{X,M}$, anaerobic plants.

Example 9.4

The design of a fixed anaerobic filter for the treatment of the wastewater used in Examples 9.2 and 9.3. The effluent concentration is assumed to correspond to that in Example 9.3. The volume of the filter is found from Expression (9.14):

$$V_2 = Q_1 \cdot (C_1 - S_{S,3})/(r_{X,M} \cdot X_2) \tag{9.14}$$

From Example 9.3 we have:

$Q_1 = 1{,}200 \text{ m}^3/\text{d}$

$C_1 = 2.7 \text{ kg COD/m}^3$

$S_{S,3} = 0.2 \text{ kg COD/m}^3$

The sludge specific removal rate is estimated at 0.5 kg COD/(kg COD · d) from Table 9.7, and the sludge concentration, X_2, is estimated at 10 kg COD/m^3 from Table 9.4.

By substitution into Expression (9.14) we find:

$V_2 = (1{,}200 \text{ m}^3/\text{d}) \cdot ((2.7 - 0.2) \text{ kg COD/m}^3)/(0.5 \text{ kg COD/(kg COD} \cdot \text{d}) \cdot 10 \text{ kg COD/m}^3)$

$V_2 = 600 \text{ m}^3$

The specific filter area is estimated at 100 m^2/m^3 from Table 9.4. Hence the total area is:

$A_{2*} = V_2 \cdot \omega = 600 \text{ m}^3 \cdot 100 \text{ m}^2/\text{m}^3 = 60{,}000 \text{ m}^2$

(which, as previously mentioned, (only) has an academic interest for the time being, but as this book is written for scientifically oriented persons, it is of course of interest to the reader).

9.4.3. Gas production, anaerobic processes

In connection with the anaerobic removal, gases are produced, including:

- methane
- carbon dioxide
- hydrogen
- hydrogen sulphide
- free nitrogen

The production of methane and carbon dioxide is absolutely dominant compared to other gasses.

The produced amount of methane can be determined by means of a COD-balance as methane represents a certain amount of COD:

$$CH_4 + 2\,O_2 \rightarrow CO_2 + 2\,H_2O$$

that is, 1 mole $CH_4 \sim 2$ mole O_2 or 16 g methane $\sim$ 64 g O_2 = 64 g COD

Hence the conversion factor between COD and methane is

$\nu_{COD,CH4} = 0.25 \text{ kg CH}_4/\text{kg COD}$.

At 25°C it corresponds to a methane gas volume of 0.38 m^3 CH$_4$/kg COD.

From the COD balance, Expression (9.1), we find the methane production, $r_{V,CH4} \cdot V_2$:

$$Q_1 \cdot C_1 - r_{V,X} \cdot V_2 = Q_3 \cdot C_3 \tag{9.1}$$

and on the assumption that $Q_1 \sim Q_3$, the result is:

$$r_{V,CH4} \cdot V_2 = Q_1 \cdot (C_1 - C_3) \, v_{COD,CH4} \tag{9.15}$$

For most anaerobic processes, 90-95 per cent of the removed COD can be recovered as methane, the rest is found in the form of sludge production. The methane gas production can be used for heating or electric power generation. In Table 9.8 we have a number of conversion factors for methane gas.

The carbon dioxide content in the gas from the anaerobic plants depends on the composition of the organic matter (its oxidation step) and the buffer characteristics of the wastewater. The CO_2 content is often 20-30 per cent. Figures found from experience are shown in Table 9.9.

$v_{VSS,COD}$ = 1.3-1.7 kg COD/kg VSS						
$v_{CH4,oil}$ = 0.84 kg oil/m³ CH_4						
$v_{CH4,oil}$ = 1.0 l oil/m³ CH_4						
$v_{fats,CH4}$ = 0.75 m³ CH_4/kg fats						
$v_{carbohydrate,CH4}$ = 0.42 m³ CH_4/kg carbohydrate						
$v_{protein,CH4}$ = 0.47 m³ CH_4/kg protein						
$v_{COD,CH4}$ = 0.25 kg CH_4/kg COD						
$v_{COD,CH4}$ = 0.38 m³ CH_4/kg COD						
1 m³ CH_4 (standard condition, 25°C, 1 atm.) weighs 0.667 kg						
1 m³ oil (standard condition, 25°C, 1 atm.) weighs 850 kg						
1 m³ CH_4 = 35,000 kJ						
1 m³ CH_4 = 9.7 kWh						
1 m³ CH_4 converted to electricity: 30% electricity = 2.9 kWh; 50% heat = 4.9 kWh; 20% loss = 1.9 kWh						

Methane content in gas	%	50	60	70	80	100
Caloriferic value	kJ/m³ gas	17,500	21,000	24,500	28,000	35,000
At electr. prod. — electricity	kWh/m³	1.5	1.7	2.0	2.3	2.9
At electr. prod. — heat		2.5	2.9	3.4	3.9	4.9
At electr. prod. — loss		1.0	1.1	1.3	1.5	1.9

Table 9.8. Conversion factors/stoichiometric coefficients in connection with gas production in anaerobic plants.

Example 9.5

How big is the methane production in the plant from Example 9.3?

The organic matter in the influent and the effluent, respectively, is as follows:

$C_1 = 3.0$ kg COD/m^3
$C_3 = 0.7$ kg COD/m^3

The volume of wastewater, Q_1, is 1,200 m^3/d.

From Expression (9.15) we find the methane production:

$r_{V,CH4} \cdot V_2 = 1{,}200$ m^3/d $(3.0-0.7)$ kg COD/m$^3 \cdot (0.25$ kg CH$_4$/kg COD$) = 690$ kg CH$_4$/d

corresponding to 1,030 m^3 CH$_4$/d (1 atm., 25°C).

0.38		m^3 CH$_4$/kg COD (removal)
0.30-0.38		m^3 CH$_4$/kg COD (input)
0.46-0.62	a)	m^3 gas/kg COD (input)
0.45-0.60	b)	m^3 CH$_4$/kg VSS (input)
0.69-0.92	a) b)	m^3 gas/kg VSS (input)

a) for 65% CH$_4$ in gas
b) 1 kg VSS = 1.5 kg COD

Table 9.9. Figures found from experience with gas production, anaerobic plants

9.4.4. Optimization, anaerobic plants

Quite a few operational problems have to be dealt with in anaerobic plants - just as the case is in other biological plants. Some of these problems as regards start-up, disturbances and toxic substances are discussed below.

9.4.5. Start-up, anaerobic plants

The slow growth of the methane producing bacteria means that the start-up of new anaerobic plants may take a long time. By seeding with anaerobic bacteria, for example from another plant, the start-up period can be reduced.

The start-up period will normally be 2-4 times the sludge age. For a plant at 35°C it means 30-60 d.

The start-up normally takes place in 2 phases:

1. Seeding, recycling of water in the plant approx. one week without any influent wastewater.
2. Step-by-step increasing load, normally starting with approx. 10 per cent of the maximum load. When the content of the volatile fatty acid in the effluent is sufficiently low (200-400 g/m^3, calculated as HAc), the load is increased by 50-100 per cent.

Figure 9.12 shows an example of the start-up of an anaerobic plant.

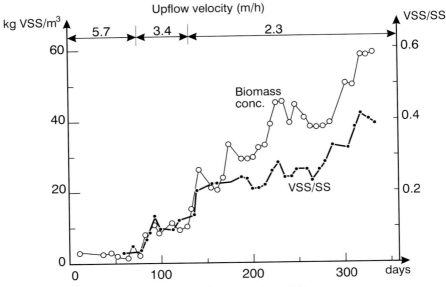

Figure 9.12. Start-up of an anaerobic plant, data from /10/

9.4.6. Disturbances, anaerobic plants

Operational parameters

The operation of an anaerobic plant should be observed by means of a number of measurements. Which and how many measurements to be applied for a given plant should be subject to an evaluation of the plant and the environmental conditions it is subjected to. As shown in Table 9.10, the measurements can be divided into measurements used for process monitoring and process control, as well as functional control.

The process control parameters give notice of problems ahead and may therefore be used to initiate measures to prevent operational problems from occurring. For example, the result of an increasing content of volatile acids may be that the organic load must be reduced, or that in more serious cases, base must be added to the tank to maintain the pH value at a sufficiently high level.

Measurement	Purpose	Normally allowable interval/variation
Process monitoring		
Temperature	Maintain constant temperature	± 1°C/d
Load with organic matter (kg COD/d)	Prevention against overload	+ 50 per cent dissolved COD/d + 100 per cent suspended COD/d
Process control		
Volatile acids, concentration	Detection of instability in the process	Total: 200-500 g as HAc/m^3 HAc: 200-500 g as HAc/m^3 HPr: 50-100 g as HAc/m^3
Gas production	Control of methane bacteria	± 20%/d (a function of influent load of organic matter)
pH	Control of instability	6-7. Variation ± 0.5/d
Functional control		
Effluent concentration, organic matter	Control of treatment efficiency	Variation ± 10%/d
Gas production/quality	Control of produced amount of methane	Variation ± 20%/d. Methane content 60-75%
Sludge quality (% volatile)	Control of sludge stabilization function	60-70% normal, variation ± 5%

Table 9.10. Operational parameters for anaerobic processes.

Operational problems

Most operational problems in anaerobic plants are in some way or other caused by the methane producing bacteria. Either the problem is that these bacteria function badly, or the problem quickly causes them to malfunction. The operational problems discussed in this section can be divided into 3 types, depending on how they intervene in the ecological system of the anaerobic tank:

– Disturbance of the balance in the 3-stage-process (see figure 3.15).
– Washout of methane bacteria.
– Inhibition of the biomass (including the methane bacteria).

Disturbances

Disturbance of the balance in the 3-stage-process, consisting of hydrolysis, acid production and methane production, is a frequent cause of problems.

A 3-stage-process with two slow stages first and last and a fast intermediate stage will not normally cause operational problems as long as the substrate passes through all three stages. It means that suspended organic matter can be applied to the process in rather varying loads (which is also the experience from the traditional anaerobic sludge stabilization). If, however, dissolved organic matter is suddenly applied, this will easily cause problems with inhibition of the methane producing bacteria due to the quick conversion into fatty acids. Such a disturbance will occur fairly quickly (within one hydraulic retention time, that is, 1-24 hours).

Washout

The washout of methane bacteria will take place when the sludge age becomes too low. This may take place, partly by extra amounts of suspended solids in the influent (both organic and inorganic) and by reduction of the sludge mass in the tank, for example due to bad retention of the sludge or increased sloughing off of the sludge from the filter media. The washout may also take place even if the organic load of the process is unchanged. Disturbances caused by washout will occur relatively slowly (within one sludge age, that is, perhaps 1-2 weeks).

Inhibition

Inhibition of the biomass may take place by means of internally produced inhibition (fatty acids, ammonia, pH) or by means of substances supplied from the outside environment (sulphate, ammonia, metals, specific organic materials). It will be characteristic for the external inhibition that the whole biomass is inhibited however, still in such a way that the greatest effect can be observed on the methane producing bacteria. External inhibition will normally occur quickly (in the course of a few hours), whereas produced inhibition occurs somewhat slower. Table 9.11 gives a brief survey of typical disturbances, and how they influence the anaerobic process. Notice that an observed abnormality in a control parameter may often be caused by several different interferences (simultaneous or isolated). For example, the reason for a quick increase in the content of volatile acids may be

– increased load of dissolved organic matter,
– temperature variations,
– (heavy) pH increase,
– pH decrease,
– inhibition caused by hydrogen sulphide (sulphate in the influent), ammonia, or other toxic substances.

It can further be seen from the table that the pH is an extremely bad control parameter (apart from the fact that it is easy to measure).

Inhibition of anaerobic wastewater processes with substances from the outside follows a pattern which deviates somewhat from what we know from anaerobic sludge stabilization. Type of substance as well as exposure time and plant layout play an essential role for the inhibition phenomenon in anaerobic wastewater processes. Table 9.12 gives a survey of limits for inhibition for a number of toxic substances.

Influence (so strong that the process is influenced)	Biological effect	Primary influence of control parameters, change/velocity		
		Volatile fatty acids	Gas production	pH
Increased load with dissolved organic matter	3-stage-process is disturbed	is increased/ quickly	is increased/ quickly	is reduced/ moderately
Increased load with suspended organic matter	3-stage-process is disturbed/ washout	is increased/ moderately	is increased/ moderately	
Increased input of inorganic suspended solids	washout	is increased/slowly	is reduced/slowly	
Increased amount of suspended solids in effluent	washout	is increased/slowly	is reduced/slowly	
Temperature variation	3-stage-process is disturbed	is increased/quickly	is reduced/quickly	is reduced/ moderately
pH increase in influent (and reactor)	(inhibition)	(is increased/ quickly)	(is reduced/ quickly)	is increased/ quickly
pH decrease in influent (and reactor)	inhibition	is increased/ quickly	is reduced/quickly	is reduced/quickly
Sulphate in influent	inhibition	is increased/ quickly	is reduced/quickly	is reduced/ moderately
Ammonium in influent	inhibition	is increased/ quickly	is reduced/quickly	is reduced/ moderately
Organic nitrogen in influent	inhibition	is increased/ moderately	is reduced/ moderately	is reduced/slowly
Toxic substances in influent	inhibition	is increased/ quickly	is reduced/quickly	is reduced/ moderately

Table 9.11. Disturbances in anaerobic processes.

For certain inhibitors, anaerobic wastewater processes can withstand much higher concentrations than anaerobic sludge processes of wastewater. The reason is that the hydraulic retention time for water is much shorter than the sludge age in the anaerobic wastewater treatment methods. The micro organisms are therefore only exposed to toxic substances for a shorter period of time. In this connection reactors with plug flow have a special advantage, for example fixed filters.

In Table 9.12 we also note the difference between single doses and continually applied inhibitors. For substances where the microorganisms can adjust (for example cyanide), they can withstand considerably higher concentrations by continual supply than by single doses.

For substances with a direct toxic effect and with little possibility of adaption (for example nickel) we see the adverse effect, where a short exposure of a rather high concentration does not yield any special effect, whereas a constant exposure at lower concentrations will cause inhibition in the long run. For almost all toxic substances in moderate doses it applies, however, that in the event of a long-term exposure of the sludge, a sludge will be developed (selected) which can better withstand the toxic substance.

Parameter	Inhibition by single dose	Inhibition by continuous supply
pH	< 6 > 8	< 5 > 8.5
Ammonia, free, NH_3	> 100 g N/m^3	> 200 g N/m^3
Hydrogen sulphide, H_2S	> 250 g/m^3	> 1000 g/m^3
Cyanide, CN^-	> 5 g/m^3	> 100 g/m^3
Trichloromethane	> 1 g/m^3	> 50 g/m^3
Formaldehyde	> 100 g/m^3	> 400 g/m^3
Nickel, Ni	> 200 g/m^3	> 50 g/m^3

Table 9.12. Anaerobic wastewater treatment. Inhibition /7/,/8/,/9/.

References

/1/ Berg, L. van den and Lentz, C.P. (1980): Effects of Film Area-to-Volume Ratio, Film Support, Height and Direction of Flow on Performance of Methanogenic Fixed Film Reactors. In: Anaerobic Filters: An Energy Plus for Wastewater Treatment, Proceedings of the Seminar/Workshop January 9-10 Florida, pp. 1-10. Argonne National Laboratory, Argonne, Illinois. (ANL/CNSV-TM-50).

/2/ Mosey, F.E. (1981): Anaerobic Biological Treatment of Food Industry Waste waters. *Water Pollut. Contr.*, **80**, 273-291.

/3/ Jewell, W.J., Switzenbaum, M.S. and Morris, J.W. (1981): Municipal Wastewater Treatment with the Anaerobic Attached Microbial Film expanded Bed Process. *J.WPCF,* **53**, 482-490.

/4/ Friedman, A.A. and Tait, S.J. (1980): Energy Recovery from Anaerobic Rotating Biological Contractor (AnRBC) treating High Strength Carbonaceous Wastewaters. In: Proceedings: First national Symposium/Workshop on Rotating Biological Contractor Technology, Pensylvania, Febr. 4-6, pp. 759-789. University of Pittsburgh, PA.

/5/ Winther, L., Henze, M., Linde, J.J. and Jensen, H.T. (1998): Spildevandsteknik. (Wastewater Engineering), Polyteknisk Forlag, Lyngby, Denmark.

/6/ Lettinga, G. et al. (1981): Anaerobic Treatment of Wastes containing Methanol and Higher Alcohols. *Water Res.,* **15**, 171-182.

/7/ Clark, R.H. and Speece, R.E. (1971): The pH Tolerance of Anaerobic Digestion. In: Advances in Water Pollution Research. Proceedings of the 5th International Conference, San Francisco and Hawaii 1970, volume 1, pp. II-27/1-14. Pergamon Press, London.

/8/ Yang, I. et al. (1981): The Response of Methane Fermentation to Cyanide and Chloroform. *Water Sci. Technol.,* **13**, (2), 977-989.

/9/ Parkin, G.F. et al. (1980): A Comparison of the Response of Methanogens to Toxicants: Anaerobic Filter vs. Suspended Growth Systems. In: Anaerobic Filters: An Energy Plus for Wastewater Treatment. Proceedings of the Seminar/Workshop, January 9-10, 1980 Florida, pp. 37-57. Argonne National Laboratory, Argonne, IL. (ANL/CNSV-TM-50).

/10/ Yoda, M., Kitagawa, M. and Miyaji, Y. (1989): Granular Sludge Formation in the Anaerobic Expanded Micro-carrier Bed Process. *Water Sci. Technol.,* **21**, (4/5), 109-120.

/11/ Switzenbaum, M.S. (ed.) (1991): Anaerobic Treatment Technology for Municipal and Industrial Wastewaters. *Water Sci. Technol.,* **24**, (8).

/12/ Henze, M. and Harremoës, P. (1983): Anaerobic Treatment of Wastewater in Fixed Film Reactors - A Literature Review. *Water Sci.Technol.,* **15**, (8/9), 1-101.

/13/ Demoulin,G.,Pitman,A.R,Sojkova,V. and Hrich,R. (1999): Betriebserfahrungen mit der optimierten Primärschlannfermentation (PSF) und Analyse der bemessungtechnichen Auswirkungen auf die biologische Nährstoffentfernung (Operational experiences with primary sludge fermentation and analysis of its effect on the design of nutrient removal processes). *Korrespondenz Abwasser,* **46**, 1076-1084.

/14/ Scheer, H. (1994): Vermehrte biologische Phosphorelimination – Bemessung und Modelierung in Theorie und Praxis – (Biological phosphorus removal – design and modelling in theory and practice). Universität Hannover, Germany. (Veröff. des Instituts für Siedlungswasserwirtschaft und Abfalltechnik, Heft 88).

/15/ Münch,E.v. and Koch, F.A. (1999): A survey of prefermenter design, operation and performance in Australia and Canada. *Water Sci. Technol.,* **39**, (6), 105-112.

Anaerobic sludge digesters, Singapore. The microbial processes in anaerobic sludge and wastewater treatment are similar, the difference is the concentration of suspended solids in the process.

10 Treatment Plants for Phosphorus Removal from Wastewater

By Erik Arvin and Mogens Henze

Phosphorus is one of the vital macro nutrients for living organisms. Phosphorus in itself is non-toxic. The problem with phosphorus arises when there is an excessive input to aqueous recipients with the wastewater together with other nutrients, such as nitrogen. The effect is eutrophication caused by a heavy growth of algae with a subsequent, unwanted disturbance of the ecological balance of the recipient. A consequence is, for example, that the living conditions for fish are deteriorated due to oxygen depletion, etc. Moreover the recreational value of the waters is reduced due to the turbid appearance and the smell of the water. If the water of the recipient is also used as raw water for water supplies, serious technical and financial complications for the treatment of drinking water usually arise due to the unwanted tastes and odours produced by the algae and the organic materials.

In order to limit or prevent eutrophication, the phosphorus discharged with the wastewater must be reduced. The discharged concentration of phosphorus can be limited by reducing or eliminating the content of phosphorus in detergents. The most widespread intervention is, however, the introduction of some kind of treatment process.

10.1. Mass balances for phosphorus removal processes

No water treatment processes make phosphorus vanish into the air. The purpose of all the processes is to convert dissolved phosphates into suspended phosphorus which can be retained in a separation process, for example in a settler.

Phosphorus in treatment plants is always phosphate in one form or the other. In figure 10.1 we see a typical distribution of phosphorus components in municipal wastewater before and after a biological treatment. Polyphosphates are hydrolyzed to orthophosphates and the major part of dissolved organic phosphates is also degraded to orthophosphates by the biological degradation of organic matter.

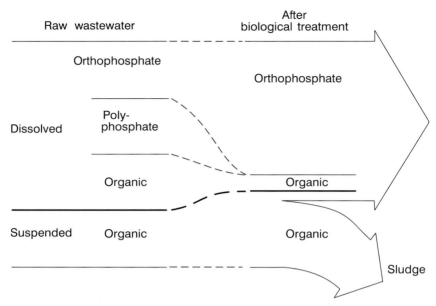

Figure 10.1. Phosphorus fractions in wastewater.

Suspended phosphates are normally considered to be organic phosphates, but they may also include chemically precipitated orthophosphates and biologically bound polyphosphates. The distribution of different phosphate fractions in the suspended phosphate is normally less interesting as the suspended solids are separated from the water by separation methods which are not sensitive to the detailed composition of the particles but rather to their size, charge, density etc.

The mass balance for a treatment plant for phosphorus removal looks as follows when using the symbols in figure 10.2:

$$Q_1 \cdot C_{TP,1} + Q_4 \cdot C_{TP,4} = Q_2 \cdot C_{TP,2} + Q_3 \cdot C_{TP,3} \tag{10.1}$$

where C_{TP} is the concentration of total phosphorus (for example g P/m^3)

The precipitant does not normally contain phosphate, that is, $C_{TP,4} = 0$.

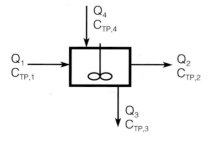

Figure 10.2. Basic layout, treatment plant for chemical phosphorus removal.

The concentration of phosphorus in the effluent, $Q_2 \cdot C_{TP,2}$, normally consists of a dissolved part, $Q_2 \cdot S_{TP,2}$, and a suspended part, $Q_2 \cdot X_{TP,2}$

$$Q_2 \cdot C_{TP,2} = Q_2 \cdot S_{TP,2} + Q_2 \cdot X_{TP,2} \qquad (10.2)$$

For the phosphorus sludge it normally applies that the amount of suspended phosphorus is much larger than the amount of dissolved phosphorus, that is,

$$Q_3 \cdot X_{TP,3} > Q_3 \cdot S_{TP,3}$$

hence

$$Q_3 \cdot C_{TP,3} \sim Q_3 \cdot X_{TP,3}$$

Example 10.1

A treatment plant treats 11,000 m³ of municipal wastewater per day. The content of phosphorus is 12 g P/m³.

The plant is an activated sludge plant with a surplus sludge production of 4,000 kg SS/d with a content of dry solids of 20 per cent.

Phosphorus is to be removed by simultaneous precipitation to an effluent concentration of 1 g P/m³.

What is the content of phosphorus in the surplus sludge production?

Phosphorus in the surplus sludge, $Q_3 \cdot C_{TP,3}$, is found from the mass balance, Expression (10.1):

$$Q_1 \cdot C_{TP,1} + Q_4 \cdot C_{TP,4} = Q_2 \cdot C_{TP,2} + Q_3 \cdot C_{TP,3} \qquad (10.1)$$

It is assumed that there is no phosphorus in the precipitant, that is, $C_{TP,4} = 0$, and that $Q_4 = 0$ ($<<Q_1$).

Phosphorus in the surplus sludge, $C_{TP,3}$, is found from the expression

$$Q_3 \cdot C_{TP,3} = Q_1 \cdot C_{TP,1} - Q_2 \cdot C_{TP,2}$$

$$C_{TP,3} = (Q_1 \cdot C_{TP,1} - Q_2 \cdot C_{TP,2})/Q_3$$

Here we know

$Q_1 = 11,000$ m³/d

$C_{TP,1} = 12$ g P/m³

$C_{TP,2} = 1$ g P/m³

Q_3 and Q_2 must be calculated.

The surplus sludge, Q_3, is found from

$Q_3 \cdot X_{SS,3} = 4,000$ kg SS/d

$X_{SS,3} = 20\%$ SS, that is, 200 kg SS/m³

$Q_3 = (4,000$ kg SS/d$)/(200$ kg SS/m³$) = 20$ m³/d

From a water balance we find Q_2

$Q_1 = Q_2 + Q_3$

$Q_2 = Q_1 - Q_3 = 11{,}000 \text{ m}^3/\text{d} - 20 \text{ m}^3/\text{d} = 10{,}980 \text{ m}^3/\text{d}$

In practice we can almost always leave out the sludge production, Q_3, in connection with the water balance - but not in sludge mass balances!

By substitution we find $C_{TP,3}$

$C_{TP,3} = (11{,}000 \text{ m}^3/\text{d} \cdot 12 \text{ g P/m}^3 - 10{,}980 \text{ m}^3/\text{d} \cdot 1 \text{ g P/m}^3)/20 \text{ m}^3/\text{d}$

$C_{TP,3} = 6{,}051 \text{ g P/m}^3$

If we assume that the content of dissolved phosphorus in the surplus sludge, $S_{TP,3}$, corresponds to the total phosphorus content in the effluent (it is definitely wrong), the content of phosphorus in the suspended solids of the surplus sludge is calculated:

$X_{TP,3}/X_{SS,3} = (C_{TP,3} - S_{TP,3})/X_{SS,3} = (6{,}051 \text{ g P/m}^3 - 1 \text{ g P/m}^3)/200 \text{ kg SS/m}^3 = 30 \text{ g P/kg SS}$ or 3%.

We see that the content of dissolved phosphorus in the surplus sludge (1 g/m³) is very low compared with the total content (6,051 g/m³), and that the assumption above that the content of dissolved phosphorus was 1 g/m³ is of no practical importance for the end result.

10.2. Mechanisms for chemical/physical phosphorus removal

For chemical and physical processes for phosphorus removal it applies that the rate of both precipitation/adsorption and ion exchange is so high that this has no importance for the achieved treatment efficiency. What happens is that equilibrium is reached in a few seconds or minutes whereas the hydraulic retention time is nevertheless much higher. The only exception is phosphate crystallization where the formation of apatite crystals is a relatively slow process. Hence the design will in this case be based on the actual precipitation rate.

Chemical removal of phosphate takes place in 4 stages (see figure 10.3)

- precipitation
- coagulation
- flocculation
- separation

The precipitation and coagulation stages take place simultaneously as they both occur quickly.

In connection with the precipitation, the dissolved phosphorus, primarily orthophosphates, is converted into solids of a low solubility by the addition of metal salts, typically of iron, aluminium or calcium. The precipitated substance is readily available after the formation of very small particles, that is, colloids which are particles of less than 1 μm.

During the coagulation, the colloids formed by the precipitation and those already present in the wastewater, are agglomerated into so-called primary particles of a diameter of 10-50 μm.

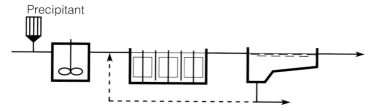

Figure 10.3. Schematic representation of a precipitation process.

The precipitation and coagulation stages are carried out in a mixing unit where the aim is a quick and efficient mixing of the precipitant in the wastewater flow. The precipitant is normally added in the form of a concentrated solution. A simple venturi flume is an excellent mixing device.

Below, the four stages of chemical precipitation and the theories on which they are based will be discussed.

10.2.1. Precipitation

In a simplified form we can say that the precipitation with iron and aluminium is very similar and is based on solubility and the molar ratio between added metal ion, Me, and the orthophosphates present. For calcium as a precipitant it applies that it is a highly pH dependent precipitation where especially the alkalinity of the wastewater is important for the determination of the chemical dosage.

Ferric iron and aluminium as precipitants

Fe^{+++} and Al^{+++} salts are frequently used as precipitants. In terms of precipitation, their behaviour is almost identical, and they can be discussed collectively as Me^{+++}.

In the literature, the following simplified precipitation model is frequently used:

Primary reaction: $Me^{3+} + H_2PO_4^- \rightarrow MePO_4 + 2\,H^+$ (10.3)

Side reaction: $Me^{3+} + 3\,HCO_3^- \rightarrow Me(OH)_3 + 3\,CO_2$ (10.4)

In both cases, the precipitation is accompanied by an alkalinity reduction and hence a pH reduction.

The essence is of course that as large an amount of the metal ions as possible is precipitated as phosphate, and we could get the impression that the hydroxide precipitation is a disadvantage. However, it does have a function in connection with the flocculation (Section 10.2.2) as the hydroxides are precipitated as voluminous particles, and on their way through the liquid these particles catch small particles which could otherwise not settle. Phosphate is also absorbed to the hydroxide particles, and thus they contribute to the overall P-removal.

The precipitation reactions shown in Expressions (10.3) and (10.4) are very simplified. It has been proved that the precipitated material contains calcium ions and in certain cases carbonate ions /1/. An empirical formula for the precipitated material is as follows:

Primary product: $Ca_k Me_m (H_2PO_4)_f (OH)_h (HCO_3)_c$ (10.5)

By-product: $Me_x (OH)_y (HCO_3)_z$ (10.6)

Whether a simple or a complicated product composition is used as a base, the calculation of the equilibrium concentration of dissolved phosphate follows the same principle. The solubility model is based on the following elements a-e:

a. An expression for the total concentration of phosphate.
b. A hypothesis of the type of precipitated phosphate compounds and by-products.
c. Solubility product relations for the precipitated phosphate compound and by-products.
d. The basic equilibrium equations of the phosphate system.
e. The equilibrium equations for the current phosphate complexes.

Examples of the solubility calculations are shown in /2/.

From a practical point of view it should be emphasized that the dominant phosphate components determining the total concentration of dissolved phosphate, S_P, in the pH range 5-11 are:

$$S_P = S_{H2PO4^-} + S_{HPO4^{--}} + S_{CaHPO4(aq)} + S_{CaPO4^-}$$ (10.7)

The last two terms are the concentrations of the phosphate complexes $CaHPO_4(aq)$ and $CaPO_4^-$. The iron and aluminium complexes are unimportant above a pH of 5.

It can be shown that it is possible to set up a generalized expression for the total concentration of dissolved phosphate at equilibrium on the assumption of the presence of the complicated precipitation products shown in the Expressions (10.5) and (10.6.), see /1/.

Ferrous iron as a precipitant

Ferrous iron, Fe^{++}, is frequently used as a precipitant due to its low price compared with ferric iron, Fe^{+++} - see figure 10.4. If it is to function efficiently for the removal of phosphorus, we can use it in two ways:

- oxidation of ferrous iron into ferric iron,
- combination precipitation with calcium.

Oxidation of ferrous iron into ferric iron
In practice, the oxidation takes place by adding Fe^{++} to an aerobic tank in a biological treatment plant:

$$Fe^{++} + 0.25\ O_2 + H^+ \rightarrow Fe^{+++} + 0.5\ H_2O \qquad (10.8)$$

The reaction consumes oxygen and acidity and hence produces alkalinity. The oxygen consumption is 0.14 g O_2/g Fe^{++}, and with the usual iron doses, 20-40 g Fe/m^3, the extra load of the aeration system is insignificant compared with the amount of oxygen which is used for the mineralization of organic matter.

If situations should occur in a biological plant with iron precipitation (simultaneous precipitation) where the concentration of oxygen is zero (overloaded plant), Fe^{+++} can once again be reduced to Fe^{++} resulting in a certain phosphorus resolution. However, this will not take place until any nitrate present has disappeared by denitrification as nitrate, especially by means of bacteria, is able to oxidize Fe^{++} to Fe^{+++}. In this connection it should be mentioned that Al^{+++} cannot be reduced, that is, other things being equal, reducing conditions cannot result in the resolution of phosphorus from aluminium precipitated sludge.

Figure 10.4. Dosing of ferrous iron for phosphorus precipitation. Sjölunda wastewater treatment plant (Malmö, Sweden).

Combination precipitation with calcium.
An efficient precipitation of phosphate with a combination of $Fe^{++} + Ca^{++}$ can be reached. Hence oxidation of Fe^{++} to Fe^{+++} is not necessarily a precondition for the use of Fe^{++} as it was earlier believed. But a relatively high Ca^{++} concentration in the wastewater is important. The precipitation product is presumably a calcium ferrous phosphate compound with ferrous carbonate as a by-product. No one has as yet characterized the precipitation product, the reason for which might be that it is a very complex material consisting of amorphous substances combined with adsorbed material. The precipitation may occur around pH 7, but the pH is frequently increased to 8-10 by addition of $Ca(OH)_2$. There is an apparent need for a more detailed documentation of the chemical conditions in connection with this precipitation process.

Calcium as a precipitant

Phosphate precipitation with calcium, added as $Ca(OH)_2$ or CaO, is misleadingly termed "lime precipitation" in normal English usage. In a simplified form, the precipitation occurs as follows:

Primary reaction: $5\,Ca^{++} + 7\,OH^- + 3\,H_2PO_4^- \rightarrow Ca_5OH(PO_4)_3 + 6\,H_2O$

Side reaction: $Ca^{++} + CO_3^{--} \rightarrow CaCO_3$

The phosphate compound produced by the primary reaction is hydroxyapatite.

However, the conditions are in reality much more complicated. First of all hydroxyapatite is never produced in practice, and secondly the apatite production is the final stage in a series of reactions in which a number of more easily soluble calcium phosphates are produced which may even frequently turn out to be the components which in practice determine the phosphate solubility.

Apatites produced in treatment plants, surface waters, the soil, and even in the enamel of the teeth and bones, are always substituted components where the hydroxyl group may be partly replaced by fluoride F^- (fluoroapatite), the phosphate group partly by carbonate CO_3^{--} (carbonate apatite), and calcium partly by Na^+, Fe^{+++}, Al^{+++}, Mg^{++} and Zn^{++}. By a relatively large substitution with iron, aluminium and carbonate we obtain the calcium iron or calcium aluminium phosphate components discussed under the precipitation of iron and aluminium. In practice, substituted apatites are always end products of the calcium phosphate precipitation.

Prior to the production of apatite we have the production of more or less amorphous and more easily soluble compounds: Dicalcium phosphate, $CaHPO_4$, octacalcium phosphate, $Ca_4H(PO_4)_3$ and tricalcium phosphate, $Ca_3(PO_4)_2$.

A relatively large content in the wastewater of magnesium, Mg^{++}, polyphosphates, and hydrogen carbonate, HCO_3^-, will inhibit the recrystallization into apatite, so that one of the above-mentioned three calcium phosphate compounds determines the phosphate solubility /3/.

10.2.2. Coagulation

The knowledge about the coagulation process is at present very limited compared to the knowledge we have about the flocculation process. This is due to the complexity of the coagulation process.

In the coagulation process, the colloids are agglomerated into larger particles, and in respect of the particle diameter, d_p, they may be characterized in comparison with other particles:

Settleable flocks	:	$d_p > 100$ µm
Primary particles	:	1 µm $< d_p < 100$ µm
Colloids	:	10^{-3} µm $< d_p < 1$ µm
Dissolved material	:	"d_p" $< 10^{-3}$ µm

Colloidal particles can remain stable in this finely dispersed condition. One of the reasons is that the particles are so small that the Brownian movements, caused by the collision of the water molecules with the colloids, dominate over the influence of the gravity on the particles. Furthermore, electric repulsive forces, caused by surface charges on the colloids, prevent the coagulation of the particles and the subsequent sedimentation.

The aggregation of particles which occurs in the coagulation process is the result of a destabilization of the colloids, usually produced by adding chemicals. The destabilization may take place through several different mechanisms depending on the chemical composition of the colloids and the type of the coagulation chemical. However, the understanding of the destabilization mechanisms is as such very incomplete.

Below three concepts are discussed which are important for the coagulation - not just in connection with the precipitation of phosphates from wastewater, but for the coagulation processes in general. These are:

- hydrophillic/hydrophobic properties
- zeta potential
- isoelectric point (isoionic pH)

Hydrophillic/hydrophobic properties
The wastewater contains hydrophillic particles - that is, colloids having a strong tendency to bind or absorb water, and hydrophobic particles - that is, water repellent colloids. Roughly speaking, the organic colloids are hydrophillic and the inorganic colloids are hydrophobic. The reason for the hydrophillic property of the organic colloids is that on their surface there are water absorbing or binding groups

Mechanisms for chemical/physical phosphorus removal

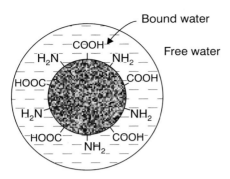

Figure 10.5. Schematic representation of a colloid hydrophillic protein particle surrounded by bound water. /4/.

such as amino groups (-NH$_2$), hydroxyl groups (-OH) and organic acid groups (-COOH), see figure 10.5. The binding or absorption of water occurs through the production of so-called hydrogen bindings. The result is that the hydrophillic particles remain enclosed by a water jacket which follows the particles in their movements. This water jacket is termed bound water. The hydrophobic particles do not have such a water jacket.

Zeta potential

The colloids in the wastewater mainly have a negative charge. This phenomenon can be verified by placing the particles in an electric field. From the field potential, the rate of movement, the viscosity etc., a potential can be calculated, termed the zeta potential, the importance of which will be further detailed below.

The charge of the particles, the primary charge, is a result of the dissociation of end groups on the surface of the particles, such as the amino, hydroxyl and acid groups mentioned above, and/or through the adsorption of ions from the water phase. The charge of hydrophillic colloids is usually caused by the first phenomenon and the charge of the hydrophobic particles by the second phenomenon. The primary charge of the colloids attracts oppositely charged ions, the so-called counter ions, which accumulate in a layer at the surface of the particles, thereby partly neutralizing the surface charge, and an electric double layer is produced.

The layer of ions closest to the particle is called the "fixed" layer, which is surrounded by the so-called "diffuse" layer in which the concentration of counter ions is gradually reduced until it reaches the concentration in the liquid as such. See figure 10.6 concerning the different layers.

Figure 10.7 shows the potential variation around the particle. The plane of shear is the interfase between the proportion of the liquid moving with the particle and the other part of the liquid. The potential at this interface is called the zeta poten-

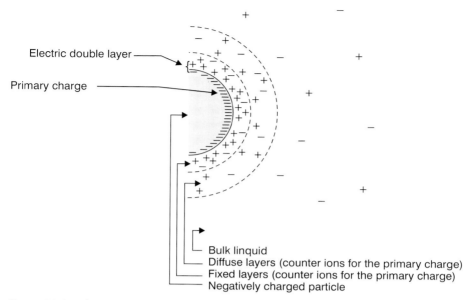

Figure 10.6. Schematic representation of the double layer of ions around a colloid particle with a negative surface charge. From /4/.

tial. And as mentioned above, this is what we measure in conjunction with the movement of the particle in the electric field.

The plane of shear of the hydrophobic colloids is close to the interface between the fixed and the diffuse layer.

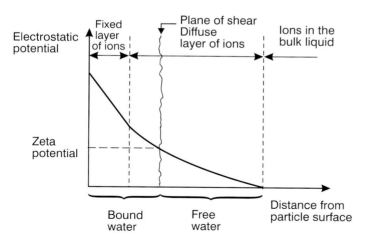

Figure 10.7. Electric potential distribution around a charged hydrophillic colloid particle. /4/.

Mechanisms for chemical/physical phosphorus removal

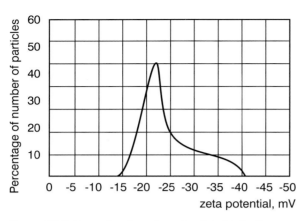

Figure 10.8. Example of a zeta potential distribution curve (and hence the charge distribution curve) for wastewater particles. /5/.

For hydrophillic colloids, the plane of shear coincides with the outer limit of the "bound" water.

The reason why the zeta potential is an important parameter in the coagulation process is that it gives an impression of the extent of the repelling electrostatic forces and hence of the colloidal stability.

However, we should not be led to believe that the conditions have been clarified solely on the basis of this parameter. The conditions are much more complicated than that. For the hydrophillic colloids, the layer of "bound" water is also important for the stability, and in practice there is not just one single zeta potential, but a distribution of potentials depending on the composition of the colloidal particles and their structure, see figure 10.8.

Isoelectric point

The surface charge of inorganic and organic particles can be changed by means of acids and bases, depending on the acid/base properties of the surface groups. Figure 10.9 shows how an organic particle with an amino and a carboxyl group on the

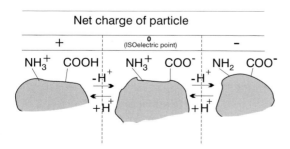

Figure 10.9. Changes of the surface charge of an organic particle by exchange of hydrogen ions.

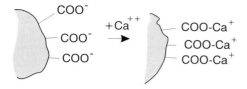

Figure 10.10. Change of surface charge with the exchange of calcium ions.

surface changes the net charge by exchange of hydrogen ions. The pH at which the net charge is zero, the isoelectric point (isoionic pH), depends respectively on the base and acid strength of the amino and carboxyl groups. Normally the isoelectric point for organic wastewater particles lies at a pH of 3-5, that is, the particles have a negative charge at a neutral pH.

The isoelectric point may to a certain extent be manipulated with by adjusting the cation content of the water. Calcium, for example, has an ability to bind more or less specifically to carboxyl groups depending on the chemical composition of the colloids. Thereby the net charge can be changed from a negative to a positive value, see figure 10.10.

Figure 10.11 correspondingly shows how the net charge on an inorganic particle can be changed by proteolytic reactions. The shown surface groups can both behave as acid and base (have amphoteric properties). Also in this case the isoelectric point is completely dependent on the specific chemical composition of the particles.

According to the double layer theory it applies that a colloidal system is stable as long as the repelling electric forces are stronger than van der Waals' attraction forces. The balance between the two forces is illustrated in figure 10.12.

By reduction of the particle surface charge, the extent of the electric field is reduced whereby the particles can come so close to each other when they collide that the van der Waals' forces dominate and the primary particle formation may begin.

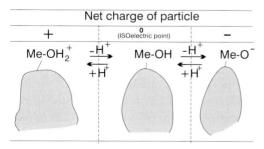

Figure 10.11. Changes of the surface charge of an inorganic particle by exchange of hydrogen ions. Me symbolizes a metal atom.

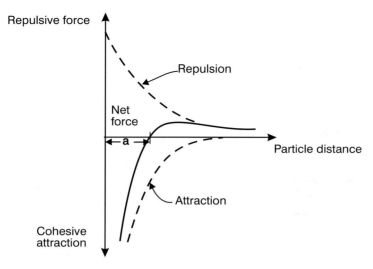

Figure 10.12. Inter-particulate forces as a function of the distance between the particles. /4/. At a particle distance less than a, there will be a net attraction between the particles and hence primary particles can be formed.

Destabilization of colloids

Destabilization of colloids is assumed to take place through one or more of the following mechanisms:

a. Bridging between the colloids via organic or inorganic polymers.
b. Entrapment, or enmeshment, of the colloids.
c. Reduction of the thickness of the electric double layer of the particles resulting in a reduction of the particle repulsion.

as to a As the colloids usually have a negative charge it could be expected that only positively charged (cationic) polymers are able to produce bridging. However, negatively charged (anionic) and nonionic polymers can also be very effective.

The effect of the anionic polymers can be explained in that a multivalent cation, for example calcium ions, is the link between polymer and colloid.

The effect of nonionic polymers may be due to the polar groups which are present in the polymer chain where a positive and negative charge is found around certain atoms. These charged areas may interact with the charged colloids.

It should be stressed that the polymers cannot just function through the bridging mechanism, but they also partly neutralize the surface charge of the colloids and hence reduce the particle repulsion. This is further discussed under item c.

as to b It is a precondition for the entrapment, or enmeshment, mechanism that apart from the colloids there are flocs in the liquid, for example of precipitated metal hydroxide, which catch the colloids.

as to c Reduction of the particle repulsion can only take place in that cations, for example Ca^{++}, Fe^{+++} or Al^{+++} ions, accumulate close to the negative colloid surface. This means a displacement of the isoelectric point for the particle. Acids and bases can exert a similar effect.

10.2.3. Flocculation

During the flocculation process, the primary particles are brought together to larger particles, flocs. The purpose of this sub-process is to increase the particle size of the precipitated material so that it can be separated which typically takes place by sedimentation or flotation.

The flocculation is often carried out in a system as shown in figure 10.13.

In each of the flocculation chambers, the water is subjected to a gentle stirring by means of mechanical flocculators, also called paddle flocculators. This causes the primary particles to collide and aggregate to flocs. The sketch shows three flocculation chambers. In practice, normally 1 to 4 chambers are used.

In some cases the aeration tank is used in activated sludge plants as precipitation and flocculation tank (simultaneous precipitation). Here the aeration system agitates the water. It may result in the problem that the turbulence caused by the aeration is so big that the chemical flocs are partly broken up so that the effluent from the activated sludge plant contains very finely dispersed material which cannot be separated in the subsequent separation unit.

The floc formation during the flocculation takes place because velocity gradients are induced in the water by stirring causing the primary particles to collide and in certain cases stick together.

During the flocculation process, two oppositely directed mechanisms apply, see figure 10.14.

– floc formation whereby primary particles are removed,
– floc break-up whereby primary particles are produced.

The kinetics for those two sub-processes will be discussed below, based on reference /6/.

The floc formation velocity, $r_{V,f}$, equals the velocity, $r_{V,p}$, at which the primary particles are removed, that is,

$$r_{V,f} = -r_{V,p} = K_F \cdot n_p \cdot \Phi \cdot G \tag{10.9}$$

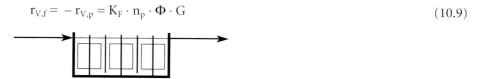

Figure 10.13. Schematic representation of a flocculation unit.

Figure 10.14. Formation of flocs from primary particles. The flocs are assumed to be settleable, whereas the primary particles cannot settle.

where $r_{V,f}$ is the floc formation velocity (unit for example number of primary particles built into flocs/(m³ water · s)),

$r_{V,p}$ is the primary particle formation velocity (unit for example number of primary particles (removed) /(m³ water · s)),

K_F is the floc formation constant, (unit for example m³ water/m³ flocs),

n_p is the number of primary particles per unit volume of water (unit for example number of primary particles/(m³ water)),

Φ is the floc volume ratio, the volume of flocs per unit volume of water (unit for example m³ flocs/m³ water),

G is the mean velocity gradient (unit for example s⁻¹).

If the volumetric number of flocs is n_f and their radius r_f (the flocs are assumed to be spherical and uniform!), we find

$$\Phi = n_f \cdot \frac{4}{3} \pi \cdot r_f^3 = n_f \cdot \alpha_f = n_{p,i} \cdot \alpha_p \qquad (10.10)$$

where α_f is the volume of a floc
α_p is the volume of a primary particle
$n_{p,i}$ is the volumetric number of primary particles in the influent to the flocculation tank

The mean velocity gradient, G, is an expression of the turbulence in the liquid, defined as:

$$G = (W/\mu_a)^{1/2} \qquad (10.11)$$

W is the power supplied per unit volume of liquid,

μ_a is the absolute viscosity of the liquid.

Example 10.2

Find the mean velocity gradient, G, in a 10 m³ flocculation tank where a power input of 8 watts is supplied. The absolute viscosity of the wastewater is assumed to be 0.001 kg/(m · s), see Table 1.10.

Expression (10.11) is used to calculate G

$$G = (W/\mu_a)^{1/2} \qquad (10.11)$$

The volumetric effect $W = 8$ watts/10 m^3 = 0.8 watt/m^3 = 0.8 kg · m^2 · s^{-3}/m^3

The absolute viscosity, μ_a, is 0.001 kg/(m · s)

W and μ_a substituted in Expression (10.11) give

$$G = (0.8 \text{ kg} \cdot \text{m}^2 \cdot \text{s}^{-3}/\text{m}^3)/(0.001 \text{ kg}/(\text{m} \cdot \text{s}))^{1/2}$$
$$G = 28 \text{ s}^{-1}$$

This value corresponds to the order of magnitude which is normally needed in flocculation chambers (10 - 50 s^{-1}).

According to Expression (10.9), the floc formation is first order with respect to n_p, Φ and G.

The concentration in terms of the number of primary particles, n_p, is an inconvenient parameter to operate with as it is difficult to measure. If the precipitated material contains a substance, such as phosphorus, which is homogeneously dispersed in the solid material, the concentration in terms of numbers can be replaced by the concentration of the substance in question:

$$X_{P,p} = M_{P,p} \cdot n_p \qquad (10.12)$$

where $X_{P,p}$ is the concentration of phosphorus in primary particles per unit volume of water (unit for example g P/(m^3 water)),

$M_{P,p}$ is the content of phosphorus in a primary particle (unit for example g P/primary particle)

The floc break-up, corresponding to the production of primary particles (or the release of phosphorus from flocs to primary particles) can be described as

$$-r_{V,f} = r_{V,p} = K_B' \cdot \Phi \cdot G^p \qquad (10.13)$$

where K_B' is a constant with a varying unit depending on the size of p,

p is a dimensionless constant.

In practice, p can be set at 2 in systems with iron, aluminium and calcium precipitation corresponding to a floc break-up taking place relatively faster than the floc formation at high turbulence levels.

By applying the kinetics for floc formation and break-up, the effluent concentration of suspended phosphate can be determined for different reactor types. As shown in figure 10.13, reactors connected in series are especially relevant.

For series-connected ideally mixed tanks of the same volume and with p = 2 in Expression (10.13), and by using the symbols from figure 10.15, we have /6/:

$$FG = \frac{X_{P,p,1}}{X_{P,p,3}} = \frac{(1+K_F \cdot \Phi \cdot G(\theta/n))^n}{1+\frac{K_B \cdot G}{K_F \cdot \Phi}((1+K_F \cdot \Phi \cdot G(\theta/n))^n - 1)} \qquad (10.14)$$

where FG   is the degree of flocculation (-)
        $X_{P,p}$ is the concentration of phosphorus in primary particles (g P/m³ water)
        G   is the mean velocity gradient (s⁻¹)
        θ   is the hydraulic retention time (s)
        n   is the number of ideally mixed tanks in series (-)
        Φ   is the floc volume ratio (m³ flocs/m³ water)
        K_F is a constant (-)
        K_B is a constant (= $K_B' \cdot \alpha_p$) (s)

We can see that $K_F \cdot \Phi$ appears in the expression. For a given dosage, Φ is constant and hence also $K_F \cdot \Phi$. Table 10.1 summarizes values of $K_F \cdot \Phi$ and K_B for different precipitants.

Precipitant	$K_F \cdot \Phi \cdot 10^4$ [-]	$K_B \cdot 10^7$ [s]	Dosage/pH
Al^{+++}	2.85 ± 0,08	3.45 ± 0.16	Al/P = 1.8
Al^{+++} + polymer	2.68 ± 0.11	0.98 ± 0.07	Al/P = 1.8, polymer 0.5 mg/l
Ca(OH)₂	5.58 ± 0.22	2.38 ± 0.15	pH = 11.2
Fe^{++} + Ca(OH)₂	7.68 ± 0.44	4.83 ± 0.40	pH = 8.0, Fe/P = 3

Table 10.1. Summary of $K_F \cdot \Phi$ and K_B for different chemicals used for post-precipitation. Data from /6/. Influent: 9.7 g P/m³. Alkalinity: 2 eqv/m³.

K_B in Expression (10.14) equals K'_B in Expression (10.13) multiplied by the constant α_p which, according to Expression (10.10), expresses the volume of a primary particle.

In order to give an impression of the effect of the type of precipitant, the number of flocculation tanks, the mean velocity gradient and the mean hydraulic retention time, the result of calculations is shown in Figs 10.16 and 10.17 based on the figures in Table 10.1.

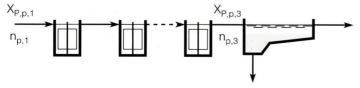

Figure 10.15. Flocculation in ideally mixed tanks connected in series.

Figures 10.16 and 10.17 show that there is a big difference in the flocculation ability of the different chemicals, that it is advantageous to divide the flocculation tank into several chambers, and that in order to obtain a given degree of flocculation, a minimum retention time is required, and that there is an optimum mean velocity gradient, G.

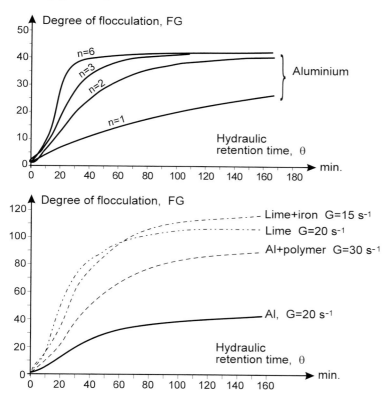

Figure 10.16. Top: Degree of flocculation (FG) as a function of the hydraulic retention time (q) and the number of flocculation chambers (n), for Al^{+++} as precipitant and $G = 20 \, s^{-1}$. Bottom: Degree of flocculation (FG) as a function of the hydraulic retention time (q) and type of precipitant. Flocculation with 2 tanks in series. /6/.

Example 10.3

Phosphorus is precipitated by adding an aluminium sulphate product, 250 g/m³. The aluminium content is 8 per cent on a weight basis. The wastewater, 180 m³/h, contains 14 g P/m³.

1. What is the molar ratio applied?

Added aluminium = $(0.08 \cdot 250 \, g/m^3)/(27 \, g \, Al/mole) = 0.74 \, mole \, Al/m^3$
Phosphate content = $(14 \, g \, P/m^3)/(31 \, g \, P/mole) = 0.45 \, mole \, P/m^3$
The molar ratio Al/P = 0.74/0.45 = 1.64 mole Al/mole P

Mechanisms for chemical/physical phosphorus removal

2. The flocculation takes place in 2 identical flocculation chambers in series. A degree of flocculation of 30 is required.

Find the necessary mean velocity gradient and the tank volume.

From figure 10.17 we can read that a degree of flocculation of 30 can be obtained at a hydraulic retention time of 55 min. and with a mean velocity gradient, G, of 20 s^{-1}. This could be found by using Expression (10.14).

The total flocculation chamber volume is

$V = \theta \cdot Q = 55$ min. $\cdot (180$ m^3/h)/(60 min./h) $= 165$ m^3

Each of the two chambers must therefore have a volume of 82.5 m^3.

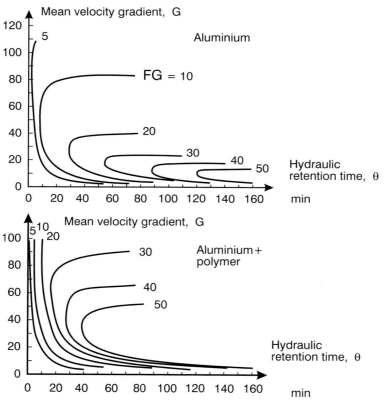

Figure 10.17. Degree of flocculation as a function of mean velocity gradient, G, hydraulic retention time, q, and precipitant. Number of chambers, n = 2, /6/.

10.2.4. Phosphorus binding in soil

There are many points of similarities between the phosphorus binding in a chemical precipitation plant and in the soil medium. Large contents of iron, aluminium and calcium in the soil give a relatively strong phosphorus fixation. However, the binding does not occur as quickly in the soil as when adding chemicals to the chemical treatment plant.

The sketch below illustrates the principle in a model which reasonably well explains the phosphorus binding in the soil.

P		P		P
in soil water	⇌	adsorbed/ion exchanged in the soil	→	precipitated

Phosphorus adsorption/ion exchange to the soil matrix occurs so quickly that equilibrium can be assumed to exist at the usual infiltration velocities. Subsequently the adsorbed/ion exchanged phosphorus will slowly be converted to a precipitated form of relatively low solubility.

The precipitation process can be described as a first order reaction, /8/:

$$\frac{-dX_{P,a}}{dt} = k_1 \cdot X_{P,a} \qquad (10.15)$$

where $X_{P,a}$ is the concentration of adsorbed phosphorus in the soil, for example mg P/kg soil,

k_1 is a first order rate constant.

The concentration of adsorbed/ion exchanged phosphorus in the equilibrium state can be expressed using an adsorption isotherm, /8/:

Langmuir isotherm:

$$X_{P,a} = \frac{B \cdot S_P}{K_1 + S_P} \qquad (10.16)$$

Freundlich isotherm:

$$X_{P,a} = K_{fr} \cdot S_P^{1/n_{fr}} \qquad (10.17)$$

where S_p is the concentration of phosphorus in the soil liquid,

B, K_1, K_{fr} and n_{fr} are constants.

A given adsorption constant may only be used within limited intervals with respect to environmental conditions such as pH, the redox potential and the content of metal in the soil.

Example 10.4

The adsorption capacity of a soil sample for phosphorus in wastewater has been measured experimentally in two ways.

The adsorption capacity was 427 mg P/kg soil, for a phosphate concentration in the soil water of 10 g P/m^3 and 135 mg P/kg soil for a phosphate concentration of 1 g P/m^3.

Within the investigated concentration interval, it is assumed that the adsorption can be described by a Freundlich isotherm where the constant $n_{fr} = 2$.

What is the constant K_{fr}, and what will the amount of adsorbed phosphorus be if the phosphorus concentration in the soil water is 5, respectively 100 mg P/l?

By using Expression (10.17), K_{fr} can be determined:

$$X_{P,a} = K_{fr} \cdot X_p^{1/n_{fr}} \tag{10.17}$$

Substitution based on one of the experiments gives

427 mg P/kg soil = $K_{fr} \cdot$ (10 mg P/l)$^{1/2}$

$K_{fr} = 135$ (mg P/l)$^{-1/2} \cdot$ mg P/kg soil

By using the second experiment we find the same value of K_{fr}.

The amount of adsorbed phosphorus at a soil water concentration of 5, respectively 100 mg P/l, is found from Expression (10.17)

$X_{P,a} = 135 \cdot 5^{1/2} = 300$ mg P/kg soil

$X_{P,a} = 135 \cdot 100^{1/2} = 1.350$ mg P/kg soil

In the second experiment we have measured the maximum adsorption capacity of 500 mg P/kg soil. What's wrong with the above calculation of the adsorption at a soil liquid concentration of 100 mg P/l?

Several things are wrong: First of all, Expression (10.17) only applies in a narrow concentration interval (1-10 mg P/l), and secondly it is unlikely in practice to have a phosphorus concentration in soil water of 100 mg P/l.

10.3. Treatment plants for phosphorus removal

10.3.1. Precipitants

In practice, relatively few types of precipitants are used, that is, salts of aluminium, iron or calcium, for example:

Al^{+++} : $Al_2(SO_4)_3 \cdot 16\,H_2O$
Fe^{+++} : $FeCl_3 \cdot H_2O$ or $FeCl(SO_4)$
Fe^{++} : $FeSO_4 \cdot 7\,H_2O$
Ca^{++} : $Ca(OH)_2 / CaO$

For financial reasons, technical grade chemicals are always used implying a certain content of impurities, soluble as well as insoluble. In many cases combinations of the mentioned chemicals, especially $Ca^{++} + Fe^{++}$, are used.

Furthermore there are polymerized iron and aluminium salts which should have the advantage of reducing the amount of metal ions which are "wasted" through side reactions in the form of hydroxides and to enhance the removal of organic matter. The polymerized aluminium salt has the formula $Al_n(OH)_n^{(3-n)+}$

Through the precipitation we obtain a conversion of dissolved phosphates into solid phosphates as well as a flocculation. The latter sub-process can in certain cases be significantly enhanced by supplementing with organic polymers. Normally, cationic (positively charged) or non-ionic (neutral) polymers are used.

Example 10.5

In an activated sludge plant, the simultaneous precipitation is carried out with ferrous sulphate. 28 g Fe^{++}/m^3 is added whereby 7.75 g P/m^3 is precipitated.

How much oxygen is used for oxidation of ferrous iron into ferric iron, and what is the change in alkalinity?

From Expression (10.8) we find the oxygen consumption:

$$Fe^{++} + 0.25\, O_2 + H^+ \rightarrow Fe^{+++} + 0.5\, H_2O \tag{10.8}$$

from which we find that 0.14 g O_2/g Fe^{++} is used, or totally:

28 g $Fe^{++}/m^3 \cdot$ 0.14 g O_2/g Fe^{++} = 3.9 g O_2/m^3

A change in alkalinity occurs in all three reactions (10.8), (10.3) and (10.4).

(10.3) and (10.4) are:

$$Fe^{+++} + H_2PO_4^- \rightarrow FePO_4 + 2\, H^+ \tag{10.3}$$

$$Fe^{+++} + 3\, HCO_3^- \rightarrow Fe(OH)_3 + 3\, CO_2$$
(10.4)

The change in alkalinity is

Expression	eqv/mole Fe	from component
(10.8)	+1	H^+
(10.3)	−2	$2\, H^+$
(10.4)	−3	$3\, HCO_3^-$

Notice that $H_2PO_4^-$ (in Expression (10.3)) and CO_2 (in Expression (10.4)) do not influence the alkalinity as both components occur in this form at a pH of 4.5.

By the precipitation, 28 g Fe^{++}/m^3 = (28 g Fe^{++}/m^3)/(56 g Fe^{++}/mole) = 0.5 mole Fe^{++}/m^3 is oxidized.

This 0.5 mole Fe^{++} is distributed between the Expressions (10.3) and (10.4). As 7.75 g P/m^3 = (7.75g P/m^3)/(31 g P/mole) = 0.25 mole P/m^3, is precipitated, it means that 0.25 mole of Fe^{++}/m^3 is removed via both expressions.

Hence the change in alkalinity is

0.5 · (+ 1) + 0.25(−2) + 0.25(−3) = −0.75 eqv/m^3.

For high alkalinity wastewater, it has no significant importance as the alkalinity is 3-7 eqv/m^3, see Table 1.10. However, in regions with poorly buffered water the reduction in alkalinity may be important for the biological processes, for example nitrification.

10.3.2. Treatment processes

Depending on where in the treatment process the chemicals are added, chemical precipitation is divided into the process types shown in figure 10.18. These can be mutually combined, and furthermore after-treatment can be added as ponds, infiltration plants, sand filters, etc. By combining precipitant and process types it is realized that in practice there are many ways of performing phosphorus removal.

Direct precipitation is used in connection with the discharge into waters where the consumption of oxygen from the organic matter of the wastewater has no significant impact. If, however, the organic matter is to be removed extensively, preprecipitation, simultaneous precipitation, or post precipitation must be chosen.

Direct precipitation
As precipitant, $Fe^{++} + Ca(OH)_2$ at a pH of 8-10 is normally used. $Ca(OH)_2$ can also be used alone, but in that case the pH must be approx. 11 to obtain effective flocculation. Finally, we can use $Ca(OH)_2$ in combination with 3-5 per cent sea water in relation to the volume of wastewater. Magnesium in the sea water functions as an auxiliary coagulant, whereby the pH only needs to be 9-10, which saves $Ca(OH)_2$.

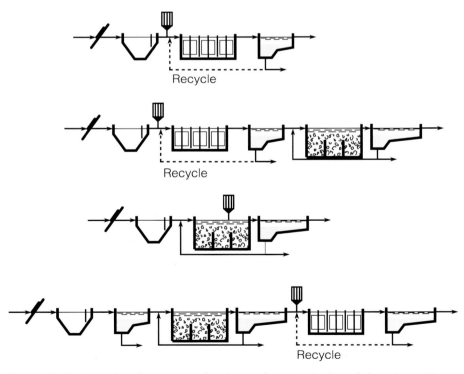

Figure 10.18. Examples of treatment plant layout for precipitation of phosphorus. Top to bottom: Direct, pre, simultaneous and post precipitation./7/.

Preprecipitation

This type of process has become popular as a method to reduce the overload of biological treatment plants. The preprecipitation increases the removal of organic matter to 50-70 per cent compared with the approx. 30 per cent which is removed by ordinary mechanical treatment.

The process may be installed in connection with existing mechanical treatment and often the flocculaton step is left out. The addition of chemicals can be made in the sand filter if it is aerated. The essential problem is a heavy increase in the sludge production and hence the need for extra sludge treatment capacity.

As precipitants, iron or aluminium salts or a mixture of $Fe^{++} + Ca(OH)_2$ are used.

Simultaneous precipitation

This type of process is the most widespread in respect of the removal of phosphorus. Here ferrous sulphate is most commonly used as precipitant which gives a small extra oxygen consumption for the oxidation of ferrous iron into ferric iron, see Expression (10.8). The precipitation process is alkalinity consuming, as appears from Expressions (10.3) and (10.4). For soft water types it may therefore be necessary to add alkalinity in the form of lime. The precipitant can be added in the sand filter or directly in the aeration tank. By simultaneous precipitation we have no direct control of the flocculation, but normally the flocculation of the biological sludge is sufficient to avoid problems. It may be an advantage to combine simultaneous precipitation with preprecipitation and/or post precipitation and finally contact filtration, see below.

Post precipitation

This type of process is normally based on the use of aluminium salts. We get a pure chemical precipitation process without much interference from organic matter or particles. On the other hand we get a chemical sludge which is both voluminous and relatively difficult to dewater.

Contact filtration

If we aim to obtain very low effluent concentrations for phosphorus, we can make a post treatment of the wastewater in a contact filter, see figure 10.20.

In principle, a contact filter is a rapid filter where the precipitant is added in the influent. Hereby a major part of the remaining dissolved phosphate is bound in particles which are caught in the filter. Furthermore, the small particles which did not settle in the chemical plant are removed. In order to prevent the running time of the filter from being too short, it is an advantage for example to use a two-media

Figure 10.19. Plant for simultaneous precipitation. The precipitation chemical is ferrous sulphate. Søholt treatment plant (Silkeborg, Denmark). Simultaneous precipitation is the most commonly used treatment method for phosphorus. An attempt has been made to reduce noise, smell and aerosol problems by silencing the motor (to the left of the bridge), by incorporating the rotor in a concrete case and by providing it with skirts to catch aerosols. The bottom picture shows the storage tank for ferrous sulphate.

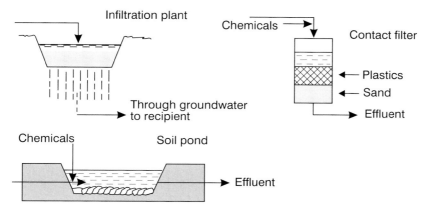

Figure 10.20. Post treatment methods to obtain low phosphorus concentrations.

filter with coarse-grained plastics at the top and sand at the bottom. Due to the relatively large pore volume in the plastics layer, a combined plastics/sand filter can retain considerably larger volumes of sludge than a traditional sand filter.

The precipitation and flocculation mechanisms in a contact filter fully correspond to those in the other precipitation processes. Ferric salts are often used as a precipitant.

Ponds

This type of process can be used for polishing or as the primary part of the phosphorus precipitation process. The function depends on the size of the pond and the chemical dosing. A soil pond as shown in figure 10.20 functions as a large secondary clarifier for the chemical flocs which are formed by the supplementary addition of chemicals. If the soil pond is made sufficiently big, it will, in fact, function as an algae pond. In that case, it may not be necessary to add chemicals to the influent of the pond. On the other hand it may be necessary to establish a chemical precipitation plant for the separation of the algae in the effluent from the pond.

As precipitant, iron and aluminium salts may be used. The chemical sludge has to be removed from the bottom of the pond at (large) intervals.

Infiltration ponds

Here it is the capacity of the soil to bind phosphate which is used. In order for the process to function, a preceding treatment is required where particles and most of the organic matter are removed. The ponds are often operated intermittently, with rest periods of 1-2 weeks. In these periods the pond remains dry, and the upper strata are oxidized which prevents clogging caused by micro-organisms which grow on the organic matter deposited.

These processes are widespread in the southern part of Europe, see figure 10.21.

Figure 10.21. Infiltration plant in Viols-le-Fort (Provence, France). Notice the maquis* in the background. In ancient times there were big forests here which were destroyed by unrestrained cutting. (*maquis is a type of vegetation that is composed of low shrubs with hard drought-resistance).

10.4. Design of plants for phosphorus removal

10.4.1. Chemical precipitation

Design can be carried out in two levels

I. Total phosphorus.

II. Dissolved plus suspended phosphorus.

Design based on total phosphorus is the easier, however, the more unreliable of the two methods.

Design based on total phosphorus

A quick - but rather unreliable - method to determine the process conditions for a given, required effluent concentration of total phosphorus is based on leaving out separate calculations for dissolved and suspended phosphorus.

Table 10.2 states relevant methods to obtain given total phosphorus concentrations and the associated metal ion dosings and requirements of pH, if any.

In the literature, we can find a number of examples illustrating the total phosphorus of the effluent as a function of the molar ratio. Figure 10.22 gives examples

Effluent concentration: 2-3 g P/m³	Biological phosphorus removal Simultaneous precipitation, Fe^{++} or Al^{+++}, MR = 0.8 Preprecipitation, Al^{+++}, MR = 1.
Effluent concentration: 1-2 g P/m³	Simultaneous precipitation, Fe^{++} or Al^{+++}, MR = 1 Preprecipitation, Ca^{++} + Fe^{++}, pH 8-9, MR (Fe) = 1 Direct precipitation, Ca^{++}, pH 10-11 Direct precipitation, Al^{+++}, MR = 1.5 Post precipitation, Al^{+++}, pH 6.5-7.2, MR = 1
Effluent concentration: 0,5-1 g P/m³	Simultaneous precipitation, Fe^{++} or Al^{+++}, MR = 1.5 Simultaneous precipitation + preprecipitation or soil ponds, Fe^{++} or Al^{+++}, MR = 1.5 Post precipitation, Al^{+++}, pH 5.5-6.5, MR = 2 Direct precipitation, Ca^{++}, pH 10-11 + sea water Preprecipitation, Ca^{++} + Fe^{++}, pH 9-10, MR (Fe) = 1.5
Effluent concentration: 0.3-0.5 g P/m³	Simultaneous precipitation, Fe^{++} or Al^{+++} + contact filtration Fe^{++} or Fe^{+++}, MR both processes = 2. Post precipitation, Al^{+++}, pH 5.5-6.0, MR = 2, + contact filtration, Fe^{+++}, MR = 2.

Table 10.2. Example of processes of technical-financial relevance to obtain given effluent concentrations for total phosphorus. The abbreviation MR (molar ratio) means: Number of moles of metal ions added per mole of total phosphorus in the influent.

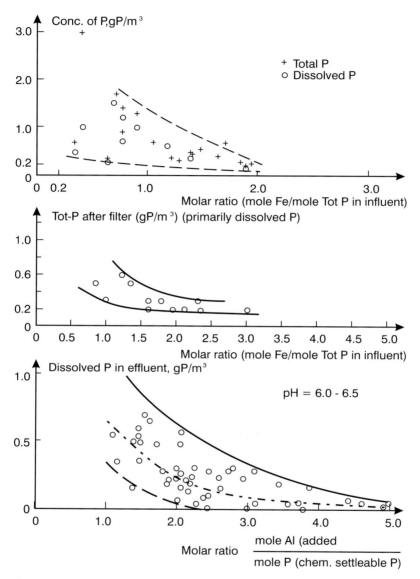

Figure 10.22. Relationship between phosphorus in the effluent and the metal/phosphorus molar ratio for simultaneous precipitation plants. Top: Haslev treatment plant (Denmark) /10/. Middle: Søholt treatment plant (Denmark) /11/. Bottom: American treatment plant, /9/.

of simultaneous precipitation. The phosphorus removal in a 2-media (plastics/sand) contact filter with iron and aluminium precipitations is shown in Tables 10.3 and 10.4.

Precipitant	Molar ratio Me/dissolved P in influent	Influent		Effluent*		Molar removal Δ mole P_{sol}/mole Me
		Tot. P g P/m³	PO_4-P g P/m³	Tot. P g P/m³	PO_4-P g P/m³	
Fe^{+++}	0.8	0.64	0.40	0.31	0.25	0.46
Fe^{+++}	1.1	0.40	0.29	0.24	0.16	0.40
Fe^{+++}	1.1	1.73	1.22	0.90	0.65	0.42
Fe^{++}	1.2	3.87	3.30	1.71	1.55	0.44
Fe^{++}	1.2	1.09	0.76	0.50	0.33	0.47
Fe^{++}	5.5	1.37	0.72	0.44	0.22	0.13
Al^{+++}	0.3	-	1.61	-	1.23	0.79
Al^{+++}	1.1	-	1.70	-	0.83	0.47
Al^{+++}	1.4	-	1.14	-	0.34	0.50
Al^{+++}	6.9	-	0.22	-	0.05	0.11
Al^{+++}	7.7	0.58	0.27	0.19	0.05	0.11

Table 10.3. Phosphorus removal in contact filter by precipitation with Fe^{+++}, Fe^{++} and Al^{+++}. The influent is the effluent from simultaneous precipitation with ferrous sulphate /1/. *pH in the effluent is approx. 7.

	Suspended solids g SS/m³	Suspended-P g P/m³
Effluent simultaneous precipitation (= influent contact filter)	10 ± 3	0.44 ± 0.18
Effluent contact filter	6 ± 3	0.21 ± 0.09

Table 10.4. Operating results in respect of suspended solids and phosphate from simultaneous precipitation with ferrous sulphate followed by contact filtration. /1/.

Design based on dissolved plus suspended phosphate
The content of total phosphorus, C_P, can be determined as the sum of dissolved phosphorus, S_P, and suspended phosphorus, X_P.

$$C_P = S_P + X_P \tag{10.18}$$

Theoretical calculations of dissolved phosphorus, S_P, after chemical precipitation processes are still uncertain /1/ and /13/. The alternative is to use existing empirical measuring results for dissolved phosphorus from chemical precipitation plants. The phosphorus concentrations are normally stated as a function of the metal/phosphorus molar ratio which means that the effects of pH, calcium and alkalinity cannot be evaluated.

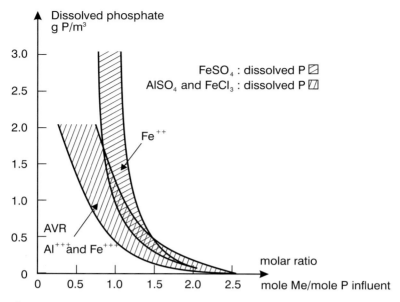

Figure 10.23. Average concentration of dissolved phosphate, S_P, in the effluent from a simultaneous precipitation plant as a function of the metal dosing (Me/P). pH and calcium variations in the wastewater have not been taken into consideration. The importance of these is shown in figure 10.24. The oxygen saturation has typically been 25 per cent. Treatment of wastewater from Lundtofte, (Denmark) /1/.

Examples of the relationship between dissolved phosphorus and the metal dosing, the pH, the calcium concentration and alkalinity, are given in Figs 10.23, 10.24 and 10.25 for different precipitation processes.

The content of suspended phosphorus, X_P, depends on the flocculation and the subsequent settling.

The flocculation is designed by means of the values in Table 10.5.

The concentration of suspended solids (particles) in the effluent from the final settling is not solely determined by the flocculation, but to a large extent also by the design of the settling tank and the properties of the suspended solids.

Based on experience, figures found for the concentration of suspended solids - see Table 10.6 - and the phosphorus content in the suspended solids, the concentration of suspended phosphorus in the effluent, X_P, can be determined.

The phosphorus content in the suspended solids can be determined by means of a sludge balance as illustrated in Example 10.1.

Example 10.6

Phosphorus has to be removed from municipal wastewater (1,200 m³/d) by simultaneous precipitation with ferric chloride. The effluent requirement is 1.5 g P/m³ and the phosphorus concentration in the influent to the plant is 13 g P/m³.

How much iron salt is going to be used for the precipitation process?

It is assumed that the effluent from the plant contains 15 g SS/m³ with a phosphorus content of 3 per cent. That is, suspended phosphorus in the effluent, X_P, is 15 g SS/m³ · 0.03 g P/g SS = 0.45 g P/m³.

From Expression (10.18) the allowable dissolved phosphorus in the effluent, S_P, can be determined.

$$C_P = S_p + X_P \qquad (10.18)$$

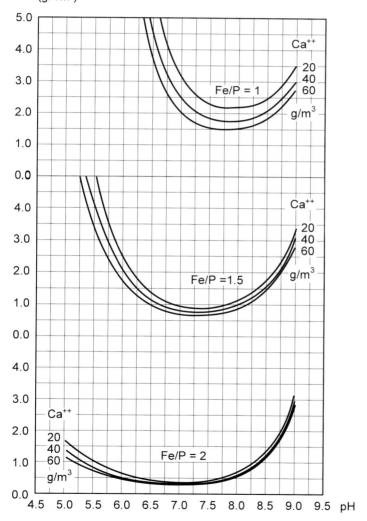

Figure 10.24. Simultaneous precipitation with ferrous sulphate, Fe(II). Dissolved phosphate as a function of pH and calcium for varying Fe/P-molar ratios: 1.0, 1.5 and 2.0. /4/

Design of plants for phosphorus removal

Precipitant	First chamber $G(s^{-1})$	Intermediate chamber $G(s^{-1})$	Last chamber $G(s^{-1})$
Al^{+++}	40-50	15-25	10
$Ca(OH)_2$	30-40	15-20	10
$Fe^{++} + Ca(OH)_2$	25-35	10-20	10
Al^{+++} + polymer	50-70	30-40	10

Precipitant	θ, total hydraulic retention time (min.)
Al^{+++}	40
$Ca(OH)_2$	20
$Fe^{++} + Ca(OH)_2$	20
Al^{+++} + polymer	30

Table 10.5. Mean velocity gradient, G, and total hydraulic retention time. /6/.

Plant type	SS g/m³	P in SS %
Preprecipitation (effluent primary settling)	30-50	2-3
Preprecipitation (effluent secondary settling)	10-20	1-2
Simultaneous precipitation	10-20	2-4
Post precipitation	5-10	10-20
Contact filtration	3-8	2-4
Soil pond after simultaneous precipitation	5-10	2-4
Biological phosphorus removal	10-20	2-4

Table 10.6. Figures based on experience for the concentration of suspended solids in the effluent from settling tanks and the phosphorus content in suspended solids.

$1.5 = S_P + 0.45$

$S_P = 1.05$ g P/m³

From figure 10.23 we find that in order to obtain $S_P = 1.05$, the molar ratio must be 0.7-1.2 mole Fe^{+++}/mole P.

Here a molar ratio of 1.1 mole Fe^{+++}/mole P, is chosen, that is,

mole Fe^{+++}/mole P = 1.1

mole Fe^{+++} = 1.1 · mole P

$= 1.1 \cdot (13$ g P/m³$)/(31$ g P/mole$)$

$= 0.46$ mole Fe^{+++}/m³

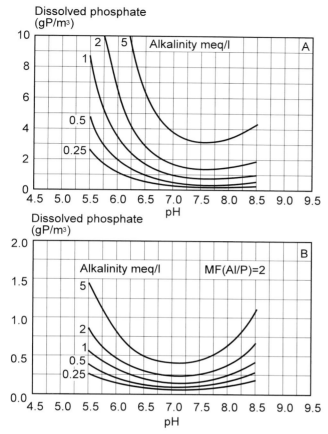

Figure 10.25. Post precipitation with Al(III). Dissolved phosphate as a function of pH and alkalinity. Calcium activity = 40 g/m³. A: Al/P molar ratio 1. B: Al/P molar ratio 2. Notice the difference between the scale of the ordinate axis of curves A and B. /4/.

The dosing of ferric chloride must be 0.46 mole = 0.46 mole · 162.5 g/mole = 75 g $FeCl_3$ (100% pure)/m³. If the technical substance contains for example 32% $FeCl_3$, the dosing is 75 g/m³/0.32 = 234 g/m³.

10.4.2. Phosphorus binding in soil

Frequently, the phopherous adsorption capacity of a soil is characterized by the maximum capacity, $X_{P,a,max}$, possibly supplemented by the capacity at a given phosphorus concentration in the soil water, for example 5 g P/m³.

There is a big difference in the ability of the different soil types to adsorb phosphorus. The maximum capacity may vary from 20 mg P/kg in sandy soil to 500 mg P/kg in clay.

In Table 10.7, average figures for 35 soil samples (top soils) in the state of New York, USA are shown. At the same time, the CEC value for the soil samples in question is stated. CEC means "Cation Exchange Capacity" and expresses the ability of the soil to bind cations, for example NH_4^+, K^+, Ca^{++}, Fe^{+++}, Al^{+++}.

By solely considering the phosphorus adsorption/ion exchange in the soil and not the associated slow chemical binding of adsorbed/ion exchanged phosphorus, we get a conservative estimate of the phosphorus binding capacity of the soil. In /8/ a phosphorus transport model is shown which includes the chemical binding. The model is based on the expressions for phosphorus binding in the soil as shown in Section 10.2.4. It should be noted, however, that the experimental basis for estimating the precipitation velocity is extremely vague.

Depth	Soil density	$X_{P,a,max.}$	$X_{P,a}$ at 5 g P/m^3	Cation Exchange Capacity, CEC
cm	g/cm^3	mg P/kg	mg P/kg	meqv/100 g
15	1.3	457	305	18.7
70	1.6	429	334	8.8
94	1.7	189	122	6.5

Table 10.7. Max. phosphorus adsorption in soil, $X_{P,a,max.}$ and phosphorus adsorption at 5 g P/m^3 in soil water, $X_{P,a}$. The average for 35 top soils in the state of New York, USA, /14/.

Example 10.7

A root zone process for the treatment of wastewater holds 180 m^3 of soil with a density of 1.7 kg/l. The maximum adsorption capacity, $X_{P,a,max.}$, is 110 mg P/kg soil.

The plant is loaded with 10 m^3 of wastewater per day with a phosphorus content of 11 g P/m^3.

How long time does it take before the maximum adsorption capacity for phosphorus is used up?

It is assumed that the effluent contains 5 g P/m^3. Adsorption per day is

10 m^3/d · (11 − 5) g P/m^3 = 60 g P/d.

The maximum adsorption capacity is

180 m^3 soil · 1,700 kg/m^3 soil · 0.110 g P/kg = 33,660 g P

corresponding to

(33,660 g P)/(60 g P/d) = 561 d.

We see that if the adsorption of phosphorus was the only possibility, the plant would only be able to function for about a year. Due to precipitation the time of operation is normally much longer.

10.5. Operation of plants for phosphorus removal

Operational problems will be discussed under the following headings:

a. Safety problems.
b. Mechanical and measuring problems.
c. Process problems.

a. Safety problems

The precipitants are either moderately strong acids, such as Fe^{+++}, Fe^{++} and Al^{+++} or, in respect of $Ca(OH)_2/CaO$, relatively strong bases. Therefore the chemicals should be handled with caution in due respect of the danger of eyes and skin effects. For hydrated lime, $Ca(OH)_2$, and unhydrated lime, CaO, it further applies that the powdered substances may give severe dust problems. To safeguard against dust problems, closed dosing systems should be used.

b. Mechanical and measuring problems

Corrosion may be a problem when using iron and aluminium salts. Plastics are suitable materials for pipes and containers. Furthermore, clogging problems should be observed. Precipitants are technical grade products with a certain content of insoluble material. These materials tend to clog pipe bends, valves etc. Besides, the insoluble material may cause damaging wear in pumps. Aluminium sulphate, for example, contains aluminium oxides which are excellent grinding agents.

In chemical precipitation systems, the measuring of pH is essential for the process monitoring and/or control. Here it should be observed that the coating of the pH electrode may lead to erroneous measurements and hence erroneous dosings. Coating problems are also found when for example measuring conductivity, turbidity, oxygen, etc.

c. Process problems

Process problems are here understood as problems in complying with the effluent requirements for phosphorus. In this respect it is important to distinguish between dissolved and suspended phosphorus as the measures to be taken for the two fractions of phosphorus are different.

In respect of **simultaneous precipitation** with iron, usually Fe^{++}, there are two factors which may cause problems, that is, lack of oxygen control in the biological-chemical reactor and excessive turbulence by the aeration.

By overloading with organic matter, anaerobic conditions will occur in the simultaneous precipitation system. The result is a reduction of Fe^{+++} to Fe^{++}, and this again may result in a certain resolution of phosphorus depending on the concentration of calcium ions in the water. The resolution is especially important around pH 7 and lower pH values.

Aerobic conditions can be reestablished by reducing the supply of organic matter or increasing the aeration capacity. The aeration capacity may for example be increased by blowing more air through the diffusors in activated sludge plants. This may, however, cause another problem in that the increased turbulence destroys the biological-chemical flocs to the effect that the content of suspended phosphorus in the effluent will increase. It appears from Table 10.5 and figure 10.17 that optimum flocculation occurs at a mean velocity gradient, G, of 10-50 s^{-1}. The G value in an aeration tank in an activated sludge plant may, however, be 100-200 s^{-1}, in other words, considerably above the optimum value. It should therefore be endeavoured to transfer the necessary oxygen with as little turbulence as possible, for example by using fine bubble aeration instead of coarse bubble aeration.

In respect of simultaneous precipitation with Fe^{++} it should finally be mentioned that in medium loaded plants, that is, B_X = 0.2-0.4 kg BOD/(kg SS · d), considerable foaming problems have been observed in the aeration tank due to the development of a certain genus of micro-organism (actinomycetes). The problem does not occur in low-loaded plants /12/.

When using Al^{+++}, there will be no problems with phosphorus resolution under anaerobic conditions.

Finally it should be mentioned that a pH control of the simultaneous precipitation process is important at low dosings, that is, a molar ratio around 1.

The biggest process problem at **post precipitation and contact filtration** with iron and aluminium salts is the great influence of the alkalinity of the water at relatively low dosings. In order to obtain low alkalinity, and hence low phosphate concentrations, acid must be added. It may be done by adding, for example, sulphuric acid or hydrochloric acid, more dosing agent (which in itself is acid), or by establishing nitrification in the biological process during which alkalinity is consumed.

Another method to lower the phosphate concentration is to recycle chemical sludge from the post precipitation to the biological process, thus combining simultaneous precipitation and post precipitation and utilizing the fact that the sludge from the post precipitation has a binding capacity in respect of phosphorus at the somewhat higher phosphate concentrations which are found in the simultaneous precipitation plant compared with post precipitation.

Post precipitation may also cause problems in relation to suspended phosphorus. One reason may be that the operation is not carried out at an optimum G value. Apart from optimizing the turbulence conditions, organic polymer may be added which enhances the mechanical strength of the flocs, see figure 10.17. Furthermore, chemical sludge from the secondary settling tank may be recycled back to the flocculation tank, see Figure 10.18, thus increasing the floc volume fraction Φ.

References

/1/ Arvin, E., Petersen, G. and Skårup, J. (1981): Fosforfjernelse fra byspildevand med jern- og aluminiumsalte. Volume 1. Teoretisk baggrund og forsøgsresultater. (Phosphorus removal from municipal wastewater with iron and aluminium salts. Volume 1. Theoretical background and experimental results). Department of Environmental Engineering, Technical Universty of Denmark, Lyngby, Denmark. (Rep. 80-67).

/2/ Harremoës, P., Henze, M., Arvin, E. and Dahi, E. (1989): Teoretisk vandhygiejne. (Water Chemistry). 3rd ed. Polyteknisk Forlag, Lyngby, Denmark.

/3/ Legeros, R.Z. (1981): Apatites in Biological Systems. *Prog. Crystal Growth Charact.*, **4**, 1-45.

/4/ Rich, G.L. (1963): Unit processes of sanitary engineering. John Wiley & Sons, New York, N.Y.

/5/ Boliden (1970): AVR Flockning. (AVR Flocculation). Boliden AB, Helsingborg, Sweden.

/6/ Ødegaard, H. (1979): Orthokinetic flocculation of phosphate precipitates in a multicompartment reactor with non-ideal flow. *Prog. Water Technol.*, **10**, (Suppl. 1), 61-88.

/7/ Eikum, A.S., Ofte, J. and Balmér, P. (1979): Kjemisk felling av kommunalt avløpsvann. PRA prosjektkomiteen for rensning av avlopsvann. (Chemical precipitation of municipal effluent. The PRA project committee for the treatment of effluent). Norsk Institutt for Vannforskning, Oslo, Denmark. (PRA User report PRA 24).

/8/ Enfield, C.G. (1982): Modelling phosphorus sorption and movement in soils in relation to septic tank leach fields. In: Eikum, A.S. and Seabloom, R.W. (eds.): Alternative wastewater treatment. Low-cost small systems, research and development. Proceedings of the Conference held in Oslo, September, 1981, pp 153-168. D. Reidel Publ. Co., Dordrecht.

/9/ EPA (1976): Process design manual for phosphorus removal. U.S. Environmental Protection Agency, Cincinnati, OH. (EPA 625/1-76-001a).

/10/ Thorsen, E., Lynggaard-Jensen, A and Sørensen, P.E. (1981): Simultanfældning og filtrering. Delrapport 1: Undersøgelser af langtidslufter med simultanfældning og kvælstoffjernelse samt tomedia-filtrering. (Simultaneous precipitation and filtration. Subreport 1: Examination of extended aeration with simultaneous precipitation and nitrogen removal and two media filtration). (Søholt treatment plant). The Water Quality Institute, ATV, Hørsholm, Denmark.

/11/ Andersson, L.G. and Sørensen, P.E. (1981): Simultanfældning og filtrering. Delrapport 2: Undersøgelse af langtidsluftere med simultanfældning. (Simultaneous precipitation and filtration. Subreport 2: Examination of extended aeration with simultaneous precipitation). Haslev treatment plant and Slagslunde treatment plant. The Water Quality Institute, ATV, Hørsholm, Denmark.

/12/ Andersson, L.G. and Sørensen, P.E. (1981): Simultanfældning og filtrering. Delrapport 3: Undersøgelse af simultanfældning i normalt belastet aktiv-slamanlæg (forsøgsanlæg i Valby). (Simultaneous precipitation and filtration. Subreport 3: Examination of simultaneous precipitation in a normally loaded activated sludge plant (test plant in Valby). The Water Quality Institute, ATV, Hørsholm, Denmark.

/13/ Arvin, E. (1978): Biological-chemical phosphorus removal. Laboratory and pilotscale experiments. Department of Environmental Engineering, Technical University of Denmark, Lyngby, Denmark. (Rep. 78-1).

/14/ Tofflemire, T.J., Arnold, R. and Chen, M. (1978): Phosphate adsorption capacity and cation exchange capacity of 35 common soil series in New York. In: State of knowledge in land treatment of Wastewater, International Symposium, New Hampshire, vol. 2, pp. 89-96. U.S. Army Crops of Engineers, Cold Regions Research and Engineering Laboratory, Hannover, NH.

Secondary settling tank. Last stage before post precipitation with aluminium sulphate, flocculation and flotation. Stacao de tratamento dé esgoto Norte (Brasilia, Brazil).

Phosphorus can also be removed in an algae/fish pond. The building to the left is a pigsty (with pigs) with a slotted floor. (AIT, Thailand). Alternative methods for the treatment of wastewater are especially used in developing countries. In recent years, however, the interest in root zone plants, plants for the treatment of grey wastewater and aquaculture plants has increased in the developed countries as well.

11 Model features, calibration and application

By Poul Harremoës

Over a period of a generation the engineering approach to design of wastewater treatment plants has developed from being very pragmatic to a science based concept. The pragmatic approach is based on the experience gained from trial and error approaches to design and operation of plants. The scientific approach is based on description of cause-effect relationships - mostly in mathematical form. The advantage of the scientific approach is that the relationships should be of a more universal nature than just experience, which is limited to the range of variability from which the specific experience has been gained. It is a general engineering experience that a more generalised scientific description contains a better approximation to reality and may better be extrapolated to circumstances outside the narrow range of experience.

11.1. Pragmatism versus theory-based models

11.1.1. Engineering craftsmanship
Pragmatism has been the successful basis for engineering development for centuries, even millennia. The craftsmanship of structures and machinery has been a sound basis for development, but the fact is that this craftsmanship in combination with science has improved practise and has expanded the perspectives in all disciplines of engineering. The fact is that the practise of wastewater treatment has been late in the transformation from sheer pragmatism to science-based approaches. This is most likely due to the complexity of the biological systems used in biological wastewater treatment. However, there are exceptions: The theory of aeration was established at the turn of the century 1900. That simple theory of proportionality has been applied since the first development of the activated sludge treatment plant. "There is nothing more practical than a good theory".

The practise of design of wastewater treatment was based on recommended loading Figures and simple rules of thump no more than a generation ago. This simple approach is described in the previous chapters for different categories of treatment plants. This will never be obsolete, because the simplicity will always appeal to the practitioner.

11.1.2. Science-based determinism

The complexity of wastewater treatment plants by to days standard is hard to incorporate in simple design rules of thump. The combination of several different processes into one plant (organic carbon removal, nitrification, denitrification, bio-P-removal) calls for a systematisation and structuring of the experience. The mathematical formulation of a model constitutes such a systematisation and structure. It creates a framework of understanding and a compounding of experience.

It has to be realised that the mathematical description of transport and processes is an idealised world based on the formalities of logic, which can be brought to fit the real world only by experience. In philosophical terminology a mathematical description is based on *deduction* by the logic rules of mathematics on the basis of certain assumptions, but the applicability to engineering practise can be achieved only by *induction*, i.e. by experience.

In its idealistic form this approach is called *determinism*. It is assumed that if the assumptions were met and the parameters of the model were true, then a known input of driving forces on the system will create only one solution. This basic assumption has to be kept in mind under all circumstances because there are many deviations from this idealisation. However, the virtue of the approach is that the deterministic structure is attempted to be universal in nature and to the extent that is the case, it is possible to expand the application into areas without experience. The practise of using models is to apply this feature and at the same time be aware of the idealisations that may in fact not fit as well to reality as well as the modeller tends to assume. The use of models creates a whole new set of approaches to the practice of engineering. We are still in the transition from one form of practice to another.

11.1.3. Model structure, variables, parameters and forcing input

The formulation of a model is based on the following components:

Model structure
The structure of a deterministic model is the mathematical formulation of all the laws of nature that is assumed to be relevant to the simulation of the performance of a treatment plant. The detailed deterministic model is based on a *reductionistic* approach, by which all the phenomena of relevance are described in detail and build into the complex model on the basis of fundamental laws of integration, like

the mass balance equations for water and for each compound of the system. These laws and mass balance equations are described in the previous pages.

Models variables
Any model contains variables, like the concentration of a compound or biomass, which are the descriptors of the performance of the model.

Model parameters
The model structure and the laws associated with the structure contain parameters, which together with the formulation of the law constitute the simulation of the performance of the system. It is attempted to choose laws of such universality that the parameters are invariant, i.e. they are constants, independent of environmental circumstances or history; but the reality is that the structure does not always feature such universality that the parameters are equally universal. In fact, some parameters are not constant at all, because they depend on circumstances or cover details not accounted for in the laws incorporated and thus in the formulation of the model structure. It is important to have these parameters well identified and the choice related to the pertinent circumstances.

Forcing input
The model is capable of simulation of the cause-effect relationship between an external load on the system, which is described by the model. The application of the model consists of analysing the effects of chosen loads on the system. These external loads have to be determined or assumed. In the deterministic formulation the model calculates the output associated uniquely with the chosen input.

The practise of model application is to choose a model structure, such that it represents the key features of relevance to the engineering application in question. This can be based on a priori knowledge and a dedicated investigation of the local circumstances at reasonable cost. The aim is to create results, which are of adequate reliability for the application.

11.2. Model applications

Models can be applied for the following purposes.

11.2.1. Planning tool

Treatment plants are just one component in a much larger system of water management. Water management is delineated by the hydrological catchment, in which system the treatment plant functions as the engineering facility that converts polluted water to purified water before discharge from an urban sub-catchment to the receiving waters. This system is very complex and difficult to manage. In recent years the analysis of managerial options have been investigated with models of the whole hydrological catchment or of the local urban catchment. That calls for models of treatment plants that are integrated with other models of the urban sewer system and of the receiving waters. Only the most relevant of features can find room in models of such scale and complexity.

11.2.2. Analysis of existing plants

The reason for investigating existing plants may be either because they do not operate as supposed or because they have to be up-graded to better performance, e.g. extended to include nutrient removal. The virtue of the analysis is that there is a lot of experience to deal with. A model can be adjusted to fit the data and then used to analyse options for improvement. However, that applies only to the extent that experience contains relevant data sets suitable for model fitting.

11.2.3. Design of new plants

New plants suffer from lack of information on the circumstances of operation in the future. Forecasting is inherently uncertain and the design has to account for that uncertainty. Therefor, it is frequent practise to use the traditional design rules approach for design. This can be justified due to the uncertainty of forecasting. However, in important cases it is relevant to test the performance of complex plants against the assumed forcing inputs and to test the sensitivity of the design to variability of assumed input.

11.2.4. Real time control of plant

The tradition of the trade is to operate wastewater treatment plants based on fixed settings and manual control. Development has demonstrated the virtues of dynamic control of the plants. Several advantages can be achieved: better effluent and savings on energy and chemicals. Real time control of wastewater treatment plants is still in its infancy and there are many approaches to real time control/28/.

Fixed rule based control
The idea is that operation is controlled automatically on the basis of fixed rules on an "if-then-basis". That is in fact how treatment plants have been run for years. The difference by to day's standard is that information from sensors can be used as the information base for fixed rules. The virtue of the simulation model is the fact that such rules can be tested with respect to performance by running the model of

the plant on the basis of information on the existing plant in an effort to find good rules.

Optimised real time control
The optimised control requires an on-line model for the optimisation. The optimisation is based on an objective function. Such a function consists of the relevant performance variables and the merit associated to that performance. It may be least cost or best effluent - or a combination of the two. The model of the plant can predict the performance on the basis of forecasted input. The operational options are analysed in order to choose the best value of the objective function (a minimum or maximum). The model can be adjusted to account for the most recent experience with the plant operation - periodically or on-line. In this case, the model is a prerequisite for the improved performance. This system is quite complex. To make the totality of the system less overwhelming there is a need to simplify the models to suit the purpose of real time control.

11.2.5. Models as research tools
There are still many features of the operation of treatment plants, which are not well understood and not well described. Models may serve as tools with which to analyse information gained from dedicated investigations. This calls for models chosen to describe specifically the processes under investigation in order to identify which interpretation fits the best - or even better: show the greatest universality.

11.2.6. Level of aggregation
"The level aggregation" means the level of detail or resolution in the description of the outcome. The level of aggregation is determined by the engineering application. In the extreme: the planning tool needs a low level of aggregation while the research tools necessarily needs a high level of aggregation on the processes to be investigated. In an integrated model of the river Rhine little detail and low resolution in time and space is required, and uncertainty is evened out in the integration of the system. On the other hand, a specific investigation of the performance of an existing treatment plant during rain may call for a high resolution in time in order to account for the processes during the fast changes associated with transient loading during rain.

The level of desired aggregation has implications on the level of complexity of the model to be chosen. Any model structure is a choice between simplicity and complexity. The key is to fit the choice to the application in question. This choice will be discussed in detail in chapter 11.3.

11.3. Model calibration and parameter estimation

11.3.1. Model structure

A model consists of a model structure, variables and parameters. The ideal of a model is achieved as an attempt to create a model with as much universality as possible. Looking at the last 500 years of scientific development it becomes clear that this choice is no simple thing. It is based on the accumulated know-how over an equal span of time. The ideal model is the model which is simple and which has been shown by induction to apply universally to the scope of engineering application in question.

Nobody with a scientific background will question the universality of the concept of mass balance as a valid tool in dealing with models. It is another question whether we can formulate the components in the equation with equal certainty. Example: The equations for bacterial growth are well established, see chapter 3.2.1, equations 3.1 and 3.2. Though these equations are recognised as the basic formulation of growth and yield, it is well known that these equations are mere simplifications of a much more complex reality, governed by enzymatic activity in a complicated interplay between the organisms and the environment. It is implicitly assumed that the equations apply and that the three parameters (max. growth rate, yield constant and saturation constant) are constants. Whether that is a good assumption depends entirely on the engineering problems in question. The art of model application is to know when such assumptions constitute a sufficiently good approximation to the reality of the problems in question. This can be achieved by induction only, i.e. by experience. The advantage of such formulations is that the model constitutes a simplified condensation of a much more complicated experience, which is complicated due to the lack of structure otherwise associated with experience. The formulation can claim universality due to induction within the range of experience only.

The issue of universality is crucial. It is not possible to make a scientific investigation into the applicability of all the components in a model for each case of application. The claim of universality rests on a-priori knowledge of the laws of nature and their applicability. The idea of this textbook is in fact to present in a condensed form the a-priori knowledge applicable to modelling of wastewater treatment plants.

It is very difficult to test the applicability of the model structure, other than by trial and error by comparison with the sum of a-priory knowledge. However, it is important always to be on the alert for signs in each particular case with respect to applicability. A few examples will illustrate this:

- Nitrifiers are sensitive to inhibitors in domestic wastewater, see chapter 3.4.4. That may change the formulation of the growth equation and/or change the value of the constants. There are several new formulations of the growth equation available and new parameters are introduced.
- The yield constant is not a constant, because the bacteria can store substrate and utilise it later. Description of these phenomena require a new formulation of the growth equation, see chapter 3.
- Change in population in the treatment plant may alter the properties of the bacteria and the growth equations and/or the parameters may change. There is no simple set of options for alternative formulations of equations in this case.

11.3.2. Parameter calibration, verification and estimation

For any application of a model it is essential to determine the parameters that fit the problems in question. Some parameters are reasonably well known based on a-priory knowledge. That applies to the acceleration of gravity, the viscosity of water, some stoichiometric constants, etc.. Nobody dealing with modelling of wastewater treatments would dream of finding such parameters on the basis of a dedicated local investigation. In other cases the parameters are known to depend on local conditions and have to be determined on the basis of local data. There are several procedures for this very important process:

Calibration
The most frequently used procedure with which to make models fit to a particular situation is called *calibration*. This is based on a time series of loading of the treatment plant and of the concentrations of the effluent from the plant. This combination of the input and output data express the transformation performed by the plant, which is precisely what the model is supposed to simulate. The procedure is now to change the parameters of the model such that the best fit to the effluent data is achieved. This is frequently done empirically on a trial and error basis. The best fit can be defined as the set of parameters that gives the minimum of the standard variation of the difference between the data and the output from the model calculations.

Verification
In an attempt to evaluate the result of the model performance with these parameters, another time series of loading of the treatment plant and of the concentrations of the effluent from the plant is used to compare the fit without adjusting the parameters. This procedure is frequently considered to be a test of the quality of the model application. The approach is called *verification*. In practise the procedure is to use one half of a time series of data for calibration and the rest of the available data series for verification.

There is reason for concern in the interpretation of the result :

- There is no guarantee that another parameter set could not have achieved a similar fit. In other words, the parameter set is not uniquely defined by the procedure. In fact, there may be little problem in getting a good fit if there are a sufficient number of parameters to manipulate, see below.
- The data set may not contain information of such kind or extent as to allow the parameters to be determined on that basis. Some parameters may not be *identifiable* on the basis of the information contained in the data series, see below.
- Suppose that the two data series have identical statistical properties (belong to the same statistical population), then there is no reason to expect a different fit for the verification, except for variations within the standard deviation of the fit. Then, why not use the whole series as a means for determination of the standard deviation of the fit and use that as a measure of the fit.
- Suppose that the two series do not have the same statistical properties, then the whole approach is false and the result may correctly be interpreted as a poor fit. However, the reason may be a poor set of parameters. It might be wiser to use the whole series for calibration because the wider range of information will create a better calibration.

The procedure of calibration/verification has one significant virtue:

- The procedure can identify *overparametrisation*, i.e. there are too many parameters to determine compared to the information available in the time series. In such a case, a good fit can be achieved by calibration, but it will fail in verification. The fact is that most models are *empirically underdetermined* by information from data series because the series from practical use do not contain sufficient information.

Each increase of complexity of a model increases the number of parameters to be calibrated. There are examples to show that increase in model complexity and detail can in fact decrease the fitness and the predictive quality of the model due to overparametrisation.

Depending on the quality of the data series, the tendency is that quite a small number of parameters and only certain parameters can be calibrated by this approach. The solution is to choose which parameters to be calibrated and consider all other parameters to be uniquely determined by a-priory knowledge. The art is to choose the right parameters for calibration, because the requirement is that they are identifiable on the basis of the data sets available - and that there is sufficient a-priory knowledge on which to fix the rest of the parameters to a predetermined value.

Parameter estimation
There exist statistical procedures for parameter estimation. The virtue of these procedures is that they can be used to determine the best parameter values, that satisfy chosen statistical criteria. The result includes the uncertainty with which these parameters have been determined and thus whether the parameter is identifiable on the basis of the data available. Included is also the standard deviation of the fit and information on the fitness of the model structure. These are the procedures to be recommended. The problem is that these procedures are quite complex and usage quite specialised.

Experimental design
The experience is that model application in practise is based on inadequate data. Without realising it, the modeller fits his model to inadequate data and assumes to have calibrated the model for practical application. The consequence may be use for extrapolations far outside the range of information covered by the data and outside a reasonable expectation of model performance.

The solution is to apply experimental design dedicated to the problem in question. The ground rule is that the data provided from experimentation shall contain data with such information that the parameter in mind can be identified. A simple example can illustrate the point, e.g. if the intention is to determine the K_s-value in the growth equation, it is quite obvious that the experiments have to include data on growth rates at concentrations in the low range of performance, say below $2 \cdot K_s$. If not, there is no way by which to determine K_s. However, in a usual calibration procedure there is no warning not to fix the K_s-value - unless the trial and error approach reveals that changing the K_s-value does not influence the fit. In a more sophisticated fashion a sensitivity analysis can be made on K_s. In a situation with no data in the proper range, there is no alternative to consider K_s an a-priory known value. The real alternative in case K_s needs to be determined is to design an experiment such that it does contain the information required to determined K_s. The approach is to make K_s identifiable by design. This reasoning applies to all determinations of parameter, e.g. determination of μ_{max}, in relation to K_s and Y_{max}.

The point is that data series should contain *excitations* created by transient loading to cover a large range of situations. This has interesting implications with respect to operation: In order to get adequate information, the treatment plant should be run with as much variation as possible without disturbing the effluent compliance. This is interesting because the standard practise of operating treatment plants is the run them as steadily as possible. Ironically, that is the way by which to get the least information about the kinetic properties of the treatment plant.

In conclusion, no model should be applied without proper attention to the requirements associated with the application in mind. No model should be applied without deciding on a proper model structure and the approach to determination of model parameters. These may be considered adequately determined by a-priori knowledge and fixed accordingly. The other parameters have to be determined for the situation in question. This requires experimental design, either by creating data time series for input and corresponding output with adequate excitations or by making dedicated experiments in the laboratory or on-line at the full scale treatment plant.

Model uncertainty
Any model prediction has an inherited uncertainty. That uncertainty can be determined by parameter estimation and by the standard deviation of fit, but care has to be taken with respect to extrapolation. Uncertainty of extrapolation can be estimated on the assumption that the model structure is a good approximation to reality and that the model parameters are constant, but subject to statistical variation. Any model prediction should in fact be described as a predicted mean performance and an estimated uncertainty. By to day's standard, the information on uncertainty is as important as the prediction of a mean.

In estimation of model uncertainty, sensitivity analysis and analysis of error propagation are important tools.

Sensitivity analysis
In sensitivity analysis each parameter is estimated with respect to importance regarding the ultimate result of performance. From such an analysis the importance of each parameter can be estimated for the range of operation in question. That is important for the selection of which parameters to select as fixed on the basis of a-priory knowledge and which to select for dedicated experimentation.

Error propagation
The error propagation is a means for determination of the uncertainty of the prediction. It is done by choosing an uncertainty for each parameter and the input data on the basis of a-priori knowledge and then combine the uncertainty in the model as it affects the ultimate prediction. This can be done by Monte Carlo simulations, which constitute many runs with random selection of parameters according to their a-priory estimated statistical distribution. In this way an output statistical distribution is generated.

11.4. Treatment plant design

11.4.1. Identification of problem

Design in this context is understood to be determination of the configuration and size of each component in the treatment train from inlet to outlet. It may be a virgin layout, because no treatment plant was there in advance or it may be the expansion of an existing plant, typically to include upgrading from a traditional treatment for BOD-removal to nutrient removal.

An important component in design is to choose the parameters characterising the influent to be treated. That applies to the flow to be designed for and the typical concentrations, see chapter 1 on "Wastewater, Volumes and Composition" for details. The existing inflow and influent concentrations can be measured accordingly. The uncertainty is the estimate of the future conditions under which the treatment plant is to be operated.

The level of design has to be chosen:

- In case of very small plants, e.g. single family houses, the plant does not have to be designed, but pre-designed plants can be bought according to size
- In case of small plants, e.g. villages and small towns with no anticipated problems from industrial wastewater, the plants can be designed on the basis of the well established loading rules.
- In case of larger plants, e.g. middle size towns or any plant with a significant industrial load, the plant can be designed on the basis of design rules, but it is wise to check the design with computer simulations of the design for expected wastewater loading.
- In case of plants for large cities, the plants should be designed and the design checked by computer simulations. This should include sensitivity analysis of the operation to various loading and parameter options, including alternative future optional development of loading. This could include pilot plant studies in order to identify problems and to check the parameter values and functionality of the simulation.

As described in chapter 1, the loading can be characterised by certain standard figures., e.g. $Q_{h,av}$ and $Q_{h,max}$. This is suitable for standard design, but in case of model simulations it is much better to run the simulation for a period based on measured (or anticipated) diurnal variations and transient variations, e.g. during rain. Similar holds true for temperature variations and concentrations variations.

The following two examples illustrate the models used for design of an activated sludge system and of a biofilm system.

Example 11.1. Model for activated sludge system

Computer programs of varying complexity are available as a support for the design of activated sludge plants. Most programs are based on or inspired from IAWQ's model for activated sludge, ASM no. 1, 2 /1, 2/. An essential feature of the model description is the process matrix shown in Table 11.1. The essential problem associated with the use of the computer models is the characterisation of the forcing input, i.e. wastewater flow and concentrations. If there are no detailed analyses available, an estimate can be made on the basis of Tables 1.7-1.10. An estimate will often be sufficient for the design for ordinary municipal wastewater. For industrial wastewater and for a-typical municipal wastewater, detailed in-

Component $j \rightarrow$ / Process $\downarrow$	1 S_I	2 S_S	3 X_I	4 X_S	5 $X_{B,H}$	6 $X_{B,A}$	7 X_P	8 S_O	9 S_{NO}	10 S_{NH}	11 S_{ND}	12 X_{ND}	13 S_{ALK}	Process Rate, ρ_j $[ML^{-3}T^{-1}]$
1 Aerobic growth of heterotrophs		$-\frac{1}{Y_H}$			1			$-\frac{1-Y_H}{Y_H}$		$-i_{XB}$			$-\frac{i_{XB}}{14}$	$\hat{\mu}_H \left(\frac{S_S}{K_S+S_S}\right)\left(\frac{S_O}{K_{O,H}+S_O}\right) X_{B,H}$
2 Anoxic growth of heterotrophs		$-\frac{1}{Y_H}$			1				$-\frac{1-Y_H}{2.86 Y_H}$	$-i_{XB}$			$\frac{1-Y_H}{14 \cdot 2.86 Y_H} - \frac{i_{XB}}{14}$	$\hat{\mu}_H \left(\frac{S_S}{K_S+S_S}\right)\left(\frac{K_{O,H}}{K_{O,H}+S_O}\right) \times \left(\frac{S_{NO}}{K_{NO}+S_{NO}}\right) \eta_g X_{B,H}$
3 Aerobic growth of autotrophs						1		$-\frac{4.57-Y_A}{Y_A}$	$\frac{1}{Y_A}$	$-i_{XB} - \frac{1}{Y_A}$			$-\frac{i_{XB}}{14} - \frac{1}{7 Y_A}$	$\hat{\mu}_A \left(\frac{S_{NH}}{K_{NH}+S_{NH}}\right)\left(\frac{S_O}{K_{O,A}+S_O}\right) X_{B,A}$
4 'Decay' of heterotrophs				$1-f_P$	-1		f_P					$i_{XB} - f_P i_{XP}$		$b_H X_{B,H}$
5 'Decay' of autotrophs				$1-f_P$		-1	f_P					$i_{XB} - f_P i_{XP}$		$b_A X_{B,A}$
6 Ammonification of soluble organic nitrogen										1	-1			$k_a S_{ND} X_{B,H}$
7 'Hydrolysis' of entrapped organics		1		-1										$k_h \frac{X_S/X_{B,H}}{K_X + (X_S/X_{B,H})} \left[\left(\frac{S_O}{K_{O,H}+S_O}\right) + \eta_h \left(\frac{K_{O,H}}{K_{O,H}+S_O}\right)\left(\frac{S_{NO}}{K_{NO}+S_{NO}}\right)\right] X_{B,H}$
8 'Hydrolysis' of entrapped organic nitrogen											1	-1		$\rho_7 (X_{ND}/X_S)$
Observed Conversion Rates $[ML^{-3}T^{-1}]$														$r_i = \sum_j \nu_{ij} \rho_j$
	Soluble inert organic matter $[M(COD)L^{-3}]$	Readily biodegradable substrate $[M(COD)L^{-3}]$	Particulate inert organic matter $[M(COD)L^{-3}]$	Slowly biodegradable substrate $[M(COD)L^{-3}]$	Active heterotrophic biomass $[M(COD)L^{-3}]$	Active autotrophic biomass $[M(COD)L^{-3}]$	Particulate products arising from biomass decay $[M(COD)L^{-3}]$	Oxygen (negative COD) $[M(-COD)L^{-3}]$	Nitrate and nitrite nitrogen $[M(N)L^{-3}]$	$NH_4^+ + NH_3$ nitrogen $[M(N)L^{-3}]$	Soluble biodegradable organic nitrogen $[M(N)L^{-3}]$	Particulate biodegradable organic nitrogen $[M(N)L^{-3}]$	Alkalinity – Molar units	Stoichiometric Parameters: Heterotrophic yield: Y_H Autotrophic yield: Y_A Fraction of biomass yielding particulate products: f_P Mass N/Mass COD in biomass: i_{XB} Mass N/Mass COD in products from biomass: i_{XP} / Kinetic Parameters: Heterotrophic growth and decay: $\hat{\mu}_H$, K_S, $K_{O,H}$, K_{NO}, b_H Autotrophic growth and decay: $\hat{\mu}_A$, K_{NH}, $K_{O,A}$, b_A Correction factor for anoxic growth of heterotrophs: η_g Ammonification: k_a Hydrolysis: k_h, K_X Correction factor for anoxic hydrolysis: η_h

Table 11.1. The now classic process matrix for computer models for activated sludge plants /2/.
($\hat{\mu} = \mu_{max}$, $X_p = X_{I,fp} = f_{XB,XI}$, $Y = Y_{max}$, $i_{XB} = f_{XB,N}$, $i_{XP} = f_{XI,N}$)

vestigation by analyses of the water or water from similar situations else where should be carried out.

The other need is to choose the parameters of the program, as discussed in the previous chapter 11.3. In the design case there is little information to be gained from calibration, because there is no plant on which to make the calibration. The only option is to estimate the parameter values on the basis of a-priori knowledge, as accumulated from large numbers of investigations. Estimates can be made for municipal wastewater on the basis of the values listed in Table 11.2.

These values are taken from four different literature references, based on four different programs. Please notice the difference in parameters, otherwise supposed to be invariant. The reason for the fact that the parameters are not the same may be:

- The investigations have been made under different circumstances (different climates; different wastewater characteristics; different measuring techniques)
- There will always be a statistical difference
- The models may have different model structures, which obviously affect the parameter values. E.g. if growth parameters are determined on the basis of operational data, the μ_{max}-values will be influenced by the number of Monod-terms in the equations.

Constant	Unit	UCTASP /25/	EFOR /24/	ASM1 /1/
Stoichiometric constants				
Y_A	g COD/g N	0.15	0.24	0.24
Y_H	g COD/g COD	0.67	0.67	0.67
f_I	-	0.08	0.08	0.08
$f_{B,N}$	g N/g COD	0.068	0.086	0.086
$f_{XI,N}$	g N/g COD	0.068	0.02	0.06
Kinetic constants				
$\mu_{max,H}$	d^{-1}	3.2	6.0	6.0
$\mu_{max,A}$	d^{-1}	0.45	0.8	0.8
$K_{S,COD}$	g COD/m^3	5	2.5	20
$K_{S,O2}$	g O$_2$/m^3	0.002	0.2	0.2
$K_{S,NO3}$	g NO$_3$-N/m^3	0.1	0.5	0.5
$K_{S,NH4,A}$	g NH$_4$-N/m^3	0.002	0.4	0.4
$K_{S,O2,A}$	g O$_2$/m^3	1.0	1.0	1.0
b_H	d^{-1}	0.62	0.62	0.62
b_A	d^{-1}	0.04	0.15	
η_g	-	0.33	0.8	0.8
η_h	-	0.35		0.4
k_h [1)	g COD/(g COD d)	1.35	5.0	3.0
K_X	g COD/g COD	0.027	0.03	0.03

[1) The constant is linked to different rate expressions in the models.

Table 11.2. Examples of standard constant sets used in computer models for activated sludge processes (20°C)

- The parameters may not be uniquely identifiable on the basis of the data available and may therefor be mutually dependent. E.g. . if growth parameters are determined on the basis of operational data, the μ_{max}-values will be influenced by the K_s-values used in the Monod-terms in the equations.

A good example of the last point is the interrelationship between the yield and the growth rate, which combine to derive the reaction rate, see equation 3.2. What is measured in practise is the reaction rate, which relates to the ratio between the growth rate and the yield. It is much more difficult to measure biomass and it is therefor difficult to measure growth and yield. None of them are identifiable unless the investigations explicitly include independent analyses of growth or yield. Frequently, the yield is taken as the most invariant of the two, see example 11.2 below.

Anyhow, the estimates listed in Table 11.2 in combination with choice of model is a good initial basis for analysis of design, e.g. with the aim to analyse the treatment plant performance under diurnal loading or transient rain loading.

11.5. Model for biofilm system

The models for biofilms are complicated by the fact that the diffusion limitation of reaction has to be taken into account. Attempts to use activated sludge models for biofilters have failed. The reason is that the phenomena typical of biofilms simply cannot be modelled without accounting for the zonation of the biofilm. Different processes take place in each zone, because the redox-conditions vary. The typical example is denitrification under aerobic conditions in the bulk water, while diffusion limitation of oxygen penetration creates anoxic conditions in a zone submerged in the biofilm, see chapter 7.1.2. Another example is the conditions for nitrification described in chapter 6.1.2. This outgrowth of the nitrifiers in biofilms is a very different mechanism, compared to the wash-out and sludge age criteria for activated sludge plants. The criteria for nitrification in biofilms described in chapter 6.1.2 is simplified compared to a much more detailed model of the phenomena in the biofilm.

The complication of the models for biofilms is, that the processes known from activated sludge have to be incorporated into a model structure. This includes the diffusion of substrates into and the products out of the biofilm, the growth of the biofilm and the population dynamics inside the biofilm; as described in chapter 5.12. Many of the parameters in the model are quite uncertain due to less consistent a-priori knowledge of model structure and parameters.

The criteria for nitrification in biofilm reactors will serve as a suitable example of the use of models of biofilms:

Example 11.2. Model of nitrification in biofilters

The following example is based on simulations performed with the biofilm model: AQUASIM /3/. The process matrix is illustrated in Table 11.3. These are the processes at any particular point in the biofilm under the circumstances of the conditions there (which are different from the conditions in the bulk). This process matrix is not different from a similar matrix for

Table 11.3. Process matrix, denitrification and nitrification in rotating disc filters /4/.

Components i	Particulate				Dissolved						Process Rate Equations P_j = Process Rate = $[M L^{-3} T^{-1}]$
	1	2	3	4	1	2	3	4	5	6	
j Processes	X_I	X_H	X_{NS}	X_{NB}	S_O	S_S	S_{NH}	S_{NI}	S_{NO}	S_A	
Heterotrophic Organisms (Degradation of organic components)											
1 Aerobic Growth of heterotrophs		1			$\frac{Y_H - 1}{Y_H}$	$\frac{-1}{Y_H}$	$-i_B$			$\frac{i_B}{14}$	$\mu_{mH} \cdot \frac{S_S}{K_S + S_S} \cdot \frac{S_O}{K_{HO} + S_O} \cdot \frac{S_{NH}}{K_{HNH} + S_{NH}} \cdot \frac{S_A}{K_{HA} + S_A} \cdot X_H$
2 Anoxic Growth of heterotrophs		1				$\frac{-1}{Y_H}$	$-i_B$	?	$\frac{Y_H - 1}{2.86 \cdot Y_H}$	$\frac{1}{14} \frac{Y_H}{Y_H} \cdot \frac{i_B}{14}$	$P_1 \cdot \frac{K_{HO}}{S_O} \cdot \frac{S_{NO}}{K_{HNO} + S_{NO}} \cdot \eta_{DEN}$
3 Decay of Heterotrophs	f_I	-1				$1 - f_I$	$i_B - f_I \cdot i_P$			nueNH/14	$b_H \cdot X_H$
Autotropic Organisms (Nitrification)											
4 Growth of Nitrosomonas			1		$\frac{Y_{NS} - 3.43}{Y_{NS}}$		$-i_B - \frac{1}{Y_{NS}}$	$\frac{1}{Y_{NS}}$		$\frac{i_B}{14} - \frac{1}{7 \cdot Y_{NS}}$	$\mu_{mNS} \cdot \frac{S_{NH}}{K_{NSNH} + S_{NH}} \cdot \frac{S_O}{K_{HO} + S_O} \cdot \frac{S_A}{K_{NSA} + S_A} \cdot X_{NS}$
5 Decay of Nitrosomonas	f_I		-1			$1 - f_I$	$i_B - f_I \cdot i_P$			nueNH/14	$b_{NS} \cdot X_{NS}$
6 Growth of Nitrobacter				1	$\frac{Y_{NS} - 1.14}{Y_{NS}}$		$-i_B$	$\frac{-1}{Y_{NB}}$	$\frac{1}{Y_{NB}}$		$\mu_{mNB} \cdot \frac{S_{NI}}{K_{NI} + S_{NI}} \cdot \frac{S_O}{K_{NBO} + S_O} \cdot \frac{S_{NH}}{K_{NBNH} + S_{NH}} \cdot X_{NB}$
7 Decay of Nitrobacter	f_I			-1		$1 - f_I$	$i_B - f_I \cdot i_P$			nueNH/14	$b_{NB} \cdot X_{NB}$
Observed Reaction Rates $[M_i L^{-3} T^{-1}]$							$n = \sum_i nue_{p,i} \cdot P_j$				
Stoichiometric Parameters	COD	COD	COD	COD	O_2	COD	N	N	N	Moles	
	Inert Particulates	Heterotrophic Biomass	Nitrosomonas Biomass	Nitrobacter Biomass	Dissolved Oxygen	Soluble Substrate	Ammonium NH_4^+	Nitrite NO_2^-	Nitrate NO_3^-	Alkalinity HCO_3^-	

Y_i = Yield coefficient $[M_i M_i^{-1}]$
f_I = Fraction of particul. decay products $[M_I M_X^{-1}]$
i_B, i_P = Nitrogen content of biomass and part. products $[M_N M^{-1}]$
Indices i Organisms: H,NS,NB

Kinetic Parameters:
μ_{mi} = Maximum specific growth rate of organism i $[T^{-1}]$
b_i = Decay rate constant of organism i $[T^{-1}]$
$K_{i,j}$ = Saturation concentration for organism i for component i' $[M_i \cdot L^{-3}]$

η_{DEN} = Denitrificationfactor [-]
Indices i (Organism)
H, NS, NB: Heterotrophs, Nitrosomonas, Nitrobacter
Indices i (dissolved components)
O, S, NH, NI, NO, A: Oxygen, Substrate, NH_4^+; NO_2^-; NO_3^-; HCO_3^-

activated sludge. The same applies to the model parameters associated with the processes, as listed in Table 11.4. What is different is that also the diffusion coefficients of each substrate and product are important parameters. They are listed in Table 11.5.

The model is applied to a sequence of four, totally mixed reactors loaded with domestic wastewater /4,5/. The biofilm thickness is assumed to be 3 mm, uniformly distributed in all four reactors. In the first reactor there is no nitrification, because the nitrifiers are outgrown by the heterotrophs. In the following reactors the nitrifiers can compete for space and nitrification at reduced rate is calculated. The calculated spatial distribution of heterotrophs and

HETEROTROPHIC ORGANISMS				
Maximum growth rate		$\mu_{max,H}$	2.00	d^{-1}
Saturation constant	COD	$K_{S,COD}$	10.00	g COD/m^3
	NH_4^+	$K_{S,NH4}$	0.10	g N/m^3
	O_2	$K_{S,O2}$	0.10	g O$_2$/m^3
	HCO_3^-	$K_{S,ALK}$	0.10	eqv/m^3
	NO_3^-	$K_{S,NO3}$	0.50	g N/m^3
Fraction denitrifiers		η_g	0.70	-
Decay constant		b_H	0.35	d^{-1}
Yield constant		Y_H	0.57	g COD/g COD
Inert fraction		$f_{x,xi}$	0.08	-
Nitrogen content of biomass		$f_{XB,N}$	0.06	g N/g COD
Inert fraction decay		$f_{I,N}$	0.05	g N/g COD
NITROSOMONAS				
Maximum growth rate		$\mu_{max,NS}$	0.35	d^{-1}
Saturation constant	NH_4^+	$K_{S,NH4,NS}$	0.70	g N/m^3
	HCO_3^-	$K_{S,ALK,NS}$	0.20	eqv/m^3
	O_2	$K_{S,O2,NS}$	0.20	g O$_2$/m^3
Decay constant		$b_{A,NS}$	0.05	d^{-1}
Yield constant		$Y_{A,NS}$	0.18	g COD/g NO_3^--N
NITROBACTER				
Maximum growth rate		$\mu_{max,NB}$	0.60	d^{-1}
Saturation constant	NH_4^+	$K_{S,NH4,NB}$	0.05	g N/m^3
	NO_2^-	$K_{S,NO2,NB}$	0.50	g N/m^3
	O_2	$K_{S,O2,NB}$	0.10	g O$_2$/m^3
Decay constant		$b_{A,NB}$	0.09	d^{-1}
Yield constant		$Y_{A,NB}$	0.06	g COD/g NO_3^--N

Table 11.4. Kinetic, stoichiometric and physical parameters in rotating discs /4/.

Oxygen, O_2		1.0-2.1
COD (glucose), $C_6H_{12}O_6$		0.1-0.7
Ammonium, NH_4^+		0.8-1.0
Nitrite, NO_2^-		0.8-1.0
Nitrate, NO_3^-		0.8-1.0
Hydrogencarbonate, HCO_3^{--}		0.4-0.8

Table 11.5. Diffusion coefficients in nitrifying biofilms, 25°. (Based on /4, 26, 27/). Unit $10^{-6}\ m^2 \cdot d^{-1}$

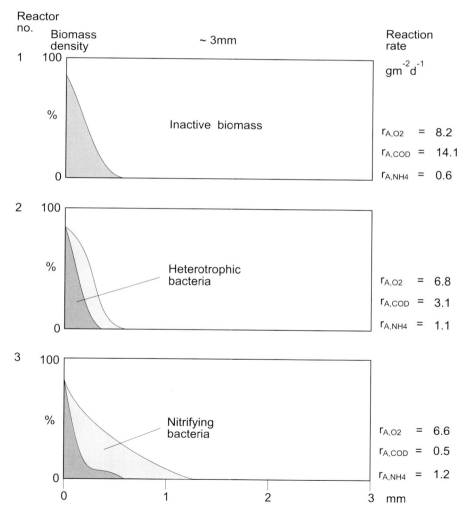

Figure 11.1. Model calculation of distribution of active heterotrophic and autotrophic bacteria in a biofilm, from /4/

nitrifiers is illustrated in figure 11.1. The surface reaction rates are indicated. Nitrifiers dominate the third reactor, because there is little organic matter left in the water.

It is clear that the active bacteria occupy the outer 0.5-1.0 mm of the biofilm. The rest is inactive due to lack of substrate. This illustrates how a parameter can be non-identifiable. The biofilm thickness is of no importance to the results at the concentration ranges involved here. Similarly, there is no way in which the biofilm thickness can be determined from the data based on operation of the biofilm reactors. The thickness can be determined experimentally only by excessive loading of the biofilters, see chapter 5.2.

Figure 11.2 illustrates the reaction rates measured and calculated from a calibrated model. The data illustrate the results from two different effluents from an existing plant without nitrification directly from the final clarifier and from the same effluent treated in an ordinary filter for suspended solids removal prior to nitrification in the biofilters. The curves illustrate two-component transition from ½-order diffusion limitation of ammonium to 0-order reaction of ammonium, because the reaction becomes limited by oxygen diffusion into the biofilm. The shapes of the curves are deceiving, because they could be described by a simple Monod-equation. However, that would be a misinterpretation, because the equation cannot account for the effects of changes in biofilm thickness if of relevance nor can it account for other effects of zonation in the biofilm, e.g. simultaneous denitrification.

The rates for filtered wastewater are higher, because the load of organic matter is smaller, making more room for nitrifiers as compared to heterotrophs.

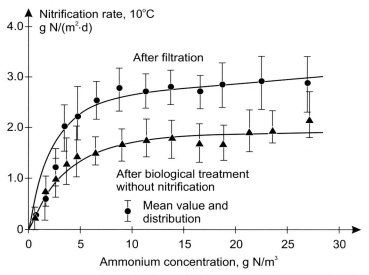

Figure 11.2. Experimentally determined ammonium flux into a biofilm, as a function of ammonium concentration in the bulk water, from /4/.

Pilot plant studies are much more expensive than model simulations. However, they do provide more substantial information about the reality in anticipation of the performance of the full-scale plant. Again, it is important to identify what the issue is. Pilot plants are mostly run for two very different reasons:

- The pilot plant study is suggested and operated entirely with the purpose to demonstrate that the configuration of the suggested design will do the job. That is the most frequent motivation. Accordingly, the pilot plant is operated for the entire period of the investigation in a fashion similar to the anticipated operation of the full-scale plant.
- The pilot plant is suggested and operated entirely in order to identify/verify the values of the design parameters. It has to be identified which design parameters are the essential ones. That is no easy task, because the period is short in most cases and the cost is high. One approach is to verify the standard loading figures, another to verify the parameters of the model used for simulation. There is most perspective in the latter, because more information is gained per unit cost of pilot plant operation. The main point is to design pilot plant operation such that the parameters looked for become identifiable. That requires loading variations, which may not comply with the loading to be experienced in full-scale operation. In the extreme, the idea is to bring the pilot plant to the brink of failure, even beyond, in order to identify the security against failure.

These principles will be illustrated in the following examples 11.3 and 11.4 associated with one of the difficult design issues: The assurance of reliable nitrification in a nitrification/denitrification operation for nitrogen removal.

Example 11.3.
Design temperature for nitrification

Nitrification is sensitive to temperature, see figures 3.7 and 6.2. Thus temperature is an important design issue in temperate climates. A difference of just one degree in design temperature gives rise to significant differences in cost, because the growth rate of nitrifiers directly influence the required sludge age, the quantity of sludge and thus the activated sludge volume.

In the upgrading of the treatment plants for the city of Copenhagen, Denmark, the design temperature was determined as follows:

The influent temperature was available from a period of five years, prior to the design of upgrading the treatment plants. The following phenomena influence the temperature in the reaction tanks: Gain of temperature from influent, by radiation from the sun, mechanically added energy and biologically liberated energy; versus loss by evaporation and convection to the air from wind, long wave radiation, melting of fallen snow and advection to the ground. These phenomena were simulated in order to anticipate the temperature in the aeration tanks during winter operation.

The effect of temperature is short term and long term:

- In the short term, the nitrification may deteriorate while the temperature is low. The performance of the plant will deteriorate for as long as the temperature is low.

- In the longer term, the growth of the nitrifiers will be lower than the wash-out rate from the plant. With time the nitrifiers will be lost from the system and it will take weeks to build up a new population of nitrifiers.

The latter situation is much more serious than the former.

Figure 11.3 shows the extreme statistics for the low temperature in the aeration tanks /6/. Each curve is a display of the lowest running mean temperature over a period of the indicated number of days for a given return period, once per year and once every 5 years. The influence of the cooling of the water in the system compared to the influent temperature was 2-3 °C, depending on the estimated roughness of the surface of the treatment plant. The shading indicates the uncertainty of that estimate. The lowest one day temperature in 5 years is 6.5 °C. The 10 day average is 8.5 °C every 5 years and 10.5 °C once a year.

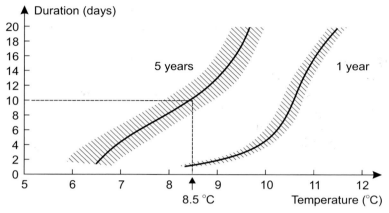

Figure 11.3. Result of statistical analysis of low temperatures in a wastewater treatment plant, illustrated as the number of days with temperature below the indicated temperature as a function of return period (the hatchings illustrate the estimated uncertainty associated with estimation of the effect of heat transfer), from /6/

It was decided to design for the lowest, average temperature for a 10 day period (~ half a sludge age) to be experienced once in five years. That is indicated to be 8.5 °C, as opposed to the initial design temperature, which was 7.0 °C, the lowest influent temperature on record. That decision constitutes a significant saving.

Changing to full aeration, including the otherwise anoxic tanks, can be used as an emergency procedure, if denitrification can be sacrificed once in five years. The rationale for that is, that loss of nitrification can be serious in the short term, because ammonia has acute effects in the receiving waters (oxygen demand and toxicity), while total nitrogen has accumulative effects (eutrofication).

Example 11.4.
Design rate of nitrification

Nitrifiers are quite sensitive to inhibiting substances, see chapter 3.4.4 for details. This was experienced at the pilot plants run for 8 years prior to the commissioning of the up-grading of the full scale plants of the city of Copenhagen /6,7,8,9/.

In order to verify the performance of nitrification, the pilot plants were run initially at the lowest temperature recorded: 7.0 °C. Based on the studies described in example 11.3, these temperatures were adjusted to design temperature for application, but the original temperature for the pilot plant operation was maintained for the sake of continuity. The pilot plants were designed for a sludge age according to figure 6.2: 17 days at 7 °C. In spite of this, the operation of the pilot plant was disrupted by loss of nitrification due to inhibition. This gave rise to detailed studies over the following years of pilot plant operation. In the first years, there was a general inhibition of nitrification, detected as a general rate of nitrification at some 80% of standard value, plus occasional acute events with complete kill of the nitrifiers, which required restart of nitrification for weeks to be re-established.

It is no easy task to operate a pilot plant (nor a full-scale plant) such that specific parameter values become identifiable. However, that is the condition for identifying inhibition to be the cause of the calamity.

The pilot plant is of the typical Danish design: The system of operation is to alternate between aerobic and anoxic conditions in the tanks, described in figure 7.13. The advantage is this form of operation is, that it involves very transient conditions, which is the optimum for gaining much information from monitoring. The system is excited all the time and in this way data are supplied for analysis.

The issue was to characterise the rate of nitrification such that no doubt was left with respect to the cause. It took a year to establish a verifiable mode of operation and consistent data. This was achieved by determination of rate of operation in three ways:

– By taking samples for batch test of rate of nitrification
– By measuring the decrease of the ammonia concentration and the increase in nitrate concentration during the phase of nitrification
– By model simulation of the pilot plant loading, operation and performance for a period. This will be described in the following example in chapter 11.4.

Batch tests

Batch test were performed in a standardised fashion, using the sludge from the pilot plant / 10/. The rate of nitrification is calculated from the decrease in ammonia and the increase of nitrate.

Monitoring pilot plant transient operation

Measuring the rate from the decrease of the concentration of ammonia during the transient operation of the treatment plant was difficult, because the decrease of the concentration of ammonia is influenced by the inflow of both ammonia and the rate of ammonification in the aeration tank. The rate of increase of nitrate gives better information, provided that simultaneous denitrification is insignificant. Because the concentrations varies in the range where the K_s-value is influential this non-linearity has to be taken into account.

The experimental procedures were improved until the point where the rates determined from each of the three approaches gave comparable results. That was the assurance that it was in fact an influence of inhibition on the rate of nitrification.

The rate is normalised with respect to the concentration of nitrifiers, which was determined by the mass balance of the pilot plant over the past sludge age:

$$f_{nit} = \frac{X_A}{X_{tot}} = Y_{nit} \frac{\theta_{tot}}{1 + b_A \cdot \theta_{tot}} \cdot \frac{N_A}{V \cdot X_{tot}}$$

where
- f_A is the fraction of nitrifiers in the activated sludge (gVSS$_{nit}$/gVSS$_{tot}$)
- X is the concentration of biomass (gVSS)
- 3 is the sludge age (d)
- b is the decay rate (d^{-1})
- M_N is the mass load of nitrified nitrogen (gN/d)
- V is total volume of sludge (m^3)

Index	A	refers to nitrifiers
	tot	refers to total biomass

This normalised rate was then compared to the expected normalised rate, which is determined as follows:

$$r_{A,ref} = \frac{\mu_{A,max}}{Y_A}$$

where $r_{A,ref}$ is the reference rate of nitrification per unit mass of nitrifiers (gN/gVSS$_A$ · h)

$$\frac{r_{A,T}}{r_{A,20}} = \exp[(T-20) \cdot \kappa_{\mu,A}]$$

$\mu_{A,max}$ is the max rate of nitrification at 20 °C (d^{-1})

Y_A is the yield of nitrifiers (gVSS/gN)

The temperature corrections is based on:

where T refers to temperatures different from reference of 20 °C

κ is the temperature correction for the growth rate of nitrifiers

The following values have been used:

$\mu_{A,max}$ = 0.86 (d^{-1})
Y_A = 0.16 (gVSS/gN)
κ = 0.088 C^{-1}

From this is derived the following normalised, reference rate of nitrification at 7 °C:

$r_{A,ref}$ = 58 (gN/gVSS$_A$ · h)

which was used as the reference rate for all results from the pilot plant studies.

Figure 11.4 shows the variation of the fraction of nitrifiers over a period of a year for two different pilot plants. The variation is essentially the result of variations in composition of the influent and the inhibition of the nitrifiers.

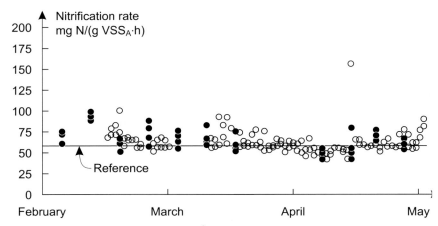

Figure 11.4. Rate of nitrification at 7 °C. Comparison between default design value, measurements with batch tests (closed circles) and measurements on-line in full scale plants (open circles).

The studies of nitrifier performance at the treatment plants were supplemented by explicit inhibition tests on the influent and upstream in the catchments /11/. It was shown that the inhibition came from the industrial areas. This gave rise to campaigns of monitoring and publicity.

Figure 11.5 shows the influence of these campaigns on the rate of nitrification over the period from identification of the problem to the commissioning of the full scale plant, when the inhibition was no longer detectable. From a value below the reference the rate has increased to above the reference at both pilot plants. In spite of statistical variation the trend over a period of 6 years is significant.

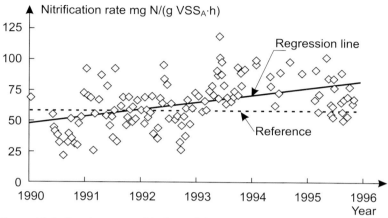

Figure 11.5. Development with time of the normalised nitrification rate in a pilot plant. Due to campaigns to reduce inhibition from industry the rate increased from below to above the default normalised value, from /6/

During this process it was decided not to change the design rate of nitrification, but to fight the cause of inhibition. The savings from this decision were considerable and by far compensated for the cost of otherwise expensive pilot plant studies. Due to the sensitivity of nitrifiers, it is anticipated that many plants in Europe will experience problems with low temperature nitrification due to inhibition, which has gone unnoticed as long as the plants were operated for BOD-removal only.

Figure 11.6 shows the normalised rate of nitrification measured by batch tests using nitrate production in samples taken at different locations (A-D), compared to results from on-line measurements of nitrate variation (1-2) in the full scale treatment plant over a period of three month during the first year testing period of the full scale plant. The final result varies around the reference value, which was maintained as the design value.

The example illustrates the comprehensive effort required to identify and determine just one parameter in the model system: the nitrification rate. Some other parameters are identified in the process (e.g. K_s, see next example). Otherwise, the default values of the other parameters are taken as true, including the yield.

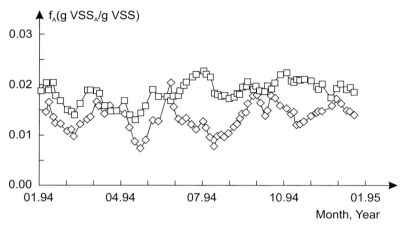

Figure 11.6. The ratio of nitrifiers to total biomass in two pilot plants, from /6/.

11.6. Analysis of existing plant/pilot plant

Model simulation of an existing plant is a valuable tool for the analysis of performance, understanding of the reasons for anormalities and investigation of means for improvement.

11.6.1. Identification of the problem

It has to be realised that the models at disposal contain many functions and many parameters, some of them of universal applicability, some subject to local variations depending on climate, influent characteristics, bacterial population and mode of operation. As mentioned in chapter 11.3 it is not feasible to monitor an existing plant with respect to all functions and parameters. Some have to be chosen as universally applicable and some have to be selected for analysis by model fitting or to be determined by dedicated experimentation. There is no clear demarcation between the two, because the operation of the plant can become part of deliberate alterations with the purpose to increase identifiability of the parameters selected for analysis. The more variation, the better.

The key is that the purpose of the analysis has to be identified and the monitoring, the operation and the external experimentation has to be decided upon as a function of that purpose. Mere model fitting may not reveal information justifying the cost.

11.6.2. Design of experimental programme

Many of the parameters can be determined by dedicated experimental procedures only. The report on ASM1 /1/ has a description of various tests to be made with respect to identification of certain important parameters. Another approach to experimentation is to monitor the performance of the plant. That can be done by taking frequent samples for chemical analysis or by on-line sensors.

Taking frequent samples have been applied as an approach. However, the cost and the effort required do not make it an attractive approach. In recent years much better monitoring has become feasible by the installation of on-line sensors. There are two different approaches to on-line measurements:

On-line, on-stream sensors which measure concentrations as they vary in the plant during operation. The spectrum was presented at the conferences on sensor technology /12/.

- On-line, off-stream sensors provide information on concentrations in the side-stream or batches taken out from the system at a certain frequency and analysed automatically. An example is to extract a sample for on-line OUR-analysis of the influent or to make side-stream batch-tests for measurements of rate of respiration in the activated sludge. The spectrum of options is described in the report /13/.
- On-line measurements give an overwhelming quantity of information, which cannot be coped with without model interpretation and automated estimation of parameters. There are new methods suitable for on-line parameter estimation, but it has to be realised that these methods can estimate a few parameters and only such parameters, which are identifiable on the basis of the data collected. Also in this case, it is important to identify the purpose of the analysis and the means by which to achieve these goals /14/.

11.6.3. Interpretation of results

Various approaches to interpretation of results is given in the following examples:

Example 11.5.
Model of phase variation

> Computer models were used to interpret the results of frequent measurements of concentrations during several phases in the pilot plants described under example 11.4. Figure 11.7 illustrates the variation of concentration during such an investigation /15/, figure 11.8 illustrates that the rate of nitrification can be determined as a function of average bulk ammonia concentration /8/. This result can be obtained because there is a lot of variation in that region of concentration in the pilot plant as a result of the mode of operation: constant variation in phases. The K_s-values can be determined from such figures. Table 11.6 illustrates the range of data obtained and the identifiability of K_s-values. It is quite striking that one of the parameters that can be determined with a reasonable degree of certainty is the K_s-value. The reason for this result is that the variation of concentration is precisely in the range of in-

ANALYSIS OF EXISTING PLANT/PILOT PLANT

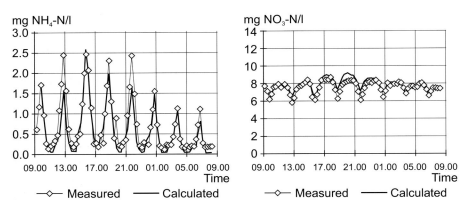

Figure 11.7. Simulation results for ammonia and nitrate in an aeration tank, from /15/.

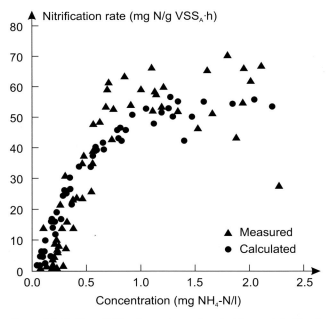

Figure 11.8. The nitrification rates estimated from data like those shown in Figure 11.7. (▲), compared to simulations in a model (●), from /8/

fluence of K_s. The value was estimated to be different for the two plants analysed. There is no explanation for this difference /8/.

(The literature contains results are mere fantasy, because a detailed analysis reveals that the range of variation simply does not contain the information required to determine a parameter. That fact may not be revealed by simple calibration by curve-fitting, but requires statistical parameter estimation).

Half-saturation constant	Pilot plant LYN	Pilot plant DAM
$K_{S,O2}$ (mgO$_2$/l)	0.4	0.1
$K_{S,NH4}$ (mgNH$_4$/l)	0.1	0.4

Table 11.6: Half-saturation constants derived from dynamic studies of pilot plant performance. The reason for the variation from plant to plant is not known /8/.

An alternative to calibration by curve-fitting is parameter estimation, based on deterministic, stochastic modelling, by which the selected parameters are determined and expressed with mean value and standard deviation /16/. This approach can be used for on-line monitoring of treatment plants and with continuous presentation of key parameters, like the nitrification rate and the denitrification rate. This is very valuable information, because it gives information directly on the key issues addressed: Do the key processes perform as could be expected. This information can be made available before the deterioration of the effluent concentrations becomes visible. Inspection of the model results and comparing model performance may help identify the cause. The model can be used to analyse options by which to rectify the calamity.

11.7. Real time control

One distinctive purpose for monitoring a plant is to improve performance by dynamic variation of operation according to information on the performance gathered on-line.

Set point operation
The approach is to select an operational variable and establish the controls by which to keep that operational variable at the desired value.

The classical example is the oxygen control in the aeration tanks. This is done on the basis of oxygen sensors that monitor the oxygen concentration in the tank. It is decided which oxygen concentration is the desired one and the supply of air is regulated such as to maintain that concentration in the aeration tank at all time. For most cases the classical PID-approach to control is used. See the classical control literature on this approach, e.g. /17/.

The role of the model is to determine the optimal set point for the oxygen concentration in the aeration tank.

Fixed rule based control
Dynamic operation is achieved automatically on the basis of fixed rules on an "if-then-else-basis" for the selected operational variables.

As an extension of the classical example, fixed rules can be chosen to vary the set-point for the oxygen concentration at certain periods of the day, or for certain ranges of loading, in an attempt to minimised the cost of aeration, when it is not required.

Another example relates to the transient loading of the treatment plant during rain. The critical issue is the transient hydraulic loading of the final clarifier. If that load persists for too long, then the activated sludge will overflow the wear and will be lost from the system. That may have such dramatic consequences for performance of the plant for weeks to come, that it must be prevented. One solution is to decrease the sludge load on the clarifier. This can be done by storing sludge temporarily in the aeration tanks. This is achieved by switching a multi-compartment train of aeration tanks from plug flow operation to step feed operation. In that way sludge will be concentrated in the up-stream tanks, eventually decreasing the sludge load on the clarifiers. An alternative is to stop aeration and achieve increased sludge quantities by temporary sedimentation in the aeration tanks. If done in the last tanks, fast transition to decrease sludge load on the clarifiers can be achieved.

Again, models can be used to optimise the rules by which to operate the plant prior to the installation of the facility.

Model predicted real time control.
The control requires an on-line model for the prediction. The idea of the prediction is to calculate the performance within the prediction horizon for different options and to choose the optimal set-points by which to achieve optimal conditions.

The past may be relevant in case the model includes stochastic elements and is adapted to the measurements registered until now. That applies to stochastic models, which adapt the parameters of the model to the data of the past, e.g. so-called ARMA-models. Such models are purely empirical, because the parameters have no physical, chemical or biological interpretation. The model of the plant can predict the performance on the basis of forecasted input, which is forecasted on the basis of observations of the past.

State-space models use only *the present* as the point of departure for prediction into the future. However, state-space models have parameters which have been adapted in each particular time step.

The basic time element is the *calculation time*. The calculation time is the time it takes to perform all calculations: collect all data from sensors, adaptation to data, prediction and optimisation plus communication of set-points to regulators. It must be short enough to allow all calculations to be made with sufficient accuracy. Very involved calculations may require very short time steps in order to be stable. Simple models may be very fast. Communication with sensors and regulators may take considerable time (minutes) and may leave little time for calculations. Varies from seconds to minutes.

The *response time* is the time it takes for the system to respond to new set-points. It may differ from seconds to hours.

The *control time* is the time between a new cycle of calculation, typically 5-15 minutes. It is essential that the control time step is sufficiently small to allow control of the most transient loading. It is essential that the calculation time is smaller that the control time step.

The *prediction time* is the horizon into the future over which the performance is predicted as a basis for optimisation of the set-points for the present time step. The prediction time varies with the problem. In case of oxygen regulation it is related to residence time in the aeration tanks. In case of storage of digester supernatant for optimal internal loading of the water treatment train the predictions is diurnal. In case of oxygen depletion in receiving waters the prediction time is days and in case of population dynamics prediction time is weeks.

The optimisation is based on an objective function. Such a function consists of the relevant performance variables and the merit associated to that performance. It may be least cost or best effluent - or a combination of the two. The operational options are analysed in order to choose the best value of the objective function (a minimum or maximum).

This system is quite complex. To make the totality of the system less overwhelming there is a need to simplify the models to suit the purpose of real time control.

11.8. Integrated modelling

Treatment plants are the engineering facilities that convert polluted water into clean water. They are merely an item in a much greater system of water supply and wastewater management in cities, which again is the urban water system that interferes with the total water circulation of a catchment and the total flow of matter in society /18/.

The first level of integration is to model the total system, involving the sewer system, the treatment plant and the receiving waters of an urban catchment, /19,20,21,22,23/. This may involve planning models, models for analysis of system performance and models for real time control. Planning models are relevant for the selection of sites for treatment plants, the layout of the sewer system and overall handling of rain runoff from the catchment. The models for analysis may be due to expected benefits from simulation of the sewer system and the treatment plant as one unit. The rationale for this approach is that increasing awareness of the sewer system not only as transport system, but also as a reactor in which processes take place which alter the characteristics of the wastewater arriving at the treatment plant. If these reactions are time-variant, it may be beneficial to incorporate the sewer system in the models. Similarly, in cases where the pollutional effects in the receiving waters are dynamically related to the effluent characteristics. This modelling approach is an alternative to the tradition, in which the boundary conditions between the system components are rules fixed once and for all. In case of rain runoff, the need to incorporate all three elements in the system is quite obvious, though not part of the tradition. In handling of runoff there are a number of options in design and operation of the system. The hydraulic loading of the treatment plant may be reduced by sewer separation or local infiltration or delayed by building detention basins. However, the tolerance of the treatment plant to hydraulic loading may also be improved by building larger final clarifiers and/or by storage of sludge in the aeration tanks during rain. The combination of options can be managed by real time control, such that water is stored in detention basins when the treatment plants can no longer cope with the inflow. The time element indicates the potential for real time control.

These different problems and options demand very different levels of aggregation and different model structures to fit the problem without unnecessary complication.

References

/1/ Henze, M, Grady, C.P.L., Gujer, W., Marais, G. v. R. and Matsuo, T. (1986): Activated sludge model No. 1. IAWPRC, London. (IAWPRC Scientific and Technical Reports No. 1).

/2/ Henze, M., Gujer, W., Mino, T., Matsuo, T., Wentzel, M.C. and Marais, G. v. R. (1995): Activated sludge model No. 2. IAWQ, London. (IAWQ Scientific and Technical Reports No. 3).

/3/ Reichert, P. and Ruchti, J. (1994): AQUASIM - Computer programme for simulation and data analysis of aquatic systems: User manual. EAWAG, Dübendorf, Switzerland.

/4/ Gujer W. and Boller M. (1990): A mathematical model for rotating contactors. *Water Sci. Technol.,* **22,** (1/2), 53-73.

/5/ Boller, M. and Gujer, W. (1986): Nitrification of tertary trickling filters following deep-bed filters. *Water Res.* **20,** 1363-1373.

/6/ Harremoës, P., Haarbo, A., Winther-Nielsen, M. and Thirsing, C. (1998): Six years of pilot plant studies for design of treatment plants for nutrient removal. *Water Sci. Technol.,* **38,** (1), 219-226.

/7/ Sinkjær, O., Yndgaard, L., Harremoës, P. and Hansen, J.L. (1994): Characterisation of the nitrification process for design purposes. *Water Sci. Technol.,* **30,** (4), 47-56.

/8/ Harremoës, P. and Sinkjær, O. (1995): Kinetic interpretation of nitrogen removal in pilot scale experiments. *Water Res.,* **29,** 899-905.

/9/ Sinkjær, O., Thirsing, C., Harremoës, P. and Jensen, K.F. (1996): Running-in of the nitrification process with and without inoculation of adapted sludge. *Water Sci. Technol.*, **34**, (1-2), 261-268.

/10/ Arvin, E. Dyreborg, S., Menck, C. and Olsen, J. (1994): A mini-nitrification test for toxicity screening, Minntox, *Water Res.*, **9**, 2029-2031.

/11/ Sinkjær, O., Bøgebjerg, P., Grüttner, H., Harremoës, P., Jensen, K.F. and Winther-Nielsen, M. (1996): External and internal sources which inhibit the nitrification process in wastewater treatment plants. *Water Sci. Technol.*, **33**, (6), 57-66.

/12/ Lynggaard-Jensen, A. and Harremoës, P. (eds.) (1996): Sensors in Wastewater Technology. *Water Sci. Technol.*, **33**, (1).

/13/ Spanjers, H., Vanrolleghem, P.A., Olsson, G. and Dold P.L. (1998): Respirometry in control of the activated sludge process: Principles. IAWQ, London. (IAWQ Scientific and Technical Report No. 7).

/14/ Harremoës, P. (1997): Transient experimentation for process understanding and control. In: Environmental Biotechnology, International Symposium, Oostende, April 21-23, pp. 1-8. Technological Institute, Oostende.

/15/ Dupont R. and Sinkjær O. (1994): Optimisation of wastewater treatment plants by means of computer models. *Water Sci. Technol.*, **30**, (4), 181-190.

/16/ Carstensen, J., Harremoës, P. and Madsen, H. (1995): Statistical identification of monod-kinetic parameters from on- line measurements. *Water Sci. Technol.*, **31**, (2), 125-133.

/17/ Olsson, G. and Piani, G. (1992): Computer systems for automation and control. Prentice Hall, London.

/18/ Harremoës, P. (1997): Integrated water and waste management. *Water Sci. Technol.*, **35**, (9), 11-20.

/19/ Harremoës, P. and Rauch, W. (1998): Optimal design and real time control of the integrated urban run-off system. *Hydrobiologia*, **410**, 177-184

/20/ Harremoës, P. and Rauch, W. (1996): Integrated design and analysis of drainage systems, indluding sewers, treatment plant and receiving waters. *J. Hydraul. Res.*, **34**, 815-826.

/21/ Rauch, W. and Harremoës, P. (1996): The importance of the treatment plant performance during rain to acute water pollution. *Water Sci. Technol.*, **34**, (3), 1-8.

/22/ Harremoës, P., Hvitved-Jacobsen, T., Lynggaard-Jensen, A. and Nielsen, B. (1994): Municipal wastewater systems, integrated approach to design, monitoring and control. *Water Sci. Technol.*, **29,** (1-2), 419-426.

/23/ Lijklema, L., Tyson, J.M. and Le Souef, A. (eds.) (1993): INTERURBA '92. *Water Sci. Technol.*, **27,** (12), 1- 236.

/24/ EFOR(1991): Computer model for treatment plants, version 2.0. EFOR Aps., Søborg, Denmark.

/25/ Dold, P.L. et al. (eds.) (1991): Activated Sludge System Simulation Programme. Water Research Commission, Pretoria, South Africa.

/26/ Perry, J. (1963): Chemical Engineers Handbook. McGraw-Hill, New York, N.Y.

/27/ Siegrist, H. and Gujer, W. (1985): Mass transfer Mechanisms in a Heterotrophic Biofilm. *Water Res.*, **19,** 1369-1378.

/28/ Olsson, G. and Newell, B. (1999): Wastewater Treatment Systems. IWA Publishing, London.

List of Symbols

Symbol	Explanation	Dimension	Unit, for example	See page
A	area	L^2	ha, m^2	
A_*	total surface area of carrier in biofilters	L^2	m^2	
AUR	Ammonia uptake rate	$M_N \cdot M_X^{-1} \cdot T^{-1}$	g N/(kg VSS· h)	
b	decay constant	T^{-1}	d^{-1}	75
B	constant in the Langmuir isotherm expression	-	-	347
b_A	decay constant/nitrifying bacteria (autotrophic biomass)	T^{-1}	d^{-1}	210
b_H	decay constant/ heterotrophic biomass	T^{-1}	d^{-1}	134
B_A	surface load	$M \cdot L^{-2} \cdot T^{-1}$	kg BOD(S)/(m^2· d)	181
B_V	volumetric load	$M \cdot L^{-3} \cdot T^{-1}$	kg BOD(S)/(m^3· d)	138
$B_{V,BOD}$	BOD volumetric load	$M \cdot L^{-3} \cdot T^{-1}$	kg BOD/(m^3· d)	152
$B_{V,COD}$	COD volumetric load	$M \cdot L^{-3} \cdot T^{-1}$	kg COD/(m^3· d)	
B_X	sludge load	$M \cdot M^{-1} \cdot T^{-1}$	kg BOD(S)/(kg SS(B) · d)	139
$B_{X,BOD}$	BOD sludge load	$M \cdot M^{-1} \cdot T^{-1}$	kg BOD/(kg VSS · d)	
$B_{X,COD}$	COD sludge load	$M \cdot M^{-1} \cdot T^{-1}$	kg COD/(kg VSS · d)	
BOD (BOD$_5$)	5-day biochemical oxygen demand	$M \cdot L^{-3}$	g O$_2$/m^3	47
BOD$_7$ (BOD$_7$)	7-day biochemical oxygen demand	$M \cdot L^{-3}$	g O$_2$/m^3	
BOD∞	total biochemical oxygen demand	$M \cdot L^{-3}$	g O$_2$/m^3	48

List of Symbols

Symbol	Explanation	Dimension	Unit, for example	See page
C	total concentration (dissolved + suspended)	$M \cdot L^{-3}$	g/m^3	43
C_I	concentration of inhibitor	$M \cdot L^{-3}$	g/m^3	87
C_P	phosphorus concentration	$M \cdot L^{-3}$	$g\ P/m^3$	
C_{TN}	total nitrogen	$M \cdot L^{-3}$	$g\ N/m^3$	57
C_{TP}	total phosphorus	$M \cdot L^{-3}$	$g\ P/m^3$	58
C/N	carbon-nitrogen ratio in denitrification process	$M_C \cdot M_N^{-1}$	kg COD/kg TN	249
$(C/N)_{optimum}$	optimum C/N-ratio for denitrification	$M_C \cdot M_N^{-1}$	kg COD/kg TN	251
$(C/N)_{practice}$	necessary C/N-ratio for a given denitrification process	$M_C \cdot M_N^{-1}$	kg COD/kg TN	251
COD	chemical oxygen demand (with potassium dichromate)	$M \cdot L^{-3}$	$g\ O_2/m^3$	48
COD_P	chemical oxygen demand (with potassium permanganate)	$M \cdot L^{-3}$	$g\ O_2/m^3$	48
COD(B)	biomass or sludge mass, measured as COD	$M \cdot L^{-3}$	$g\ O_2/m^3$	73
COD(S)	substrate (dissolved), measured as COD	$M \cdot L^{-3}$	$g\ O_2/m^3$	73
D	diffusion coefficient	$L^2 \cdot T^{-1}$	cm^2/s	167
D_{BOD}	diffusion coefficient for BOD in biofilm	$L^2 \cdot T^{-1}$	cm^2/s	
D_H	degree of hydrolysis	-	-	199

List of Symbols

Symbol	Explanation	Dimension	Unit, for example	See page
D_{O2}	diffusion coefficient for oxygen (frequently in biofilm)	$L^2 \cdot T^{-1}$	cm^2/s	194
d_p	particle diameter	L	mm, mm	319
D_{COD}	diffusion coefficient for (organic) matter in biofilm	$L^2 \cdot T^{-1}$	cm^2/s	234
E	treatment efficiency	-	-	137
f	floc	-	-	
f(pH)	growth kinetics contribution from pH	variable	variable	84
f(S)	growth kinetics zero order, first order or Monod kinetics	variable	variable	73
f(S_{O2})	growth kinetics for oxygen	variable	variable	84
f(T)	growth kinetics contribution from temperature	variable	variable	84
$f_{C/N}$	efficiency factor for denitrification	-	-	251
$f_{d,max}$	diurnal constant in max. month (infiltration)	-	-	23
$f_{h,max}$	constant for max. hour	-	-	23
$f_{h,min}$	constant for min. hour	-	-	23
F_{ESP}	excess sludge production	$M \cdot T^{-1}$	kg SS(B)/d, kg VSS/d, kg COD/d	123
F_{O2}	oxygen consumption	$M \cdot T^{-1}$	kg O$_2$/d	
$F_{O2,V}$	oxygen consumption	$M \cdot L^{-3}$	kg O$_2$/m^3	

List of Symbols

Symbol	Explanation	Dimension	Unit, for example	See page
$f_{S,SI}$	fraction of inert matter in dissolved matter	$M_I \cdot M_{SI}^{-1}$	kg COD/kg COD	
$f_{s,\,max}$	constant for max. second in maximum month (infiltration)	-	-	23
$f_{SI,N}$	fraction of nitrogen in inert dissolved matter	$M_N \cdot M_{SI}^{-1}$	kg TN/kg COD	57
F_{SP}	sludge production	$M \cdot T^{-1}$	kg SS(B)/d, kg VSS/d, kg COD/d	139
$F_{SP,V}$	volumetric sludge production	$M \cdot L^{-3}$	kg SS(B)/m^3	
$f_{X,N}$	fraction of nitrogen in suspended solids	$M_N \cdot M_X^{-1}$	kg TN/kg COD	57
$f_{X,P}$	fraction of phosphorus in suspended solids	$M_P \cdot M_X^{-1}$	kg TP/kg COD	282
$f_{X,XI}$	fraction of inert organic matter in suspended solids	$M_{XI} \cdot M_X^{-1}$	kg COD/kg COD	207
$f_{XB,XI}$	fraction of inert organic matter in biomass	$M_{XI} \cdot M_{XB}^{-1}$	kg COD/kg COD	
$f_{XS,N}$	fraction of nitrogen in X_S	$M_N \cdot M_{XS}^{-1}$	kg TN/kg COD	49
$f_{XB,N}$	fraction of nitrogen in biomass	$M_N \cdot M_{XB}^{-1}$	kg TN/kg COD	231
$f_{XI,N}$	fraction of nitrogen in suspended inert matter	$M_N \cdot M_X^{-1}$	kg TN/kg COD	57
FG	degree of flocculation	-	-	327
G	mean velocity gradient	T^{-1}	s^{-1}	326
G°(W)	Gibb's free energy at 25°C, pH = 7 and an activity of 1	$M \cdot L^2 \cdot T^{-2} \cdot \text{e-eqv}^{-1}$	kJ/e-eqv	101

List of Symbols

Symbol	Explanation	Dimension	Unit, for example	See page
h	transfer coefficient	$L \cdot T^{-1}$	m/d	169
I	pH Monod constant	-	-	87
K_B	floc degradation constant for iron, aluminium or calcium phosphorus precipitation	varying (often: T)	varying (often: s)	327
K_B'	floc degradation constant	varying (often: $T \cdot L^{-3}$)	(number of primary particles sloughed off from flocs $\cdot$ s)/m^3 flocs	327
k_E	hydrolysis constant			199
K_{fr}	Freundlich isotherm constant	varying	varying	331
k_h	hydrolysis constant (first order kinetics)	T^{-1}	d^{-1}	74
k_p	acetate pick-up constant	$M \cdot L^{-3} \cdot T^{-1}$	kg COD/(m$^3 \cdot$ d)	274
k_{hX}	hydrolysis constant (saturation kinetics)	$M \cdot M^{-1} \cdot d^{-1}$	kg COD(S)/(kg COD(B) $\cdot$ d)	74
K_I	inhibition constant, acetic acid, methane step	$M \cdot L^{-3}$	kg COD/m^3	122
K_l	Langmuir isotherm constant	$M^{-1} \cdot L^3$	m^3/g	331
K_{pH}	pH constant	-	-	87
K_S	saturation constant for substrate	$M \cdot L^{-3}$	g/m^3, g COD(S)/m^3	84
$K_{S,COD}$	saturation constant for substrate	$M \cdot L^{-3}$	g COD/m^3	
K_S'	saturation constant by inhibition	$M \cdot L^{-3}$	g/m^3	87

405

LIST OF SYMBOLS

Symbol	Explanation	Dimension	Unit, for example	See page
$K_{S,HAc}$	saturation constant for acetic acid pick-up	$M \cdot L^{-3}$	kg HAc/m^3	286
$K_{S,I}$	inhibition constant	$M \cdot L^{-3}$	g/m^3	87
$K_{S,M}$	saturation constant, Monod kinetics for methane converting biomass or sludge	$M \cdot L^{-3}$	kg/m^3	122
$K_{S,NH4}$	saturation constant, ammonium nitrogen	$M \cdot L^{-3}$	g NH$_4^+$ \ N/m^3	88
$K_{S,NH4,A}$	saturation constant (K_S) for ammonium (NH$_4^+$) by nitrification (A)	$M \cdot L^{-3}$	g NH$_4^+$ \ N/m^3	93
$K_{S,NO2,A}$	saturation constant (K_S) for nitrite (NO$_2$) by nitrification (A)	$M \cdot L^{-3}$	g NO$_2^\backslash$ \ N/m^3	93
$K_{S,NO3}$	saturation constant, nitrate	$M \cdot L^{-3}$	g NO$_3^\backslash$ \ N/m^3	104
$K_{S,O2}$	saturation constant for oxygen	$M \cdot L^{-3}$	g O$_2$/m^3	86
$K_{S,O2,A}$	saturation constant (K_S) for oxygen (O$_2$) for autotrophic bacteria (A)	$M \cdot L^{-3}$	g O$_2$/m^3	197
$K_{S,O2,H}$	saturation constant for oxygen by denitrification	$M \cdot L^{-3}$	g O$_2$/m^3	242
$K_{S,PO4}$	saturation constant, phosphate	$M \cdot L^{-3}$	g P/m^3	88
$K_{S,S}$	saturation constant, Monod kinetics, - anaerobic acid step	$M \cdot L^{-3}$	g COD/m^3	122

List of Symbols

Symbol	Explanation	Dimension	Unit, for example	See page
K_T	flocculation constant	$L^3 \cdot L_f^{-3}$	m³ water/m³ flocs	325
K_X	hydrolysis saturation constant (saturation kinetics)	$M \cdot M^{-1}$	kg COD(S)/kg COD(B)	74
k_{0Vf}	zero order rate constant in biofilm	$M \cdot L^{-3} \cdot T^{-1}$	kg COD/(m³ **biofilm**· d)	161
$k_{0Vf,M}$	zero order rate constant in anaerobic methane producing biofilm	$M \cdot L^{-3} \cdot T^{-1}$	kg COD/(m³ biofilm· d)	289
k_{0A}	zero order area specific rate constant	$M \cdot L^{-2} \cdot T^{-1}$	kg COD/(m² biofilm· d)	163
$k_{1/2A}$	half order area specific rate constant	$M^{1/2} \cdot L^{-1/2} \cdot T^{-1/2}$	$g^{1/2} \cdot m^{-1/2} \cdot h^{-1}$	192
$k_{1/2A,NH4}$	half order area specific rate constant for ammonium	$M^{1/2} \cdot L^{-1/2} \cdot T^{-1}$	$(g\ NH_4^+ \setminus N) \cdot m^{-1/2} \cdot d^{-1}$	200
$k_{1/2A,NO3}$	half order area specific rate constant for nitrate	$M^{1/2} \cdot L^{-1/2} \cdot T^{-1}$	$(g\ NO_3^{\setminus} \setminus N)^{1/2} \cdot ^{-1/2} \cdot d^{-1}$	234
$k_{1/2A,O2}$	half order area specific rate constant for oxygen	$M^{1/2} \cdot L^{-1/2} \cdot T^{-1}$	$(g\ O_2)^{1/2} \cdot m^{-1/2} \cdot d^{-1}$	179
$k_{1/2V,NO3}$	half order specific volumetric rate constant for nitrate	$M^{1/2} \cdot L^{-3/2} \cdot T^{-1}$	$g^{1/2} \cdot m^{-3/2} \cdot d^{-1}$	235
k_1	first order rate constant	T^{-1}	h^{-1}, d^{-1}	
k_{1Vf}	first order rate constant in biofilm	T^{-1}	h^{-1}, d^{-1}	159
k_{1A}	first order area specific rate constant	$T^{-1} \cdot L^{-2}$	$h^{-1} \cdot m^{-2}$	160

407

List of Symbols

Symbol	Explanation	Dimension	Unit, for example	See page
l	litre	L^3	l!	
L	biofilm thickness	L	mm, µ	159
m	metre	L	m!	
M_N	mass of nitrogen	M_N	kg TN	260
$M_{P,p}$	mass of phosphorus in one primary particle	M	g P/primary particle	327
M_X	sludge mass	M_X	kg SS, kg COD, kg COD(B)	138
$M_{X,aerobic}$	aerobic sludge mass	M_X	kg SS(B), kg COD, kg COD(B)	
MR	molar ratio (P precipitation)	$mole \cdot mole^{-1}$	mole precipitation chemical/mole P	340
MW_{biom}	molar weight for biomass or sludge (often set at 113 g/mole)	$M_X \cdot mole^{-1}$	g/mole	81
MW_{org}	mole weight for organic (not specified) matter (often set at 393 g/mole)	$M \cdot mole^{-1}$	g/mole	81
n	number of flocculation chambers	-	-	327
N	number (persons)	-	-	17
N	transport through cross section	$M \cdot L^{-2} \cdot T^{-1}$	$g/(m^2 \cdot d)$	158
n_f	number of flocs per volume of water	L^{-3}	number of flocs/m^3 water	326
n_{fr}	Freundlich adsorption isotherm constant	-	-	331
n_p	number of primary particles per volume of water	L^{-3}	number of primary particles/m^3 water	325

Symbol	Explanation	Dimension	Unit, for example	See page
OUR	oxygen uptake rate	$M_{O_2} \cdot M_X^{-1} \cdot T^{-1}$	g O_2/(kg VSS · h)	60
P	power in flocculation expression	-	-	327
p	primary particle	-	-	342
PE	population equivalent for (0.2 m³/d, 60 g BOD/d)	$L^3 \cdot T^{-1}$ or $M \cdot T^{-1}$	m³/d or g/d	24
PE_{BOD}	population equivalent for BOD (= 60 g BOD/d)	$M \cdot T^{-1}$	g BOD/d	24
PE_{water}	population equivalent for water (= 0.2 m³/d)	$L^3 \cdot T^{-1}$	m³/d	24
pK_a	the negative logarithm for the acid dissociation constant	-	-	120
Q	volume rate of water flow	$L^3 \cdot T^{-1}$	m³/h, l/s	132
$Q_{d,av}$	average volume of water per day	$L^3 \cdot T^{-1}$	m³/d	12
$Q_{d,av,max}$	average infiltration per day in maximum month of the year	$L^3 \cdot T^{-1}$	m³/d	23
$Q_{d,max}$	maximum volume of water per day	$L^3 \cdot T^{-1}$	m³/d	
$Q_{h,av}$	average volume of water per hour	$L^3 \cdot T^{-1}$	m³/h	12
$Q_{h,max}$	average volume of water of maximum hours in the days of the year (household + industry) or average hourly infiltration in maximum month of the year (infiltration)	$L^3 \cdot T^{-1}$	m³/h	12

List of Symbols

Symbol	Explanation	Dimension	Unit, for example	See page
$Q_{h,min}$	average volume of water of minimum hours in the days of the year	$L^3 \cdot T^{-1}$	m^3/h	
$Q_{s,A}$	average infiltration per second per unit area in periods with infiltration	$L^3 \cdot T^{-1}$	$l/(ha \cdot s)$, $m^3/(ha \cdot s)$	23
$Q_{s,av}$	average volume per second of water in an average hour over the year	$L^3 \cdot T^{-1}$	m^3/s	
$Q_{s,max}$	average volume of water per second in maximum month of the year (infiltration) or average maximum second (household + industry)	$L^3 \cdot T^{-1}$	m^3/s	18
Q_{year}	annual volume of water	$L^3 \cdot T^{-1}$	$m^3/year$	18
$Q_{year,pers}$	annual wastewater production per person	$L^3 \cdot T^{-1}$	$m^3/(person \cdot year)$	16
R	recycle ratio	-	-	138
$r_{A,M}$	removal rate for methane per unit area	$M \cdot L^{-2} \cdot T^{-1}$	$g\ CH_4/(m^2 \cdot h)$	289
$r_{A,NH4}$	removal rate for ammonium per unit area	$M \cdot L^{-2} \cdot T^{-1}$	$g\ NH_4^+ - N/(m^2 \cdot d)$	223
$r_{A,NO3}$	removal rate for nitrate per unit area	$M \cdot L^{-2} \cdot T^{-1}$	$g\ NO_3^- - N/(m^2 \cdot d)$	233
$r_{A,S}$	removal rate per unit area	$M \cdot L^{-2} \cdot T^{-1}$	$kg\ BOD/(m^2 \cdot d)$	192
r_f	radius of flocs	L	mm	158

List of Symbols

Symbol	Explanation	Dimension	Unit, for example	See page
r_V	volumetric reaction rate	$M \cdot L^{-3} \cdot T^{-1}$	kg BOD/(m$^3 \cdot$ d), g COD/(m$^3 \cdot$ h)	134
$r_{V,CH4}$	volumetric methane production rate	$M \cdot L^{-3} \cdot T^{-1}$	g CH$_4$/(m$^3 \cdot$ h)	303
$r_{V,f}$	volumetric floc formation rate	$L^{-3} \cdot T^{-1}$	number of primary particles incorporated in flocs/(m^3 water $\cdot$ s)	325
$r_{V,HAc}$	volumetric acetic acid uptake rate	$M \cdot L^{-3} \cdot T^{-1}$	kg HAc/(m$^3 \cdot$ d)	286
$r_{V,ki}$	volumetric removal rate for matter in - phase k	$M_i \cdot L^{-3} \cdot T^{-1}$	g/(m$^3 \cdot$ h)	204
$r_{V,p}$	volumetric primary particle formation rate	$L^{-3} \cdot T^{-1}$	number of primary particles sloughed off from flocs/(m^3 water $\cdot$ s)	325
$r_{V,S}$	volumetric substrate (dissolved) removal rate	$M \cdot L^{-3} \cdot T^{-1}$	kg COD(S)/(m$^3 \cdot$ d)	73
$r_{V,XB}$	specific volumetric removal rate of biomass or sludge	$M \cdot L^{-3} \cdot T^{-1}$	kg COD/(m$^3 \cdot$ d), kg VSS/(m$^3 \cdot$ d)	73
$r_{V,XS}$	specific volumetric removal rate of particulate substrate (X_S), for example hydrolysis	$M \cdot L^{-3} \cdot T^{-1}$	kg/(m$^3 \cdot$ d)	74
r_{Vf}	specific biofilm volumetric rate	$M \cdot L^{-3} \cdot T^{-1}$	kg/(m$^3 \cdot$ d)	144
$r_{X,S}$	removal rate per amount of biomass or sludge	$M_S \cdot M_X^{-1} \cdot T^{-1}$	kg COD(S)/(kg COD(X) $\cdot$ d)	

411

List of Symbols

Symbol	Explanation	Dimension	Unit, for example	See page
$r_{X,S(NO3)}$	nitrate removal rate per amount of biomass or sludge	$M_{NO3} \cdot M_X^{-1} \cdot T^{-1}$	g NO_3^--N/(kg VSS · h)	106
$r_{X,X}$	removal rate (of sludge) per amount of sludge	$M_X \cdot M_X^{-1} \cdot T^{-1}$	kg SS/(kg SS · d), kg COD/(kg COD · d)	
s	second	–	s	
s	dispersion	–	–	
s_{Vf}	dimensionless concentration of dissolved matter in biofilm	–	–	
s_g	dimensionless interface concentration	–	–	43
S	concentration of dissolved matter (substrate concentration)	$M \cdot L^{-3}$	kg/m³, kg BOD(S)/m³, kg COD/m³	43
S_D	diffusible matter	$M \cdot L^{-3}$	g/m³	200
S_E	enzyme concentration	$M \cdot L^{-3}$	g/m³	199
S_{Vf}	concentration of dissolved matter in biofilm	$M \cdot L^{-3}$	kg/m³, g N/m³	43
S_g	interface concentration	$M \cdot L^{-3}$	g/m³	169
S_{HAc}	dissolved biological, very easily degradable organic matter	$M \cdot L^{-3}$	kg COD(S)/m³	
S_I	dissolved biological inert organic matter	$M \cdot L^{-3}$	kg COD(S)/m³	49
$S_{I,N}$	dissolved organic inert nitrogen	$M_N \cdot L^{-3}$	g N/m³	57
S_{NH4}	concentration of ammonium and ammonia nitrogen	$M_N \cdot L^{-3}$	g N/m³	57

List of Symbols

Symbol	Explanation	Dimension	Unit, for example	See page
S_{NOX}	concentration of nitrite and nitrate nitrogen	$M_N \cdot L^{-3}$	$g\ N/m^3$	57
S_{NO3}	concentration of nitrate nitrogen	$M_N \cdot L^{-3}$	$g\ NO_3^- \text{-} N/m^3$	104
$S_{org.P}$	dissolved organic phosphorus	$M \cdot L^{-3}$	$g\ P/m^3$	58
S_{O2}	oxygen concentration	$M \cdot L^{-3}$	$g\ O_2/m^3$	
$S_{O2,2}$	oxygen concentration in reactor	$M \cdot L^{-3}$	$g\ O_2/m^3$	86
S_{PO4}	dissolved inorganic orthophosphate	$M \cdot L^{-3}$	$g\ PO_4\text{–}P/m^3$	58
$S_{p\text{-}P}$	dissolved inorganic polyphosphate	$M \cdot L^{-3}$	$g\ P/m^3$	58
S_S	dissolved biological, easily degradable organic matter	$M \cdot L^{-3}$	$kg\ COD(S)/m^3$	49
S_2	substrate concentration in reactor	$M \cdot L^{-3}$	$kg\ BOD(S)/m^3$, $g\ COD/m^3$	
SDI	sludge density index	$M \cdot L^{-3}$	$g/l, kg/m^3$	59
SVI	sludge volume index	$L^3 \cdot M^{-1}$	$ml/g\ SS$	59
SS	suspended solids (dry solids)	M	$kg\ SS$	139
SS(B)	biomass or sludge measured as suspended solids, SS	M	$kg\ SS$	
t	time	T	h, d	
t	tonne	M	tonnes	
T	temperature	-	°C	
$t_{d,year}$	diurnal factor	$T \cdot T^{-1}$	d/year	22
$t_{h,d}$	time factor	$T \cdot T^{-1}$	h/d	

List of Symbols

Symbol	Explanation	Dimension	Unit, for example	See page
TAL	alkalinity	$eqv \cdot L^{-3}$	eqv/m^3	58
TOC	total organic carbon	$M \cdot L^{-3}$	$g\,O_2/m^3$	54
TOD	total oxygen demand	$M \cdot L^{-3}$	$g\,O_2/m^3$	54
u	interface velocity	$L \cdot T^{-1}$	mm/d	205
V	volume	L^3	m^3, l	
V_{TOT}	total volume	L^3	m^3	147
VS	volatile solids	M	kg VS	
VSS	volatile suspended solids	M	kg VSS	139
W	effect	$M \cdot L^2 \cdot T^{-3}$	watt or $kg \cdot m^2/s^3$	326
X	suspended solids concentration	$M \cdot L^{-3}$	kg/m^3	43
$\overline{X}$	mean value	-	-	
X_B	biomass (or sludge) concentration	$M \cdot L^{-3}$	$kg\,SS(B)/m^3$, $kg\,VSS/m^3$, $kg\,COD(B)/m^3$	73
$X_{B,A}$	concentration of autotrophic (nitrifying) biomass or sludge	$M \cdot L^{-3}$	$kg\,COD(B)/m^3$	138
$X_{B,H}$	concentration of heterotrophic biomass or sludge	$M \cdot L^{-3}$	$kg\,COD(B)/m^3$	138
$X_{B,M}$	biomass or sludge mass of methane producers	$M \cdot L^{-3}$	$kg\,SS(B)/m^3$	122
$X_{B,S}$	biomass or sludge mass of acid producing bacteria	$M \cdot L^{-3}$	$kg\,SS(B)/m^3$	121
X_{PAO}	biomass of phosphorus accumulating bacteria	$M \cdot L^{-3}$	$kg\,SS(B)/m^3$	287

List of Symbols

Symbol	Explanation	Dimension	Unit, for example	See page
X_{COD}	sludge concentration (biomass concentration) measured in COD	$M \cdot L^{-3}$	kg COD(B)/m^3	
X_I	suspended biological inert organic matter	$M \cdot L^{-3}$	kg COD(S)/m^3	49
$X_{I,N}$	suspended inert nitrogen	$M_N \cdot L^{-3}$	kg N/m^3	57
$X_{org,P}$	suspended organic phosphorus	$M_P \cdot L^{-3}$	kg P/m^3	58
X_{PHA}	polyhydroxy-alkanoate	$M_{PHA} \cdot L^{-3}$	kg COD/m^3	49
$X_{P,a}$	adsorbed phosphorus	$M_P \cdot M$	mg P/kg soil	330
$X_{P,p}$	amount of phosphorus in primary particles per volume of water	$M \cdot L^{-3}$	g P/m^3 water	327
X_R	non-diffusible matter	$M \cdot L^{-3}$	g/m^3	199
X_S	suspended biologically, slowly degradable organic matter	$M \cdot L^{-3}$	kg/m^3, kg COD(S)/m^3	49
$X_{S,N}$	suspended organic, slowly degradable nitrogen	$M_N \cdot L^{-3}$	kg N/m^3	57
$X_{0.5}$	sludge concentration in activated sludge after ½ hour's settling	$M \cdot L^{-3}$	g/l, kg/m^3	59
Y	yield constant	$M \cdot M^{-1}$	kg VSS/kg substrate or kg COD(B)/kg COD(S)	
Y_{COD}	yield constant per amount of COD	$M_B \cdot M_S^{-1}$	kg COD(B)/ kg COD(S)	79
Y_{max}	maximum yield constant	$M_B \cdot M_S^{-1}$	g COD(B)/g COD(S)	73

415

LIST OF SYMBOLS

Symbol	Explanation	Dimension	Unit, for example	See page
$Y_{max,A}$	maximum yield constant, autotrophic bacteria or sludge (nitrifying)	$M \cdot M_N^{-1}$	g COD/g N	197
$Y_{max,H}$	maximum yield constant for heterotrophic bacteria or sludge	$M_X \cdot M_S^{-1}$	kg COD(X)/ kg COD(S)	133
$Y_{max,M}$	maximum yield constant, methane producing bacteria or sludge	$M_X \cdot M_S^{-1}$	kg COD(B)/kg COD(S)	119
$Y_{max,NO3}$	maximum yield constant for denitrification	$M_B \cdot M_N^{-1}$	kg COD(B)/kg NO_3^- - N	105
$Y_{max,P}$	maximum yield constant, phosphorus accumulating bacteria	$M_B \cdot M_S^{-1}$	kg COD(B)/kg COD(S)	
$Y_{max,S}$	maximum yield constant, acid producing bacteria or sludge in anaerobic processes	$M_B \cdot M_S^{-1}$	kg COD(B)/kg COD(S)	118
Y_{NH4}	yield constant for ammonium oxidizers	$M_B \cdot M_N^{-1}$	kg COD(B)/kg NH_4^+ - N	
Y_{NO3}	yield constant per amount of nitrate	$M_B \cdot M_N^{-1}$	kg COD(B)/kg NO_3^- - N	
Y_{obs}	observed yield constant	$M_X \cdot M_S^{-1}$	kg COD(X)/kg COD(S)	79
$Y_{obs,NH4}$	observed yield constant for ammonium oxidizers	$M_B \cdot M_N^{-1}$	kg COD(B)/kg NH_4^+ - N	
$Y_{obs,NO2}$	observed yield constant for nitrite oxidizers	$M_B \cdot M_N^{-1}$	kg COD(B)/kg NO_2^- - N	

List of Symbols

Symbol	Explanation	Dimension	Unit, for example	See page
$Y_{obs,PAO}$	observed yield constant for phosphorus accumulating bacteria	$M_B \cdot M_S^{-1}$	kg COD(B)/kg COD(S)	
α	dimensionless biofilm constant	-	-	159
α_f	volume per floc	L^3	m^3, mm^3	326
α_p	volume per primary particle	L^3	m^3, mm^3	326
β	metabolic efficiency	-	-	90
β	degree of penetration	-	-	162
ε	efficiency factor for biofilm	-	-	160
ε_k	volume fraction	-	-	203
η	dimensionless depth in filter	-	-	177
η_g	fraction of denitrifying bacteria	-	-	60
η_h	hydrolysis reduction by denitrification	-	-	381
θ	hydraulic retention time	T	d, h	142
θ_X	sludge age, solids retention time	T	d	141
$\theta_{X,A}$	sludge age (θ_X) for nitrifying bacteria (A)	T	d	198
$\theta_{X,aerobic}$	aerobic sludge age	T	d	143
κ	temperature constant for μ_{max} and r_S	-	$°C^{-1}$	85
κ	r_A/k_{0A} (biofilm)	-	-	

List of Symbols

Symbol	Explanation	Dimension	Unit, for example	See page
λ	dimensionless ratio between hydraulic and biofilm diffusion	-	-	170
μ_a	absolute viscosity	$M \cdot L^{-1} \cdot T^{-1}$	$kg/(m \cdot s)$	326
μ_{max}	maximum specific growth rate	T^{-1}	h^{-1}, d^{-1}	73
$\mu_{max'}$	maximum specific growth rate by inhibition	T^{-1}	d^{-1}	87
$\mu_{max,H}$	maximum specific growth rate for heterotrophic organisms	T^{-1}	d^{-1}	133
$\mu_{max,M}$	maximum specific growth rate for methane producing bacteria	T^{-1}	d^{-1}	122
$\mu_{max,PAO}$	maximum specific growth rate for phosphorus accumulating bacteria	T^{-1}	d^{-1}	115
$\mu_{max,S}$	maximum specific growth rate for acid producing bacteria	T^{-1}	d^{-1}	122
μ_{obs}	observed specific growth rate	T^{-1}	h^{-1}, d^{-1}	84
$\mu_{obs,A}$	observed specific growth rate for autotrophic bacteria	T^{-1}	d^{-1}	197
$\mu_{obs,A,net}$	observed specific net growth rate for nitrifying bacteria (A)	T^{-1}	d^{-1}	197
ν	stoichiometric coefficient	$M \cdot M^{-1}$	kg/kg	

LIST OF SYMBOLS

Symbol	Explanation	Dimension	Unit, for example	See page
$\nu_{COD,CH4}$	stoichiometric coefficient	$M \cdot M^{-1}$	kg CH_4/kg COD	318
$\nu_{HAc,P}$	stoichiometric coefficient	$M_P \cdot M^{-1}$	kg PO_4-P/kg HAc	113
$\nu_{NH4,O2}$	stoichiometric coefficient	$M \cdot M_N^{-1}$	kg oxygen/kg NH_4^+ - N	215
$\nu_{NO3,COD}$	stoichiometric coefficient	$M \cdot M_N^{-1}$	kg COD/kg NO_3^- - N	244
$\nu_{O2,S}$	stoichiometric coefficient	$M \cdot M^{-1}$	kg substrate/kg oxygen	
$\nu_{ox,red}$	stoichiometric coefficient	$M \cdot M^{-1}$	kg red.mean/kg ox.mean	174
$\nu_{S,HAc}$	stoichiometric coefficient	$M \cdot M^{-1}$	kg HAc/kg COD (easily degradable matter)	
$\nu_{X,S}$	stoichiometric coefficient, conversion of suspended solids (X) into dissolved matter (S)	$M_S \cdot M_X^{-1}$	kg COD/kg COD	75
ξ	dimensionless depth of biofilm	-	-	49
ξ'	dimensionless depth for the efficient part of the biofilm	-	-	162
ω	specific filter area	$L^2 \cdot L^{-3}$	m^2/m^3	177
Φ	floc volume fraction, volume of flocs per volume of water	$L^3 \cdot L^{-3}$	m^3 flocs/m^3 water	326
Φ	the Thiele module	-	-	167

List of Symbols

Greek Symbols

α		alpha	ν		ny
β		beta	ξ	Ξ	xi
γ	Γ	gamma	ο		omikron
δ	Δ	delta	π	Π	pi
ε	E	epsilon	ρ		rho
ζ		zeta	σ	Σ	sigma
η		eta	τ		tau
θ	Θ	theta	υ	Y	ypsilon
ι		jota	φ	Φ	phi
κ		kappa	χ		chi
λ	Λ	lambda	ψ	Ψ	psi
μ		my	ω	Ω	omega

Index

A
Absolute viscosity 30
Acetate 110, 112, 168, 286
Acetic acid 52, 106, 168, 254, 263
Acetoclastic bacteria 116
Achromobacter 67
Acid production 116
Acid step 117
Acid-forming bacteria 116, 125
Activated sludge 81, 131, 133, 135, 137, 139, 141, 143, 145, 147, 149, 151, 153, 155
Activated sludge flocs 136
Activated Sludge Model 114
Activated sludge plants 144
 – process matrix 134
 – single tank 146
Activated sludge processes
 – design 151
Adsorption 347
Adsorption capacity 347
Aerobic biofilters 182
Aerobic conditions
 – PAOs 114
Aerobic sludge age 143, 212, 223, 230
Aerobic storage 113
AIT 368
Algae 66
Alkalinity 30, 58, 92, 106, 113
Alternating denitrification 263
alternating operation 224
Alternating plant 274
Aluminium 29, 331
Aluminium sulphate 345
Ammonia 324
 – fish 209
Ammoniaoxidation 99

Anaerobic bacteria
 – nutrients 120
Anaerobic contact plant 303
Anaerobic filter area 315
Anaerobic filter plants
 – design 315
Anaerobic filter processes 309
Anaerobic phosphorus tank 286
Anaerobic plants
 – design 310, 312, 314, 316, 318, 320, 322
 – disturbances 321
 – mass balances 301
 – optimazion 319
 – sludge load 313
 – start-up 319
 – technical data 312
Anaerobic processes 123 116, 118, 120, 122, 123, 124
 – alkalinity 120
 – gas production 317
 – inhibition 124
 – kinitics 121
 – nutrients 119
 – operational parameters 321
 – pH 123
 – plant types 305-306, 308
 – process matrix 302
 – reaction rate constants 125
 – reactions 118
 – toxic substances 124
Anaerobic sludge blanket plant 311
Anaerobic treatment 301
Anaerobic wastewater processes
 – inhibitors 323
Anaerobic wastewater treatment 300
anoxic 100

ASM1 381
ASM2 266
ASM2d 287
ASM3 77
assimilation 77
Athens 129
ATP 54
Autotrophic biomass 28, 138

B
Bacteria 66
Bangkok 41
Bath & wash 34
Bilbao 136
Biocarbone 198
Biochemical oxygen demand 46
Biodenipho 295
Biodenitro 255
Biofilm
— adhesion 196
— alkalinity 281
— bubble formation 197
— concentration distribution 160
— degree of hydrolysis 199
— detailed model 203
— efficiency factor 160
— external hydrolysis 199
— nitrification 214
— particle adsorption 199
— particulate organic matter 197
— penetration 162
— redox-zones 279
— sloughing 195-196
— thickness 185
— treatment efficiency 178
— biofilm control 191
Biofilm kinetic parameters 167-168
Biofilm kinetic user guide 175
Biofilm kinetics 157-158, 160, 162, 164, 166
Biofilm plants
— anaerobic 304

Biofilm system
— models 382, 384, 386, 388, 390
Biofilters 157, 159, 161, 163, 165, 167, 169, 171, 173, 175, 177, 179, 181, 183, 185, 187, 189, 191, 193, 195, 197, 199, 201, 203, 205, 207
— aeration 195
— carrier material 155
— design 188, 190-192, 194, 234
— half order constants 193
— mass balances 179-180
— nitrogen gas bubbles 267
— recycle 180
Biological growth 73
Biological phosphorus removal 51, 109-110, 112, 114, 285, 287, 289, 291, 293, 295, 297
— design 291-292, 294, 296
— kinetics 114
— mass balances 285-286
— nitrate 288
— plant types 288, 290
— reactions 111
Biomass 81
Bio-P bacteria 109
Biosorption plants 150
Bisubstrate model 72
Black wastewater 35
BOD 28, 45, 55
BOD5 46
BOD20 55
BOD-analysis 47
Bording 224
Bound water 336
Breweries 20

C
C/N ratio 259
— practice 262
Cadmium 29
Calcium 334
Calibration 377

Canneries 20
Car repair/wash 21
Carbohydrate 53, 79
Carbon source 254
Cascade aeration 144
Cation Exchange Capacity 362
Cell internal storage products 51
Chemical oxygen demand 48
Chemical precipitation 355
Chemical/physical phosphorus removal 330, 332, 334, 336, 338, 340, 342, 344, 346
Chironomus 67
Chromium 29
Cl.perfringens 30
Clogging 182
Coagulation 335
COD 28, 45, 48
COD fractions 275
COD/BOD ratio 31
COD/N-ratio 35
COD/TN-ratio 31, 225
COD-fractionation 49
COD-variations 276
Colloids 335
   – bridging 340
Competitive reversible inhibition 87
Conductivity 30
contact filter 357
Contact filtration 351
Contact process 306
Contact stabilization 223
Contact stabilization plants 148
Copenhagen 190
Copenhagen Pectin 307
Copper 29
Corrosion 363
Cow 80
Cryophilic proces 124
Cyanide 87, 324

D
Dairies 20
Decay 75
Decay constant 88, 109
Decay, nitrifying bacteria 211
Degasification 307
Degree of flocculation 345
Denitrification 100, 102, 104, 106, 108, 193, 239, 241, 243, 245, 247, 249, 251, 253, 255, 257, 259, 261, 263, 265, 267, 269, 271, 273, 275, 277, 279, 281, 283
– alkalinity 103, 270
– alternating process 255
– biofilters 252, 258
– combined sludge 251, 254
– design 270, 277
– energy sources 105
– hydrolyzate 274
– inhibition with oxygen 264
– kinetics 103
– mass balances 240
– nutrients 103
– oxygen 264
– oxygen inhibition 107
– oxygen saving 269
– pH 108
– plants 251-252, 254, 256, 258
– process matrix 241
– reaction rate constants 108
– reactions 101
– recycle plant 255
– separate sludge 251, 253
– settling tanks 266
– simultaneous 257
Denitrification rate 107
Denitrifying bacteria 61
Denitrifying biofilm
   – kinetic constants 245
Denitrifying biofilter 243
Denitrifying biomass 28

Denitrifying plants
 – design 259-260, 262, 264, 266, 268, 270, 272, 274, 276, 278
Design 374
Design rate of nitrification 388
Destabilization of colloids 340
Detergents 28
Dicalcium phosphate 334
Diffusion coefficients 168
Diffusion path 160
Diffusional limitation 166
Digester 305
Diluted SVI 59
Dinitrogen oxide 100
Direct precipitation 350
Discs
 – design 190
Dissimilatory nitrate reduction 100
Dissolved solids 43
Domestic wastewater 27
Double layer 337
DSVI 59

E
Easily degradable organic matter 53, 291
Efficiency factor 262
EFOR 381
Electrical circuit industries 21
Electron acceptor 69, 75
Endogenous energy sources 106
Energy 312
Energy production 80
Engineering craftsmanship 369
Enmeshment 340
Enterovirus 30
Entrapment 340
Environmental factors 85, 93
EPS 76, 80
Error propagation 378
Estimates 16
Eutrophication 327

Excess sludge 131
 – phosphorus 287
Excess sludge production 140
Exfiltration 23
Exo Polymeric substances 76
Exoenzymes 66
Expanded filter 186, 309
Experimental design 377
External carbon source 290
External substrate 103

F
Fats 79
FCF 81
Fecal streptococcae 30
Fermentable organic matter 54
Fermentable readily biodegradable organic material 50
Fermentation 293, 299
Fermentation products 50
Ferric iron 331
Ferrous iron
 – oxidation 333
Ferrous iron 332
Ferrous sulphate 359
Filamentous bacteria 136
Filter flies 67
Filter kinetics 175-176, 178
Filter medium 312
Filter wash water 33
First order reaction 84, 158
Fish 41, 80
Fixed anaerobic filter 309
Fixed rule based control 372, 395
Floc break-up 343
Flocculation 70, 341, 364
chambers 341
 – tanks 344
 – Flocs 342
Flotation 305
Fluidized filter 186, 258, 309, 314
Food Chain Factor 81

Formaldehyde 324
Fractile diagrams 13
Frederikssund 256, 284
Free nitrogen 317
Free water 337
Freundlich isotherm 347
Fungi 66

G
Galvanic industries 21
GAO 67, 76, 296
GAOs 111
Garbage grinders 24
Gas flux 312
Gas production 122
  – conversion factors 318
Gas stripping 306
Gas/sludge separator 314
G-bacteria 296
Gel filtration 53
GF/A filter 43
GF/C filter 43
Giardia 30
Giovanni Lorenzo Bernini 99
Glucose 168
GLY 76
Glycogen 76, 110, 112
Glycogen accumulating bacteria 296
Glycogen accumulating organisms 67
Gordonia 67
Granules
  – anaerobic 308
Grazing 75
Grey wastewater 35
Growth rate 70

H
Hammel 227
Haslev 356
Hegel 279
Hen 80
Heraklion 16

Heterotrophic aerobic conversions 88
Heterotrophic biomass 28, 138
Heterotrophic micro-organisms 85
Heterotrophic organisms 50
Hospitals 22
Hourly factor 18
Household wastewater 34
Hydraulic erosion 197
Hydraulic film diffusion 169-170
Hydraulic retention time 360
Hydraulic surface load 314
Hydraulic surface loading rate 182
Hydrogen 317
Hydrogen carbonate 168
Hydrogen sulphide 317, 324
Hydrolysis 74, 211, 286, 299, 301, 303, 305, 307, 309, 311, 313, 315, 317, 319, 321, 323, 325
Hydrolysis constant 74, 88, 109, 125
Hydrolysis process 291
Hydrolysis yield 299
Hydrolysis/fermentation
  – design 300
Hydrolyzate 254
Hydrophillic/hydrophobic properties 335
Hydroxyapatite 334
Hydroxylamine 92
Hyper-filtration 306, 314

I
Ideal mix aeration 144
Idealized biological film 157
Industrial wastewater 301
Inert particulate organic material 51
Inert soluble organic material 50
Infiltration 23
Infiltration ponds 353
Inhibition 96, 322
Inhibition constant 125
Inhibition constant, oxygen 109

Integrated modelling 399
Internal loadings 34
Inter-particulate forces 340
Ion exchange 347
Iron 83
ISAH, Johannesburg 295
Isoelectric point 338

J
Johannesburg 238
Joseph Heller 131

K
Kinetic constants 381
Kinetics 84
 – nitrification 93
Kirkeskoven 64
Kitchen 34

L
Lamella sedimentation 306, 314
Langmuir isotherm 347
Laundries 21
leachate 32
Lead 29
Lime precipitation 334
Low temperature 388
Lundtofte 358
Lynetten 190

M
Malmö 333
Mass balances, activated sludge plants 131-132, 134, 136
Max. hourly constant 18
Maximum daily flow 11
Maximum hourly flow 12
Maximum specific growth rate 73, 88
Maximum yield constant 73, 80, 88, 109
Mercury 29
Mesophilic process 124

Metals 27, 29
Metamorfosis 129
Metazoa 66
Methane 281, 302, 317
Methane bacteria 116, 303
Methane production 116
Methane step 117
Methane-forming bacteria 125
Methanol 106, 168, 246, 263
Meyzieu 198
Micrococcus 67
Microsensors 282
Microtrix 67, 111
Model
 – application 369, 371, 371-373, 375, 377, 379, 381, 383, 385, 387, 389, 391, 393, 395, 397, 399
 – calibration 374, 376, 378
 – parameters 371
 – simulation 392
 – structure 370, 374
 – uncertainty 378
Model for activated sludge 380
Model predicted real time control 396
Models
 – analysis of existing plants 372
 – variables 371
Models as research tools 373
Modified BOD 48
Molar ratio 331, 356
Monod expression 84
Monod kinetics 73
Morigosaki 64
Movable filter 187
Movable filter medium 186
MUCT 295
Municipal wastewater 27

N
Nitrate 29
 – uptake rate 164

– utilization rate 60
Nitric oxide 100, 108
Nitrification 89-90, 92-94, 96, 98, 189, 193, 209, 211, 213, 215, 217, 219, 221, 223, 225, 227, 229, 231, 233, 235, 237
  – alkalinity 270
  – design temparature 387
  – filters 226
  – inhibiting substances 97
  – pH-inhibition 217
  – plants 220, 222, 224, 226, 228
  – rate 214, 235
  – temperature 212
Nitrification and denitrification
  – combined 248
Nitrification in biofilters 382
Nitrification-denitrification
  – alkalinity 270
  – computer models 273
Nitrifying activated sludge 211
Nitrifying bacteria 90
  – decay constant 210
Nitrifying biofilm plant 211
Nitrifying biofilms
  – diffusion coefficients 385
Nitrifying filter 214
Nitrifying organisms 51
Nitrifying plants 210
  – design 229-230, 232, 234, 236
  – mass balances 210, 212, 214, 216, 218
  – optimizing 233
  – process matrix 211
Nitrite 29, 92
Nitriteoxidation 99
Nitrobacter 67, 90
Nitrogen 57
Nitrogen balance 272
Nitrogen dioxide 100
Nitrogen fractions 57
Nitrogen in the biomass 210

Nitrosomonas 67, 90
Nocardia 67
Non-competitive reversible inhibition 87
Non-settleable solids 43
Nonylphenoles 28
NUR 60
Nutrients 83

O
Odense 289
Oils 79
Optimised real time control 373
Optimum (C/N) ratio 263
Organic matter 45-46, 48, 50, 52, 54, 56, 81
  – energy 78
Organic surface loading rate 181
Other bacteria 296
OUR 60
Oxidation ditch 146
Oxygen 86, 94
Oxygen uptake rate 60

P
PAH 28
Panduro 65
PAO 50, 67, 76
PAOs, denitrifying 111
PAOs, non-denitrifying 110
Parameter estimation 374, 376-378
Partial penetration 173
Partially penetrated biofilm 170
Particulate COD 50
PE 24
Person Load 24
pH 30, 86, 95
PHA 51, 76, 110, 112, 286
  – storage rate 115
PHB 76
Phosphate
  – equilibrium 332

Phosphate complexes 332
Phosphate-accumulating organisms 50
Phosphoros removal process lay-out 295
Phosphorus 58
 – release 296
Phosphorus accumulating bacteria
 – anaerobic/aerobic conditions 115
 – nitrate 115
 – reaction rate constants 115
Phosphorus binding
 – soil 346, 361
Phosphorus fractions 328
Phosphorus removal
 – design 355-356, 358, 360, 362
 – mass balances 327-328
 – operation 363-364, 368
 – treatment plants 327, 329, 331, 333, 335, 337, 339, 341, 343, 345, 347-355, 357, 359, 361, 363, 365, 367
Phosphorus resolution 333
Photolabs 21
Phtalates 28
PHV 76
Piazza Minerva 99
Pig 80
Pilot plant studies 386
PL 24
Plug flow aeration 144
Poly-b-hydroxybuturate 76
Poly-b-hydroxyvalerate 76
Polyhydroxyalkanoates 76
Polymerized aluminium 349
Polymerized iron 349
Polyphosphate 29, 76, 110, 287, 327
Ponding 184
Ponds 353
Population Equivalent 24
Porosity 312
Post denitrification 254, 263
Post precipitation 351

Post treatment 353
Potentially limiting factor 175
PP 76
PP storage rate 115
Pre-anaerobic tank 293
Precipitants 348
 – primary reaction 331
Precipitation 331
 – side reaction 331
Pre-denitrification 254
Preprecipitation 351
Pretreatment 301
Primary particles 335, 341
Primary settling 190
Printing houses 21
Propionate 286
Protein 53, 79
Protozoa 66
Pseudomonas 67
Pure oxygen 228

R
Radioactivity 27
Reaction rate constants 88
Reaction rate per surface area 160
Readily biodegradable substrate 50
Real time control 372, 395-396
Recycle rate 138
Recycle ratio 138, 181, 312
Reject water 33
Respiration rate 60, 64
Respiration test 51
Restaurants 22
Return period 390
Rotating disc 188, 228, 235, 309
Rotifera 67

S
Safety problems 363
Salmonella 30
Saturation constant for oxygen 88
SBR 147

SBR-plant 257
Schools 22
Science-based determinism 372
Seeding 319
Selection 68
Selection in activated sludge plants 69
Selection in biofilters 69
Selector 144
Sensitivity analysis 378
Septic sludge 32
Sequencing Batch Reactor 146
Set point operation 395
Settleable solids 43-44
Settling 70, 305, 314
Settling tank 147
  – nitrogen gas 266
Simultaneous denitrification 263
Simultaneous nitrifica-
  tion-denitrification 265
Simultaneous precipitation 351
Singapore 326
Slagelse 300
Slaugtherhouses 20
Sloughing 183
Slowly biodegradable substrates 50
Sludge
  – characterization 275
Sludge age 141, 152-153, 313
  – design 154
Sludge blanket plant (UASB) 307
Sludge concentration 138, 152
Sludge hydrolysis 274
Sludge index 138
Sludge load 230
Sludge loading 139, 152
Sludge mass 138
Sludge percentage 138
Sludge production 139, 182, 314
  – estimated 140
  – measured 139
Sludge separation 314
Sludge surface load 314

Sludge volume index 59
Soil matrix 347
Soluble COD 50
Specific filter surface area 316
Sphaerotilus natans 67
Stationary filter medium 185
Step feeding 144, 149
Step-feed 225
Stoichiometric constants 381
Storage 76, 80
Submerged filter 176, 185
Sugar production. CSM 308
Sulphide 30, 87
Sulphur 83
Supernatant 33, 233, 254
Surface charge 339
Surface tension 30
Surplus sludge
  – phosphorus 294
Surplus sludge production 182
Suspended biofilm reactor 187
Suspended inert organic matter 54
Suspended solids 30, 43-44
SVI 59
Swimming baths 22
Søholt 356
Søllerød 64

T
TAL 58
Tanneries 20
Temperature 70, 85, 94, 107
  – variation 37
Textile industries 20
Thermophilic 85
Thermophilic process 124, 310
TOC 45, 54
TOD 45, 54-55
Toilet 34
Tokyo 64
Total energy yield 78
Total nitrogen 29

Total organic carbon 28, 54
Total oxygen demand 54
Total phosphorous 29
Total sludge age 143
Toxic substances 87
Treatment efficiency 137, 152, 181
Treatment plant design 381-382
Tricalcium phosphate 334
Trickling filters 81, 183
 – design 189
 – plastics 228
 – ventilation 183
Trisubstrate model 72
Tubifex 67
Two sludge activated sludge 221
Two-component diffusion 172, 174

U
UASB 300
UCT 295
UCTASP 381
Upflow filter 186

V
van't Hoff 85
Velocity gradient 360
Verification 375
VFA 52, 293
Vienna-Blumental 257
Viols-le-Fort 354

Volumetric load
 – design 311
Volumetric loading 138, 151-152
Volumetric loading rate 181
Vorticella 67
VSS/SS 31

W
Washout 322
Wastewater 11-12, 14, 16, 18
 – characterization 275
 – components 20, 22, 24, 26-28, 30, 32, 34, 36
 – industrial 21
 – institutions 22
 – microorganisms 30
 – particles 44
 – per capita 17
 – pretreatment 305
 – prognoses 25

Y
Yield 80
Yield constant 79, 113, 152
 – units 82

Z
Zero order reaction 161
Zeta potential 336
Zoogloea ramigera 67

Printing: Mercedes-Druck, Berlin
Binding: Stein+Lehmann, Berlin